# Food Microbial Sustainability

Arun Karnwal
Abdel Rahman Mohammad Said Al-Tawaha
Editors

# Food Microbial Sustainability

## Integration of Food Production and Food Safety

Springer

*Editors*
Arun Karnwal
Department of Microbiology
Lovely Professional University
Punjab, India

Abdel Rahman Mohammad Said Al-Tawaha
Department of Biological Sciences
Al Hussein Bin Talal University
Ma'an, Jordan

ISBN 978-981-99-4783-6 ISBN 978-981-99-4784-3 (eBook)
https://doi.org/10.1007/978-981-99-4784-3



This Springer imprint is published by the registered company Springer Nature Singapore Pte Ltd.
The registered company address is: 152 Beach Road, #21-01/04 Gateway East, Singapore 189721, Singapore

# Preface

Food is essential to our daily lives, nourishing and sustaining us. However, behind the scenes, there is a complex and fascinating world of microorganisms at work, shaping our food supply's safety, quality, and sustainability. As our global population continues to grow and our environment faces increasing challenges, it has become imperative to explore innovative ways to integrate food production and food safety while addressing the pressing issue of microbial sustainability. In this book, *Food Microbial Sustainability: Integration of Food Production and Food Safety*, we delve into the intricate relationship between microorganisms, food production systems, and the safety of our food. We aim to shed light on the crucial role of microbial sustainability in ensuring a secure and sustainable food supply for present and future generations.

The field of food microbiology has evolved significantly over the years. It has transcended from a discipline primarily focused on controlling and preventing the growth of harmful microorganisms to a holistic approach that recognizes the potential benefits and functions of microorganisms in various aspects of food production and safety. This paradigm shift has led to a greater understanding of the complex interactions between microorganisms and food, paving the way for innovative strategies to enhance our food systems' efficiency and sustainability.

This book brings together a collection of diverse perspectives from experts in the field, including scientists, researchers, policymakers, and industry professionals. Through their contributions, we explore a wide range of topics, including microbial ecology in food systems, the role of beneficial microorganisms in food production, the impact of food processing technologies on microbial sustainability, and the latest advancements in microbial risk assessment and food safety management. Furthermore, we examine the challenges posed by climate change, population growth, and globalization and how these factors impact the microbial sustainability of our food supply. We delve into the potential risks associated with emerging pathogens and antimicrobial resistance, exploring innovative solutions such as biopreservation, biocontrol, and utilizing microbial metabolites.

The integration of food production and food safety is not a simple task. It requires a multidisciplinary approach, collaborative efforts, and continuous learning. This book serves as a comprehensive resource for researchers, professionals, and students seeking a deeper understanding of the intricate relationship between food production, food safety, and microbial sustainability. It aims to inspire further exploration and foster dialogue among stakeholders from various disciplines, enabling us to develop sustainable and resilient food systems for a healthier future.

We extend our heartfelt gratitude to the contributing authors who have shared their expertise and insights, enriching this book with their valuable contributions. We also express our appreciation to the editorial team and publishers who have supported and facilitated the realization of this project.

We hope this book serves as a catalyst for new ideas, research endeavors, and practical interventions, contributing to the ongoing efforts toward a more sustainable, safe, and resilient global food supply. Together, let us embrace the challenge of integrating food production and food safety while nurturing the delicate balance of microbial sustainability.

Phagwara, Punjab, India — Arun Karnwal
Ma'an, Jordan — Abdel Rahman Mohammad Said Al-Tawaha

# Contents

# Chapter 1
# Role of Microorganisms in the Food Industry

**Navneet Kaur and Anjuvan Singh**

**Abstract** Microorganisms play a crucial role in the food industry as they contribute to the production, spoilage, and preservation of various food products. Dairy products like cheese, yogurt, and fermented milk rely on microorganisms to transform milk into delicious and nutritious foods. Similarly, fermented foods like sourdough bread and meats like salami also require microorganisms for production. In addition to these traditional uses, microorganisms are now being utilized to make wine, beer, and other beverages. They are also used to produce chocolate, culinary colors, and probiotics that promote human health. With the world's population increasing, it is crucial to find innovative ways to use microbes to generate wholesome food and deal with the current food supply crisis. However, while microorganisms are essential for food production, they can also cause spoilage, rendering the food unfit for human consumption. This is particularly true for bakery goods, which are prone to physical, chemical, and microbiological deterioration. Chemical and physical deterioration can decrease the shelf life of intermediate and low-moisture bread goods, while high-moisture products are more susceptible to microbiological spoilage by molds, bacteria, and yeasts. Therefore, it is vital to understand the role of microorganisms in food production, spoilage, and preservation. This chapter will discuss the significance of microorganisms in dairy and bakery goods and how they benefit the food industry. By studying their role in food production, we can develop effective strategies to preserve and extend the shelf life of bakery products, ensuring they remain safe and healthy for human consumption.

**Keywords** Food industry · Probiotics · Spoilage · Bakery products · Bacteria · Yeast · Mold · Biopreservatives

N. Kaur · A. Singh (✉)
Department of Biotechnology, School of Bioengineering and Biosciences, Lovely Professional University, Phagwara, Punjab, India



A. Karnwal, A. R. Mohammad Said Al-Tawaha (eds.), *Food Microbial Sustainability*, https://doi.org/10.1007/978-981-99-4784-3_1

## 1.1 Introduction

Since the beginning of time, the actions of microorganisms in food have either harmed or benefited human civilization. The development of human civilization is a continuous process. Around 8000 BC, the ancient human society discovered agriculture and animal husbandry. As a result, there was an excess of food, making the storage, preservation, and prevention of food deterioration became their crucial duties. Ancient people were unaware of using microbes for food preservation, so they used ice and fire to preserve the food. Ancient people adopted different methods for food preservation, i.e., sugaring, sun drying, freezing, cooking, salting, and smoking. History illustrates that ancient people were aware of using microbes in food processing but not in food preservation. During ancient civilizations, the production of alcoholic beverages, fermented food, and bread was practiced. However, ancient people were unaware of using microorganisms to preserve food. In 1837, Louis Pasteur became the first person to understand the role of microorganisms in food. He used heat to destroy the unwanted microorganisms in beer and wine; around 1860, this method was known as pasteurization. Since his invention, food microbiology has significantly developed (Mishra, 2013). The ancient civilization used the production of fermented food. From ancient times onward, the methods for producing fermented meat, vegetables, and milk were known to society (Fox, 1993). Although these methods were artisanal, no importance was given to the microorganisms' role. However, cultures developed that processing and storing particular raw materials in a particular way led to forming foods with desirable and organoleptically attractive properties and preserving qualities far better than those of the original substrate. Most of the time, small communities, monasteries, and feudal estates passed on their manufacturing techniques and knowledge from generation to generation. These communities typically produced small amounts of the product sold in or in their immediate region (Caplice & Fitzgerald, 1999).

Nowadays, there are two main techniques used for the fermentation of food. First of all, the technology used is "wild fermentation" or "spontaneous fermentation" in this, foods are naturally fermented, and it generally occurs when the microorganisms are present in the processing environment naturally, or they are present in raw food. Some of the foods are sauerkraut, Kimchi, and fermented soy products. Secondly, some fermented meals such as yogurt, Kombucha, and natto can be made by adding starter cultures that are used to start the fermentation process, and this type of fermentation is known as culture-dependent ferments (Rezac et al., 2018). Similarly, back-slopping is a technique for carrying out a culture-dependent ferment in which a small quantity of a previously fermented batch is introduced to uncooked food, such as sourdough bread (Marco et al., 2017). Starters can be used as commercial or natural starters to standardize the finished product's organoleptic properties (such as back-slopping) (Yann & Pauline, 2014).

Fermented foods are nowadays widely used in every cuisine in some or the other form. Fermented foods have seen a rise in popularity in the West in recent years, primarily due to claims that they provide health benefits and the growing interest in gastrointestinal health. Fermented food can exhibit positive effects on health through several processes. First, they contain microorganisms that may be probiotic, like lactic acid bacteria (Marco et al., 2017). According to research, most fermented products contain at least $10^6$ microbial cells per gram, with concentrations varied based on the region, age, and time of analysis and consumption of the product (Rezac et al., 2018). Through its buffering and protective action against intestinal circumstances (such as low pH and bile acids), the surrounding food matrix appears to significantly influence the survival of probiotic strains (Bove et al., 2013). Several investigations have demonstrated that bacteria from fermented foods can enter the gastrointestinal tract; however, this is likely to vary among products, and their presence in the gut appears temporary (Zhang et al., 2016). First, these microorganisms may exert a physiological advantage in the gut by competing with pathogenic bacteria and producing immune-regulatory and neurogenic fermentation byproducts (Derrien & van Hylckama Vlieg, 2015). Second, a positive effect on health can be exerted by metabolites during fermentation. For instance, lactic acid bacteria generally produce polyamines and peptides that may impact immunological and cardiovascular health and metabolites applicable to dairy and non-dairy products (Pessione & Cirrincione, 2016). Third, some substances may further undergo fermentation and become physiologically active metabolites. For instance, phenolic substances (like flavonoids) may undergo fermentation by lactic acid bacteria to active metabolites (Filannino et al., 2015). Fourth, positive health effects can be gained from fermented foods because it includes probiotics and vitamins (Salazar et al., 2016). Last but not least, the level of toxins and anti-nutrients can be lowered by fermentation. For example, the fermentation of sourdough may lower the levels of fermentable carbohydrates (Abu-Salem et al., 2014), e.g., fermentable oligosaccharides, disaccharides, monosaccharides, and polyols or FODMAPs, the tolerance of these products can be increased in people having functional bowel disorders like irritable bowel syndrome (Laatikainen et al., 2016).

The role of microorganisms in the fermentation process is crucial, as is their role in preserving the finished product. From the beginning of time, food preservation has been essential for human survival by improving the stability and safety of different types of food. The food industry still uses several traditional methods to preserve food, including fermenting, salting, drying, and heating. Since then, there has been a significant gain in knowledge regarding the primary causes of degradation, and losses from food spoilage and contamination have decreased (Gram et al., 2002).

Understanding the composition, pH, origin, water activity ($a_w$), and storage conditions (such as pressure, atmosphere, and temperature) of the food as well as the characteristics of the most prevalent and resistant microorganisms can help predict the composition of the microflora during processing and storage. The proper application of technology (such as pasteurization and sterilization), sanitary measures (such as good hygiene practices and good manufacturing practices), and traceability (to prevent and reduce the distribution of poor quality and unsafe food) that can stop and delay the colonization of spoilage and pathogenic microorganisms in food. However, food is still susceptible to degradation throughout the food chain, including in grocery stores, restaurants, and consumers' homes, and hazardous microbial infection may occur (Aung & Chang, 2014; Gram et al., 2002).

Food deterioration and contamination harm consumers and the food industry, resulting in financial loss, reputational damage, and legal Punishments (Akkerman et al., 2010). Customers reject the food when its sensory qualities decline due to several intricate mechanisms involved in food deterioration caused by microbial activity. Microorganisms (such as bacteria and molds) may have spoiled food if there is visible development in the form of slime and colonies, changes in texture brought on by the breakdown of carbohydrates, lipids, and proteins, and the sense of off-flavors and off-odors (Gram et al., 2002).

Conversely, the spread of pathogenic microbes, cross-contamination, and toxin production endanger consumer health and threaten food safety. The ailment known as gastroenteritis, marked by intestinal cramps, diarrhea, nausea, fever, and vomiting, is typically brought on by the ingestion and colonization of pathogen microorganisms in the human digestive tract (Flint et al., 2005). Although bacteria are the primary cause of gastroenteritis, other agents such as parasites, viruses, yeasts, and molds are also responsible and have become more common during the past 10 years.

This chapter mainly focuses on the role of different microorganisms in fermented food and the use of different microorganisms as probiotics. The use of microorganisms produces different types of dairy and bakery products. Spoilage is caused to bakery products due to the action of microorganisms and the effect of different physical factors that affect the growth of microorganisms and last different preservation technique to reduce the growth of microorganisms that causes spoilage.

## 1.2 Role of Microorganisms

### *1.2.1 Fermented Food*

Since ancient times, people have ingested fermented foods, and numerous health advantages have been linked to them. However, it was not until the 1970s that considerable scientific interest was generated regarding the potential probiotic effects associated with ingesting fermented items (Pessione & Cirrincione, 2016). In most of the world, fermented products from animal and plant sources are permissible and crucial to the diet. They use various raw materials, such as substrates, and technologies ranging from the most basic to the most sophisticated, producing finished goods with excellent sensory and textural properties (Bove et al., 2013).

#### 1.2.1.1 Wide Use of Fermented Food

Food preservation improved nutritional value and increased flavor and aroma. The improvement of materials for relatively high-quality products and improved health benefits are the key sources behind the development and use of fermented foods. Nowadays, originality and trendiness play an essential role. These characteristics are brought about by the activity of a group of bacterial strains, especially lactose bacteria or microbes mixed with probiotics and probiotic potential in different organic substrates (Gram et al., 2002).

Because it enables us to select and use new sources, boost the yield of already-existing sources, enhance the bioavailability of nutrients, nutritional value, and flavor, and introduce specific functional characteristics in raw materials, biotechnology has made a significant contribution in recent years. Aerobic or semi-aerobic Lactobacillus strains comprise most of the microflora in these goods (Korhonen, 1994).

#### 1.2.1.2 Advantages Related to Health

The nutritional value of a food depends on how easily it can be digested and how much of the necessary nutrients it contains. Fermentation can potentially improve both the nutrient density and the digestibility of food. Fermentation is a process in which the microbial culture's enzymatic activity may predigest the macronutrients. In healthy people having normal gastrointestinal function, predigesting proteins have a negligible positive effect. It is known that the curd clots that form in the stomach after ingesting yogurt are much smaller than those that form after ingesting milk which increases proteolysis in the stomach (Gram et al., 2002).

Carbohydrate predigesting can help people tolerate certain foods much better. For example, it is widely known that yogurt is more tolerable than milk due to its low lactose content and the lactase activity of the microbial culture for people who are lactose intolerant. Lactase activity in *Streptococcus thermophilus* and *Lactobacillus bulgaricus* strains prevents lactose from passage through the stomach and aids the digestion of lactose in the intestine. *Lactobacilli* strains include *Lactobacillus fermentum, Lactobacillus salivarius, Lactobacillus buchneri, Lactobacillus brevis,* and *Lactobacillus cellobiosis*, produce alpha-galactosidase activity, which can then be used to break down alpha-galactosides limits in raffinose and stachyose. Both soybean milk and legumes can be processed to remove the carbohydrates that cause flatulence. Older people those are having low stomach acid output, the acid produced during carbohydrate fermentation may help them lower the gastric fluid's pH (Filannino et al., 2015).

Although many molds, bacteria, and yeasts used in fermentation exhibit lipolytic activity when it comes to fats, the nutritional importance of this is probably minimal. Plant foods containing non-digestible substances, including cellulose, hemicelluloses, polygalacturonic acid, and glucuronic acid, may have less lipolytic activity after fermentation. Mineral and trace element bioavailability may be increased due to the breakdown of these molecules. Additional vital nutrients, including vitamins, amino acids, and proteins, may be produced during fermentation. Many physiologically active substances are said to be present in fermented milk, which may benefit human health. These substances include elements originating from milk, fermentation-related microorganisms, and their metabolic byproducts. While some lactose bacteria strains have been identified as probiotics in vivo, fermented milk's probiotic qualities may also be attributed to peptides made from hydrolyzed milk proteins. On proteolysis using starter bacteria or digestive enzymes, several milk proteins have been found to include peptides of varied functions (Caplice & Fitzgerald, 1999). These peptides contain mineral carriers, opioids, antibacterial, anti-cancer peptides, and anti-hypertensive. It is known that whey proteins and casein fractions serve as precursors for the fore mentioned bioactive peptides (Perdigon et al., 1988).

The *Lactobacillus* GG strain has been demonstrated in numerous clinical investigations to have positive health effects, such as the ability to treat gastrointestinal illnesses. Although the variables or mechanisms causing these effects are not entirely understood, mucosal immune system stimulation has been suggested. Therefore, it was intriguing to find out if milk fermentation with *Lactobacillus* GG would result in the formation of immunoreactive peptides (Gram et al., 2002). These preliminary findings imply that lactic acid bacteria in fermented milk alter the immunogenic characteristics of milk proteins. The creation of functional foods and probiotic lactic acid bacteria strains, among other things, may be impacted by this discovery (Perdigon et al., 1988).

## *1.2.2 Probiotics*

In recent years, probiotics have attracted both consumer and academic attention. A growing body of clinical evidence supports some potential health benefits of probiotic supplementation, notably in treating specific diarrheal illnesses. Probiotics are generally "Live microorganisms that, when administered sufficiently, provide a health benefit to the host." The term "probiotic," derived from a Greek phrase meaning "for life," was first used in the 1960s. Despite being a relatively recent concept, some meals' health benefits contain live bacteria that have been understood for ages. However, it was not until the early twentieth century that researchers proposed that the gut flora may be altered by swapping out the harmful bacteria for beneficial ones (Huis in't Veld et al., 1989).

### 1.2.2.1 Role of Probiotics in the Food Industry

Probiotics are yeast or bacteria permitted as dietary supplements and food additives. They come in capsules, tablets, packets, powders, or other forms and are found in various fermented foods, most frequently yogurt or dairy products (Valerio et al.,

2009). Probiotic products may contain single species of bacteria or a blend of several. The bacteria, specifically the *Bifidobacterium* and *Lactobacillus* species, are the most commonly utilized probiotics. *Saccharomyces boulardii*, yeast, also seems to offer health advantages. It is important to remember that probiotic benefits often only apply to a specific strain, so even within a single species, a health benefit attributed to one strain may not necessarily apply to another. Therefore, it is best to avoid broad generalizations regarding prospective health advantages (Korhonen, 1994).

#### 1.2.2.2 Some Properties of Probiotic-Rich Fermented Milks

Only a few fermented dairy products have been linked to a probiotic effect on people in Europe (yogurt). Milk with probiotic characteristics is referred to as "fermented milk," which benefits people by restoring the balance of the microbes in the intestines and the healthy microbial ecology of the gut. The effect of cultured products on the ecology of the gastrointestinal system in humans and other animals has been extensively studied. Probiotic solutions are currently available in a variety of *Lactobacillus* and *Bifidobacterium* species. Lactobacilli that are currently used as probiotics include (Adams & Hall, 1988) *L. brevis, L. cellibiosus, L. acidophilus, L. lactis, L. fermentum, L. casei, L. plantarum,* and *L. reuteri* and *B. infantis, B. thermophilum, B. longum, B. animalis,* and *B. adolescentis* are the *Bifidobacteria* now used as probiotics.

#### 1.2.2.3 The Following Is the List of the Ideal Qualities of a Good Probiotic Strain (Valerio et al., 2009)

1. Simple in vitro proliferation
2. Surviving during storage, processing, and incorporation into feed or food
3. Proliferation and colonization of the gastric-intestinal tract by forming lactose and bacteriocins
4. Noninvasive, nonpoisonous, nonpathogenic, non-tissue residues, and non-absorbed in the digestive tract
5. Genetic stability, the absence of mutations, and the incapacity to produce harmful germs
6. Fighting off harmful bacteria and preventing intestinal infections
7. Lactose intolerance (pre-digestion)
8. Effects on cholesterol levels
9. Antitumor action

Probiotics' positive effects are primarily related to their direct antagonistic action against particular subgroups of microbes (Enteropathogens), their impact on gut metabolism, or their promotion of systemic or mucosal immunity.

#### 1.2.2.4 List of Microorganisms Used as Probiotics in the Food Industry (Valerio et al., 2009)

| | Species of *Bifidobacterium* | Species of *Lactobacillus* | Others |
|---|---|---|---|
| 1. | *B. infantis* | *L. reuteri* | *Clostridium butyricum* |
| 2. | *B. breve* | *L. acidophilus* | *Saccharomyces boulardii* |
| 3. | *B. lactis* | *L. bulgaricus* | *Escherichia Coli* |
| 4. | *B. longum* | *L. gasseri* | *Streptococcus thermophilus* |

## 1.3 Diary and Bakery Products (Fig. 1.1)

### *1.3.1 Fermented Milk Made with* Bifidobacteria

Several *Bifidobacterium* strains can live in the intestine and enter the colon. *Bifidobacteria*, one of the primary species in the colon, can alter the native intestinal microflora, which has physiological implications such as regulating gut transit and changing the microbiota's enzyme activity (Corthier et al., 1985).

In some circumstances, *Bifidobacteria* can treat constipation and diarrhea. *Bifidobacteria* have been demonstrated in the improvement of parameters of digestion and have a minor anti-pathogen impact in humans (Kotz et al., 1990). They also have an impact on some elements of the immune system. Different strains inhibit the activity of colon enzymes necessary for transforming procarcinogens into human carcinogens (Corthier et al., 1985). In vitro studies with viable *L. helveticus* or *Bifidobacterium* sp. yogurt strains and fermented milk but not *L. acidophilus* revealed a positive effect on colon cancer differentiation and cell differentiation proliferation. *Bifidobacterium* sp. and *L. helveticus* inhibited the development of human HT-29 cells, and the activity of dipeptidyl peptidase IV increased, a particular marker of cell differentiation. These results imply that various lactic acid bacteria exhibit some degree of selection (Duval-Hflah et al., 1983). *Bifidobacteria* struggle in the presence of oxygen because they are not well suited to fermented milk. Growth and survival in acidified and partially aerobic environments are thus crucial selection criteria for particular strains. In certain investigations, milk that has

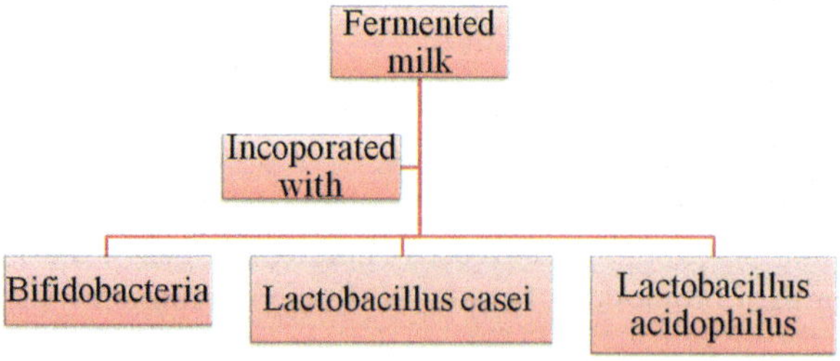


**Fig. 1.1** Figure showing different strains of bacteria used in fermented milk

been fermented with both *L. acidophilus* and *Bifidobacterium* is used. It is unknown whether this pairing has different or additional health advantages.

### *1.3.2 Fermented Milk Made with* Lactobacillus casei

The most frequent location of *Lactobacillus casei* in humans is the small intestine. Certain strains of *L. casei* can pass through the ileum and stomach in sufficient numbers to trigger a physical response. According to a recent in vivo study (Dewitt, 1985) survival rate in the ileum of many strains of *L. casei* was about 5–10% (except one strain at 0.5%) if *L. casei* was combined with *Bifidobacteria* around (10–30%) or *S. thermophilus* (0.5%).

The most researched lactic acid bacterium is *L. casei,* which has been shown to impact the frequency and length of different types of diarrhea and the immune system. It can help prevent and treat diarrhea brought on by antibiotics and traveler's diarrhea and can significantly impact infantile diarrhea. The ability of *Lactobacillus casei* to treat or prevent the potential of some types of diarrhea to change the activity of the intestinal microflora appears to be related, hence limiting the growth of potentially harmful bacteria. However, the actual methods for particular activities are intricate and poorly understood. Animal studies have shown that the general and specialized immune systems are stimulated. When children with acute diarrhea were treated, it was discovered that many immunoglobulin levels had increased using the *L. casei* strain. The result was not seen in fermented milk that had been heated (Denariaz, 1994).

A recent research program examined the impact of *L. casei* on *Salmonella.* According to the results of the random investigation, milk only had a 10% survival yield after *Salmonella* infection, while milk had a 90% survival yield after it was fermented with bacteria of yogurt and multiple *L. casei* strains. This finding suggests some strain specificity and that combining different strains can produce positive effects. It has been hypothesized but not proven that adherence of bacteria cells is required for diarrhea or immune system control. It has been demonstrated that *Lactobacillus casei* inhibits mutagenicity and reduces the enzyme activity linked to colon cancer risk (Ballongue et al., 1993). It was found to lower human superficial bladder cancer recurrence. The production of carcinogenic chemicals in the intestines and the immune-modulating impact may be part of the mechanism of action. Further research is required to learn how *L. casei*, *Bifidobacteria*, and lactose bacteria reduce the cancer risk (Yasui et al., 1992).

### 1.3.3 *Using* Lactobacillus acidophilus *to Ferment Milk*

One strain of *Lactobacillus* that does well in the gastrointestinal tract is *Lactobacillus acidophilus*. Similar *to L. casei*, it usually lives in the small intestine. Additionally, it makes up most of the flora in women's urinary tracts. *L. acidophilus* may play a role in human health in several ways, including lowering blood cholesterol and preventing vaginal Candida infections.

### 1.3.4 *Yogurt*

The most tangible health benefit of consuming yogurt is reducing lactose malabsorption. Clinical and observational studies support this effect, but the exact mechanism remains unknown. There are some organisms among *Lactobacillus*, with *L. bulgaricus* showing the most activity and *Bifidobacteria* showing the least. Yogurt with heat treatment will not have the same digestion of lactose action as yogurt with live bacteria. Isaac Carasso, a pharmacist, started the first large-scale yogurt production to try to treat intestinal illnesses. The common knowledge of the Balkan populations was the basis for this concept. Several research findings (Attaie et al., 1987) provide statistics on the yogurt effect on male diarrhea. A recent data test was conducted with *Bifidobacterium bifidum* and *Streptococcus thermophilus*, and the analysis reveals that it is impossible to predict which bacteria if any caused the observed effect. Yogurt microorganisms influence a variety of immunological variables. Yogurt consumption has been linked to reduced allergy symptoms in old age people. According to present research, yogurt and a particular *L. casei* strain found in products after fermentation have an adjuvant effect on mice's anti-choleric immune reaction to vaccination. Unlike milk, heat-treated yogurt had the same effect, implying that cell constituents or byproducts of fermentation, other than living organisms, are more important for this function.

### 1.3.5 *Cheese*

Cheese is produced by introducing a microbial culture containing a specific strain of bacteria into milk. Cheese is a food that can be made with the milk of many different animals, but the milk of the cow is typically chosen. We can also use milk from goats, reindeer, sheep, and water buffalo. Lactic acid is produced during milk fermentation due to the inoculation of bacterial cultures, and this acid gives the milk its sour flavor. Casein, or milk protein, coagulates as a result of it. The liquid portion of coagulated milk is known as whey, while the solid portion is known as cheese (Marteau et al., 1990). The cheese becomes ripe by adding bacteria, fungi, or both inoculums. The pH is lowered, the texture changes and a flavor develops when these fungal or bacterial inoculums are poured (De Simone et al., 1992). The renin enzyme, which is present in rennet pills, can be used to regulate coagulation. Calve's stomachs contain the renin enzyme. However, genetically modified microorganisms are used today to produce them (Romond et al., 1989). Lemon juice and vinegar can also be used to create coagulation.

The cheese has various flavors (Zoppi et al., 1982). Cheese comes in the following varieties, depending on the bacteria that are added:

(a) To enhance the flavor and texture of cheddar cheese, bacteria are used in the preparation process.
(b) Mold fungi are used to make Roquefort and blue cheese.
(c) Bacteria and fungus are used in conjunction to create Camembert cheese.
(d) *Propionibacterium shermanii* is used in the production of Swiss cheese.

Most cheeses range in natural form from off-white to yellow. The cheese can also be flavored with herbs and spices (Havenaar et al., 1994). Different milk fat concentrations, processing methods, aging duration variations, and mammal breeds are other factors that greatly influence the flavors and styles of cheese.

### 1.3.6 Bread

Yeast is a saprophytic, single-celled fungus. By secreting enzymes, yeast cells may digest foods, including sugar and minerals. In order to make bread, yeast is used (Kaila et al., 1992). When flour is mixed with yeast culture, water, and other ingredients, carbon dioxide is produced. This carbon dioxide is then trapped in the flour-based dough. $CO_2$ raises the dough to create the bread. Most often, starch-containing wheat flour is used. Starch serves as the yeast's energy source. Additionally, the gluten protein in the flour causes the stretchy and sticky threads to form as the inoculum of yeast reacts with the starch. The dough is lifted by these threads, which capture $CO_2$ (Ling et al., 1994).

Yeast has been used as a leavening agent in baking for a long time. *Saccharomyces cerevisiae* is the yeast species most frequently used because of its capacity to develop quickly and ferment sugar in the dough quickly (Aso et al., 1992). The rising of the dough is caused by the $CO_2$ produced during fermentation. Baker's yeast is produced in large quantities under controlled pH, humidity, and temperature conditions (Hilton et al., 1992).

### 1.3.7 Chocolate

Microbes are used in the production of chocolate. Cacao tree seeds are used to make chocolate. The cacao tree's white, fleshy pods contain these seeds (Valerio et al., 2009). To extract the seeds from the pods, with naturally occurring microbes, they are first fermented, primarily yeasts and bacteria, specifically *Acetobacter* and *Lactobacilli.* The compounds created by these microorganisms are the ethanol made by yeast, which causes a rise in temperature during the fermentation process, killing the beans and contributing to the chocolate's flavor. The fermentation process in chocolate is the process that creates fragrances, flavors, and vibrant colors. Acetic acid fermentation and alcoholic fermentation are the two steps that make up this process. In the first step of this procedure, yeast activity in the cocoa pulp transforms the sugar into alcohol. Acetic acid is then created by microorganisms oxidizing the alcohol.

## 1.4 Benefits of Microorganisms in the Food Industry (Mazhar et al., 2022)

(a) Microbes grow quickly and require less space.
(b) Microbes have a lot of protein in their cells. They have a high content of protein of roughly 40–50% in microbes and 20–40% in algae, high yielding capacities, and are less impacted by environmental factors, such as weather which does not impact them.
(c) The microorganism's protein contains all the essential amino acids.
(d) Among different microorganisms, yeast generally contains more vitamin content.

Despite the benefits of microbes in the food industry, they may also have many side effects, as mentioned below,

## 1.5 Deterioration of Bakery Products by Microbes

Microbiological deterioration is the most common cause of baked goods' shelf-life reduction. Manufacturers and consumers suffer financial losses due to spoilage brought on by microbial development. These losses may result from various specific circumstances, including product packaging, sanitary manufacturing procedures, storage, product turnover, and storage conditions.

Investigated whether lipoxygenase-induced enzymatic spoiling and microbiological deterioration caused by yeast, fungi, and bacteria could be discriminated against each other and unspoiled bread counterparts 48 h after baking (Fig. 1.2).

The volatile compounds produced by the various types of bread decomposition and their analogues were identified using gas chromatography-mass spectrometry. A microbial study revealed that the concentrations of each microbe used increased over time. *Leuconostoc mesenteroides, Leuconostoc citreum, Weissella cibaria, Lactococcus lactis,* and *Weissella confusa* were identified as the species by Valerio et al., (2009) with the use of repeated extragenic palindromic PCR and sequencing

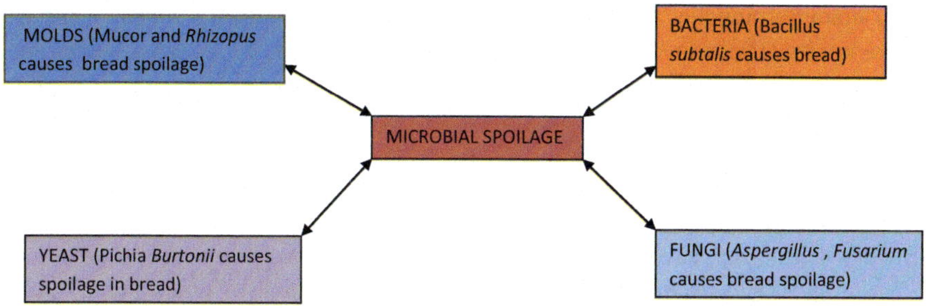


**Fig. 1.2** Figure showing spoilage of bread by different Microorganisms

of the 16S rRNA gene analysis. The REP-PCR results identified 17 different patterns.

### *1.5.1 Bacteria*

Moreover, the growth of bacteria is constrained by lowering the water activity and pH, and bacteria can contaminate baked goods. *Bacillus subtilis* spores are a good example of heat-resistant organisms; after 20 min at 65 °C, 55% are still active in amylase. This microbe, found in raw yeast, flour, sugar, and other substances, creates rope in bread. Ropey bread is distinguished by a stringy, incredibly wet crumb, a coloration ranging from brown to black, and a rotten fruit stench discharge. Typically, this issue arises during the summer when the weather is warm and muggy. Under warm, muggy conditions, ripeness can appear exceptionally quickly. Consequently, it is a widespread issue in the climate that is warm of Africa, the Mediterranean region, and Australia (Rosenkvist & Hansen, 1995).

It will be advantageous for bakers to employ ingredients with low contamination levels because raw ingredients are a primary source of *Bacillus* contamination. *Bacillus subtilis* is the principal reason behind the ropey bread; however, other Bacillus species, such as *Bacillus licheniformis, Bacillus megaterium,* and *Bacillus cereus,* are also responsible (Voysey & Hommond, 1993). The first signs of rottenness in bread include a smell of pineapple, and later on, the crumb turns discolored, squishy, and touchable sticky, rendering the bread inedible. The degradation of bread texture is caused by the development of slime, which is caused by the interaction of the amylolytic and photolytic enzymes generated by a few *Bacillus* strains. Strict hygienic standards and suitable production methods are needed to prevent problems with ropes caused by *Bacillus* spp. spores.

One sort of bacterium that might contaminate pie fillings is *Staphylococcus aureus*. Food poisoning incidents linked to cream-filled pastries have also been reported. In addition, cakes and pastries were responsible for 3% of Britain's 323 food poisoning incidents between 1969 and 1972. Other substances used in baking, including desiccated coconut, cocoa powder, and chocolate, have been discovered to contain Salmonella. For instance, *Salmonella typhimurium* significantly impacts frozen pizza (Dickson, 2001).

### 1.5.2 Yeast

Bakery items can have yeast issues. Species of *Trichosporon, Saccharomyces, Pichia*, and *Zygosaccharomyces* are examples of wild yeast. “chalk bread” originated from the white blotches *Saccharomyces* sp. produces on bread. According to research by Legan & Voysey, (1981), yeast issues in products of a bakery can also be divided into two categories: (a) visible yeast that appears as white or pinkish patches on the surface of the bread and (b) fermentative spoilage linked to essence odors and alcoholic and thus osmophilic yeasts. Yeasts, primarily *Pichia burtonii*, cause bread to become surface-spoiled (“Chalk Mold”). Osmophilic yeast contamination of products typically happens due to dirty tools and equipment. Therefore, osmophilic yeast contamination will be reduced by maintaining proper manufacturing methods.

### 1.5.3 Mold

Using preservatives is a desirable technique to lessen spoilage and ensure the food is safe because spoilage by mold is a large and expensive bakery problem. Because modern consumers dislike chemicals as preservatives, the baking industry is pressured to use fewer of them (Membre et al., 2001). Mold development is the most significant factor affecting how long high and intermediate-quality bread products will last. Although most molds require high $a_w$ values (>0.8), some xerophilic molds can grow at $a_w$ values as low as 0.65. Mold growth in bakery products is a serious problem that costs businesses money. Additionally, based on the product type, season, and processing method, losses from mold deterioration range from 1 to 5% (Malkki & Rauha, 2000). According to Hickey (1998), the food industry loses an average of 200 million pounds of production annually owing to mold degradation. Mold spores are frequently destroyed when fresh bread and other prepared foods are baked (Knight & Menlove, 2006). Therefore, for bread to develop mold after baking, contamination from the bakery surfaces, food handlers, equipment, the air, or raw materials must happen while chilling, slicing, or wrapping procedures. This implies that any mold-related degradation issues must manifest themselves after baking. The mold particle counts are higher in the summer than in the winter due to airborne particles from warmer temperatures and more humid storage conditions. Additionally, moisture condensed from the packaging before a product has wholly cooled on its surface may promote mold growth. Unpleasant odors are produced by mold rotting, which is frequently seen on a product’s surface (Jarvis, 2001). The molds most frequently found in baked goods are *Rhizopus* sp., *Monilia* sp., *Mucor* sp., *Aspergillus* sp., *Penicillium* sp., *and Eurotium* sp.

## 1.6 Microbial Growth Is Affected by Physical Factors (Effects of pH, Temperature, and Activity of Water)

The key aspect affects how long bakery products may be stored without developing microbial contamination. It is critical when molds and bacteria compete to contaminate high-moisture foods (Ponte et al., 1993). Unlike bacteria, molds are typically less particular about how pH affects them. Molders prefer an acidic pH and tolerate acidic environments (3.5–5.5). Therefore, mold is more likely than bacteria to destroy meals with a pH of 4.5. Abellana et al., (1999) established a technique to determine the effect of water activity ($a_w$) and the growth of xerophilic fungi on baked goods and isolates, temperature, and their interactions on the mycelial growth of *Eurotium* sp. Water activity ($a_w$), isolate, temperature, and two or three-way interaction all affected intra-isolate variation. All isolates performed best in the water activity ($a_w$) temperature range at 0.90 ($a_w$) and 30 °C, with an interval of the growth rate of 3.8–5.1 mm $day^{-1}$ at 0.75 ($a_w$), while growth was less than 0.15 mm $day^{-1}$. The temperature has a significant impact on how spores germinate and how mold grows. The temperature range between 18.3 and 29.4 °C is ideal for most molds. By lowering the storage temperature from 27 °C to 21 °C, (Chamberlain, 1993) discovered that the shelf life of cake free of mold doubled. He also stressed the significance of exercising prudence when distributing and keeping food.

Abellana et al., (2001) investigated how temperature and water activity affected mycelial growth rates in *Penicillium chrysogenum, Penicillium aurantiogriseum, Aspergillus flavus,* and *Penicillium corylophilum* (Chamberlain, 1993). Growth rates exhibited the expected $a_w$ and temperature dependence on an analogous sponge cake. However, no discernible variations in the growth rates of various isolates were found. The minimum $a_w$ values required for *Penicillium* sp. growth ranged from 0.85 to 0.90.

At 0.90 $a_w$, *Aspergillus flavus* could grow when the temperature was higher than 15.8 °C. They demonstrated that if the $a_w$ is maintained at 0.85, fungal development by these species on sponge cake analogues with a composition similar to typical bakery goods is avoided. The xerophilic fungi *Eurotium chevalieri, Eurotium rubrum, Eurotium herbariorum, Eurotium amstelodami,* and *Wallemia sebi* were discovered, isolated, and described by Vytrasova et al., (2002). Investigations were made into the fungi's tolerance to high temperatures and preservatives. *Eurotium* species were discovered to be more resistant than *Wallemia sebi*. Sorbic acid was more effective than calcium propionate for preservation against xerophilic fungus. A sponge cake analogues was used by Guynot et al., (2003) to study the effects of pH, water activity ($a_w$), and carbon dioxide ($CO_2$) levels on the growth of seven fungal species that frequently lead to the spoilage of baked goods, i.e., *Eurotium amstelodami, Eurotium herbariorum, Eurotium repens, Eurotium rubrum, Aspergillus niger, Aspergillus flavus,* and *Penicillium.* The primary elements significantly influencing fungal development were water activity and $CO_2$. The concentration of $CO_2$ required to avoid analogues cake deterioration was determined by the effect of water activity at levels of 0.80–0.90 on fungal growth. When the amount of $CO_2$ in the headspace grew from 0 to 70%, the lag phase level of $a_w$ was doubled to 0.85.

In general, samples packaged with 100% $CO_2$ did not exhibit any fungal growth up to 28 days of incubation at 25 °C, regardless of the $a_w$ level. Seven molds that are frequently involved in the spoilage of bakery goods were used in an investigation by Elena into the growth of mold on the fermented bakery product analogues (FBPA) of two different pHs (4.5 and 5.5), different water activity ($a_w$) levels (0.80–0.90), and different potassium sorbate concentrations (0.0–0.3%). Ten-term polynomial models were developed to explain how the variables $a_w$, pH, potassium sorbate concentration, and fungal growth interacted. It is helpful to simulate storage spoilage concerning the factors influencing fungal development. Small amounts of $a_w$ could only slightly diminish potassium sorbate concentrations at pH 4.5; however, fungal growths were still detectable at pH 5.5 despite adding 0.3% of potassium sorbate. In a 2008 study by Mariona Arroyo et al., four food spoilage fungi (*Penicillium roqueforti, Eurotium repens, Cladosporium herbarum, Penicillium corylophilum*) were examined concerning biotic factors like pH, water availability, and temperature and the presence or absence of a preservative. These included two mycotoxigenic food spoilage fungi, *Penicillium.* These studies aimed to get the potential for nutritional partitioning, niche exclusion, and relative coexistence in bread-based matrices. The niche's size significantly decreased as water availability, pH, and temperature decreased. There were some noticeable relationships between the preservative and pH. The data determined the relative niche overlap indices between the ochratoxigenic species and competing fungi. These showed that *Aspergillus ochraceus* and *Penicillium verrucosum* performed better than other species in terms of nutrition.

### *1.6.1 Salt Tolerance*

At 22 °C, *Penicillium roqueforti* and *Aspergillus niger* were cultivated with NaCl and other NaCl replacers ($MgCl_2$, $CaCl_2$, $MgSO_4$, and KCl). Challenge tests on white bread were also carried out to determine the effects of a reduction in NaCl with or without replacement in partial on the growth of *Penicillium roqueforti.* With NaCl and $MgCl_2$ having the most substantial inhibitory effects on the growth of *Penicillium roqueforti* and *Aspergillus niger,* respectively, the obtained data demonstrated that the isolates displayed varying sensitivities to the salts investigated at equivalent water phase concentrations. $MgSO_4$ was the least effective antifungal. At comparable molalities, $CaCl_2$ showed the most excellent antifungal action overall. $MgCl_2$ was found to have a more negligible inhibitory effect on *Aspergillus niger* at equivalent water phase concentrations than expected based on its $a_w$ depressing effect. This is even though the chemicals studied have significant effects that reduce water activity ($a_w$), which are essential in explaining the observed trends. *Penicillium roqueforti* growth on conventional white bread, bread with 30% less sodium chloride, and bread with 30% sodium chloride partially replaced by a potassium chloride and sodium chloride solution did not differ according to the challenge tests.

## 1.7 Microbial Growth Control in Bakery Products

### 1.7.1 Product Reformulation to Lower the Average $a_w$

Reformulation involves lowering the amount of readily available water to increase the bread product's shelf life. The $a_w$ of the foodstuff can be decreased through evaporation or freeze-drying, dehydration, or by adding highly osmotically active ingredients such as sugars and salts directly into the food. The amount of $a_w$ decrease practically impacts whether a food is considered non-perishable. The way that microorganisms respond to a particular level of $a_w$ varies significantly depending on their surroundings (Guynot et al., 2003); the increasing concentration of crystalloids renders water in solutions of carbohydrates and salts inaccessible to microorganisms. Furthermore, large quantities of these chemicals directly harm bacteria osmotically. Since all metabolic functions require an aqueous environment for chemical reactions to occur, this effect may be caused by the negative impact of decreased water availability on all of these activities. Maintaining a sufficiently low $a_w$ level is typically necessary to control mold growth in bakery items. For instance, $a_w$ of 0.75 can extend the shelf life of food without mold by 6 months. Higher $a_w$ values, such as those above 0.77, will only briefly extend the shelf life. However, caution must be exercised while reducing product $a_w$ because low levels can have a negative impact on the product's quality and result in changes to its shape and texture (Arroyo et al., 2008).

### 1.7.2 Freezing

The long-term preservation of bakery goods, particularly those with cream fillings, has been accomplished via freezing. Controlling the growth of ice crystals requires rapid freezing. When the rate of freezing is slower, huge ice crystals are generated, which have the potential to damage internal cellular structures and membranes (Samapundo et al., 2010). Pancakes, shortcakes, cookies, and other baked goods are frequently frozen and sold frozen. Bread can be stored at −22 °C and kept fresh for several months (Kyzlink, 2001). Frozen bread ages relatively slowly compared to fresh bread, which does so in less than a week. Therefore, it ages more slowly at lower temperatures. Desrosier, (2006) claimed that bread preserved at −18 °C for a year after being frozen shortly after baking was as soft as fresh bread kept at 20 °C for 2 days.

### 1.7.3 Preservatives

Most frequently, preservatives are employed to prevent mold formation in baked foods. Preservatives are "an antimicrobial agent used to preserve food by reducing the growth of germs and consequent decomposition" by the Code of Federal Regulations (CFR). Preservatives can be divided into two categories: chemical and natural. Chemical mold inhibitors allowed in bread include sorbic, propionic, and acetic acids and their salts. On the ingredient statement, natural food preservers such as raisins, cultured products, and vinegar are known by their common names.

### 1.7.4 Biopreservatives Effect

Due to growing customer demand, biopreservatives using microbes and their metabolites to stop food from spoiling and increase its shelf life have recently attracted more attention. Particularly interesting as biopreservation organisms are lactic acid bacteria (LAB).They can produce various bioactive compounds, including organic acids, fatty acids, hydrogen peroxide, and bacteriocins and have been employed for millennia as starter cultures in the food business. It is known that LAB has antifungal properties.

Mentes et al., (2005) investigated the effects of two sourdoughs created with *Lactobacillus alimentarius* and *Lactobacillus plantarum*, which have antibacterial activity on inhibiting Bacillus strains that form ropes in wheat bread. By adding 15% or 20% low pH (pH 3.5–4.0) sourdough starter to bread dough manufactured separately using two strains of *Bacillus, Bacillus subtilis* and *Bacillus licheniformis*, visible rope production was prevented *(Lactobacillus plantarum* and *Lactobacillus alimentarius*). A 10% sourdough addition, though, was insufficient to prevent the development of visual rope. When the experiment was performed with sourdoughs with a pH higher than 4 (pH > 4), additives at 10% or 15% did not prevent the production of rope, but additions at 20% prevented the production of visible rope caused by both *Bacillus licheniformis* and *Bacillus subtilis.*

Thirty fermented wheat dough microflora samples were gathered from various Tunisian bakeries (M'hir et al., 2007). Mesophilic aerobic bacteria accounted for roughly $10^6$ cfu/g in 40% of the samples (MAB). The microbiota of these samples was dominated by yeasts and lactic acid bacteria (LAB). The differences were from $10^5$ to $10^8$ cfu/g. From microbial counts, the LAB/yeast ratio ranged from 1:1 to 200:1. More than 50% of the analyzed samples lacked *Enterococcus* sp. They had $10^2$ and $10^4$ cfu/g of contaminating micrflora, including coliforms and mesophilic bacteria. About 26% of the examined samples were judged to contain LAB and coliforms. The ratio between LAB and mesophilic Bacillus is more significant. Approximately $10^4$ was the LAB/mesophilic Bacillus ratio for 45% of the samples that underwent analysis. However, no Micrococcaceae were found in any of the samples. Discovered that sourdough fermentation with antifungal strains of *Lactobacillus plantarum* inhibits ordinary bread-spoiling fungus growth. In both the sourdough wheat bread and in vitro systems, the antifungal sourdoughs primarily inhibited the formation of *Penicillium expansum, Fusarium culmorum,* and *Aspergillus niger* spores. On wheat bread, however, there was no change in the proliferation of *Penicillium roqueforti* spores. Calcium propionate (CAP) was coupled with antifungal sourdoughs to limit the quantity of chemical additives in bread, and potential mutual effects were evaluated. Including 3000 ppm CAP in the bread did not affect the development of *Penicillium roqueforti,* but it slowed down the growth of the other fungi.

When antifungal sourdoughs and CAP were added to the bread recipe, a potent synergistic effect inhibited *Penicillium roqueforti* expansion. The effectiveness of lactic acid bacteria in preventing the principal bread contaminants, *Aspergillus, Fusarium,* and *Penicillium*, was assessed by Carla Luciana. Of 95 examined strains, only three (*Lactobacillus plantarum*, *Lactobacillus reuteri*, and *Lactobacillus brevis*) exhibited antifungal activity. Acetic and phenyl lactic acids were the main antifungal substances. Sourdough containing *Lactobacillus brevis,* and yeasts had higher fermentation quotients (FQ = 2.0) and leave volumes (80 cm) than dough without *Lactobacillus brevis*. A reduction of 50% in the content of the calcium propionate chemical propionate was made possible by the starter culture's addition of antifungal LAB strains.

## 1.8 Conclusion

The current chapter emphasizes the significance of microorganisms in the food industry, as they play a crucial role in the production of various food products. However, the presence of microorganisms also leads to food spoilage, which poses a significant challenge to the food industry. Therefore, various techniques and methods are adopted to overcome this challenge and ensure the safety and longevity of food products. One of the primary techniques used to prevent spoilage is the use of preservatives in bakery foods. These include freezing, chemical preservatives, and biopreservatives, which help to extend the shelf life of high and intermediate-moisture bread items. Despite these measures, however, mold deterioration remains a severe issue that impacts the shelf life of many bakery products, resulting in significant financial losses for baking businesses. To combat this problem, it is essential to implement strategies to prevent mold growth and extend the shelf life of baked goods. This is especially important given the growing demand for baked goods globally. These strategies include maintaining high standards of bakery hygiene, which can help to prevent the growth of harmful microorganisms. Additionally, post-packing or modified environment or heat treatment packaging may be necessary to ensure that the food remains safe from spoilage by microbes. In conclusion, while microorganisms are beneficial in the food industry, they also pose a significant risk to food safety and longevity. To ensure the safety and longevity of bakery products, it is crucial to adopt effective strategies to prevent spoilage, including the use of preservatives, maintaining bakery hygiene, and implementing post-packing or modified environment or heat treatment packaging techniques as necessary. These efforts can help to minimize financial losses for baking businesses and meet the rising demand for baked goods globally.

## References

Abellana, M., Magri, X., Sanchis, V., & Ramos, A. J. (1999). Water activity and temperature effects on growth of Eurotium amstelodami, E. chevalier and E. herbariorum on a sponge cake analogue. *International Journal of Microbiology, 52*, 97–103.

Abellana, M., Sanchis, V., & Ramos, A. J. (2001). Effect of water activity and temperature on growth of three Penicillium species and Aspergillus flavus on a sponge cake analogue. *International Journal of Food Microbiology, 71*, 151–157.

Abu-Salem, F. M., Mohamed, R., Gibriel, A., & Rasmy, N. M. (2014). Levels of some antinutritional factors in Tempeh produced from some legumes and jojobas seeds. *International Scholarly and Scientific Research & Innovation, 8*, 296–301.

Acheson, D., Bresee, J. S., Widdowson, M. A., Monroe, S. S., & Glass, R. I. (2002). Foodborne viral gastroenteritis: Challenges and opportunities. *Clinical Infectious Diseases, 35*(6), 748–753.

Adams, M. R., & Hall, C. J. (1988). Growth inhibition of food-borne pathogens by lactic and acetic acids and their mixtures. *International Journal of Food Science & Technology, 23*(3), 287–292.

Akkerman, R., Farahani, P., & Grunow, M. (2010). Quality, safety and sustainability in food distribution: A review of quantitative operations management approaches and challenges. *OR Spectrum, 32*(4), 863–904.

Arroyo, M., Aldred, D., & Magan, N. (2008). Environmental factors and weak organic acid interactions have differential effects on control of growth and ochratoxin A production by Penicillium verrucosum isolates in bread. *International Journal of Food Microbiology, 98*, 223–231.

Aso, Y., Akazan, H., & BLP Group. (1992). Prophylactic effect of a Lactobacillus casei preparation on the recurrence of superficial bladder cancer. *Urologia Internationalis, 49*, 125–129.

Attaie, R., Whalen, P. J., Shahani, K. M., & Amer, N. A. (1987). Inhibition of growth of staphylococcus aureus during production ofacidophilus yogurt. *Journal of Food Protection, 50*(3), 224–228.

Aung, M. M., & Chang, Y. S. (2014). Traceability in a food supply chain: Safety and quality perspectives. *Food Control, 39*, 172–184.

Ballongue, J., Grill, J. P., & Baratte-Euloge, P. (1993). Action sur la flore intestinale de laits fermentes au Bifidobacterium. *Le Lait, 73*, 249–256.

Bove, P., Russo, P., Capozzi, V., Gallone, A., Spano, G., & Fiocco, D. (2013). Lactobacillus plantarum passage through an oro-gastro-intestinal tract simulator: Carrier matrix effect and transcriptional analysis of genes associated to stress and probiosis. *Microbiological Research, 168*(6), 351–359.

Caplice, E., & Fitzgerald, G. F. (1999). Food fermentations: Role of microorganisms in food production and preservation. *International Journal of Food Microbiology, 50*(1–2), 131–149.

Chamberlain, N. (1993). Mold growth on cake. *Biscuit Maker and Plant Baker, 14*, 961–964.

Corthier, G., Dubos, F., & Raibaud, P. (1985). *Applied and Environmental Microbiology, 40*, 250.

De Simone, C., Clardi, A., Grassi, A., Lambert, G. S., Tantzoglou, S., Trinchieri, V., Moretti, S., & Jirillo, E. (1992). Effect of Bifidobacterium bifidum and Lactobacillus acidophilus on gut

mucosa and peripherical blood B lymphocytes. *Immunopharmacology and Immunotoxicology, 14*, 331–340.
Denariaz. (1994). Fermented milks and health benefits. *Nutrition Newsletter IDF, 141*, 32–35.
Derrien, M., & van Hylckama Vlieg, J. E. (2015). Fate, activity, and impact of ingested bacteria within the human gut microbiota. *Trends in Microbiology, 23*(6), 354–366.
Desrosier, N. W. (2006). *The technology of food preservation* (pp. 110–148). Avi Publishing.
Dewitt O. (1985). *Thesis in Medecin Paris No* 162.
Dickson, J. S. (2001). Survival of selected indicator and pathogenic bacteria in refrigerator pizzas. *Journal of Food Protection, 50*, 59–86.
Duval-Hflah, Y., Chappuis, J. P., Ducluzeau, R., & Raibaud, P. (1983). *Progress in Food & Nutrion Sciences, 129*, 213.
Filannino, P., Bai, Y., Di Cagno, R., Gobbetti, M., & Gänzle, M. G. (2015). Metabolism of phenolic compounds by Lactobacillus spp. during fermentation of cherry juice and broccoli puree. *Food Microbiology, 46*, 272–279.
Flint, J. A., Van Duynhoven, Y. T., Angulo, F. J., DeLong, S. M., Braun, P., Kirk, M., et al. (2005). Estimating the burden of acute gastroenteritis, foodborne disease, and pathogens commonly transmitted by food: An international review. *Clinical Infectious Diseases, 41*(5), 698–704.
Fox, P. F. (1993). Cheese: An overview. In P. F. Fox (Ed.), *Cheese: Chemistry, physics and microbiology - 1, 2* (pp. 1–36). Chapman and Hall.
Gram, L., Ravn, L., Rasch, M., Bruhn, J. B., Christensen, A. B., & Givskov, M. (2002). Food spoilage—Interactions between food spoilage bacteria. *International Journal of Food Microbiology, 78*(1–2), 79–97.
Guynot, M. E., Marin, S., Sanchis, V., & Ramos, A. J. (2003). Modified atmosphere packaging for prevention of mold spoilage of bakery products with different pH and water activity levels. *Journal of Food Production, 10*, 1864–1872.
Havenaar, R., Marteau, P., & Huis in't Veld, J. H. J. (1994). Survival of lactobacilli strains in a dynamic computer controlled in vitro model of the gastrointestinal tract poster session. *Lactic, 94*, 7–9.
Hickey, C. S. (1998). Sorbate spray application for protecting yeast-raised bakery products. *Baker's Digest, 54*, 4–7.
Hilton, E., Isenberg, H. D., Alperstein, P., France, K., & Borentein, M. T. (1992). Ingestion of yoghurt containing Lactobacillus acidophilus as prophylaxis for Candida vaginitis. *Annals of Internal Medicine, 116*, 353.
Huis in't Veld, J. H. J, Hose H., Schaafsma G. J., Silla H., & Smith J. E. (1989). Health aspect of food biotechnology. In *91 bis Final Seminar COST subgroup 2 "Food Biotechnology"*, October 3–6, 1989.
Jarvis, B. (2001). Mold spoilage of food. *Process Biochemistry, 7*, 11–14.
Kaila, M., Isolauri, E., Soppi, E., Virtamen, E., Laine, S., & Arvilommi, H. (1992). Enhancement of the circulating antibody secreting cell response in human diarrhea by a human Lactobacillus strain. *Pediatric Research, 32*, 141–144.
Knight, R. A., & Menlove, E. M. (2006). Effect of the bread baking process on destruction of cer tain mold spores. *Journal of the Science of Food and Agriculture, 10*, 653–660.
Korhonen, H. (1994). Bioactive compounds in fermented milks. *Nutrition Newsletter IDF, 3*(141), 10–11.
Kotz, C. M., Peterson, L. R., Moody, J. A., Saviano, D. A., & Levitt, M. D. (1990). In vitro antibacterial effect of yogurt on escherichia coli. *Digestive Diseases and Sciences, 35*, 630–637.
Kyzlink, V. (2001). *Principles of food preservation* (pp. 247–370). Elsevier.
Laatikainen, R., Koskenpato, J., Hongisto, S. M., Loponen, J., Poussa, T., Hillilä, M., & Korpela, R. (2016). Randomised clinical trial: Low-FODMAP rye bread vs. regular rye bread to relieve the symptoms of irritable bowel syndrome. *Alimentary Pharmacology & Therapeutics, 44*(5), 460–470.
Legan, J. D., & Voysey, P. A. (1981). Yeast spoilage of bakery products and ingredients. *Journal of Applied Bacteriology, 70*, 361–371.

Ling, W. H., Korpela, R., Mykkanen, H., Saiminen, S., & Hanninen, O. (1994). Lactobacillus strain GG supplementation decreases colonic hydrolytic and reductive enzyme activities in healthy female adults. *The Journal of Nutrition, 124*, 18–23.

M'hir, S., Mondher, M., & Hamd, M. (2007). Microflora distribution and species ratio of Tunisian fermented doughs for bakery industry. *African Journal of Food Microbiology, 6*(18), 2122–2129.

Malkki, Y., & Rauha, O. (2000). Mold inhibition by aerosols. *Baker's Digest, 52*, 47–50.

Marco, M. L., Heeney, D., Binda, S., Cifelli, C. J., Cotter, P. D., Foligné, B., et al. (2017). Health benefits of fermented foods: Microbiota and beyond. *Current Opinion in Biotechnology, 44*, 94–102.

Marteau, P., Pochart, P., Flourie, B., & Rambeau, J. C. (1990). Effect of chronic ingestion of fermented dairy product containing Lactobacillus acidophilus and Bifidobacterium bifidum on metabolic activities of the colonic flora of humans. *The American Journal of Clinical Nutrition, 52*, 685–688.

Mazhar, S., Yasmeen, R., Chaudhry, A., Summia, K., Hussain, I., Amjad, S., & Ali, E. (2022). Role of microorganisms in modern food industry. *International Journal of Food Science & Technology, 4*, 65–77.

Membre, J. M., Kubaczka, M., & Chene, C. (2001). Growth rate and growth-no-growth interface of Penicillium brevicompactum as functions of pH and preservative acids. *Food Microbiology, 18*, 531–538.

Mentes, O., Ercan, R., & Ajcekuj, M. (2005). Inhibitory activity of two Lactobacillus strain, isolated from sourdough, against rope-forming Bacillus strain. *Journal of Food Microbiology, 18*, 359–363.

Mishra, L. K. (2013). Role of microorganisms in food. Recent advances in microbiology, 215.

Perdigon, G., Macias, M. E. N., Albarez, S., Oliver, G., & Ruiz Holando, A. P. (1988). *Immunology, 63*, 17.

Pessione, E., & Cirrincione, S. (2016). Bioactive molecules released in food by lactic acid bacteria: Encrypted peptides and biogenic amines. *Frontiers in Microbiology, 7*, 876.

Ponte, J. G., Payne, J. D., & Ingelin, M. E. (1993). The shelf life of bakery foods. In G. Charalambous (Ed.), *Shelf life of foods and beverages* (pp. 1143–1197). Elsevier Science.

Rezac, S., Kok, C. R., Heermann, M., & Hutkins, R. (2018). Fermented foods as a dietary source of live organisms. *Frontiers in Microbiology, 9*, 1785.

Romond, M. B., Romond, C., Beerens, H., & Bourloux, P. (1989). The in vivo effect of dietary buffering capacity on thebifidogenic activity of human milk oligosaccharides. *Microbial Ecology in Health and Disease, 2*(1), 29–36.

Rosenkvist, H., & Hansen, A. (1995). Contamination profiles and characterization of bacillus species in wheat bread and raw materials for bread production. *International Journal of Food Microbiology, 26*, 353–363.

Salazar, N., Gueimonde, M., De Los Reyes-Gavilán, C. G., & Ruas-Madiedo, P. (2016). Exopolysaccharides produced by lactic acid bacteria and bifidobacteria as fermentable substrates by the intestinal microbiota. *Critical Reviews in Food Science and Nutrition, 56*(9), 1440–1453.

Samapundo, S., Deschuyfteleer, N., & Van Laere, D. (2010). Effect of NaCl reduction and replacement on the growth of fungi important to the spoilage of bread. *Journal of Food Microbiology, 27*, 749–756.

Valerio, F., Favilla, M., De Bellis, P., Sisto, A., de Candia, S., & Lavermicoca, P. (2009). Antifungal activity of strains of Lactic acid bacteria isolated from semolina ecosystem against Penicillium roqueforti, Aspergillus niger and Endomyces fibuliger contaminating bakery products. *Systematic and Applied Microbiology, 32*, 438–448.

Voysey, P. A., & Hommond, J. C. (1993). Reduced additive bread-making tech. In J. Smith (Ed.), *Technology of reduced, additive foods* (pp. 80–94). Blackie, Academic and Professional.

Vytrasova, J., Pribanova, P., & L. Mar vanova. (2002). Occurrence of Xerophilic fungi in bakery production. *International Journal of Food Microbiology, 72*, 91–96.

Yann, D., & Pauline, G. (2014). Usefulness of natural starters in food industry: The example of cheeses and bread. *Food and Nutrition Sciences, 2014*, 1679.

Yasui, H., Nagaoka, A. A., Mike, K., Hayakawa, K., & Ohwaki, M. (1992). Detection of Bifidobacterium strains that induce large quantities of logA. *Microbial Ecology in Health and Disease, 5*, 155–162.

Zhang, C., Derrien, M., Levenez, F., Brazeilles, R., Ballal, S. A., Kim, J., et al. (2016). Ecological robustness of the gut microbiota in response to ingestion of transient food-borne microbes. *The ISME Journal, 10*(9), 2235–2245.

Zoppi, G., Deganello, A., Benoni, G., Saccomani, F., & Benoni, G. (1982). Oral bacteriotherapy in clinical practice: I. The use of different preparations in infants treated with antibiotics. *European Journal of Pediatrics, 139*(1), 18–21.

# Chapter 2
# Farming Microbes for Sustainable Food Production

**Abdel Rahman Mohammad Said Al-Tawaha, Arun Karnwal, Abhijit Dey, Iftikhar Ali, and Abdel Razzaq Al-Tawaha**

**Abstract** The large number and variety of microbiomes that live on and in plants are a great way to increase crop yields and adaptability in agriculture. With improvements in biological technology, microorganisms and their metabolites have become a reliable and effective tool for reducing plant stress, increasing nutrient absorption, product yield, and pest control. However, there is a big need to improve how well and reliably microbial technology works in the real world. Because of this, there is a growing interest in making the best microbial biofertilizers and biopesticides that are safe for the environment and can be used on different soils and crop species. To make "microbe-optimized plants," crop breeding projects should take into account the positive interactions between plants and microbes. Also, microbiome engineering can be used to find possible groups of microorganisms that are best suited to help plants grow. In the end, combining crop breeding programs with microbiome engineering is the only way to greatly increase agricultural yields, get the most out of them, and solve problems with food security.

**Keywords** Biological technology · Microbial inoculant · Microbial farming · Crop breeding · Microbiome engineering

---

A. R. Mohammad Said Al-Tawaha (✉)
Department of Biological Sciences, Al Hussein Bin Talal University, Ma'an, Jordan
e-mail: Abdel-al-tawaha@ahu.edu.jo

A. Karnwal
Department of Microbiology, Lovely Professional University, Phagwara, Punjab, India

A. Dey
Department of Life Sciences, Presidency University, Kolkata, India

I. Ali
School of Life Science, Chinese University of Hong Kong, Hong Kong, Hong Kong

Centre for Plant Science and Biodiversity University of Swat, Charbagh, Pakistan

A. R. Al-Tawaha
Department of Crop Science, Faculty of Agriculture, University Putra Malaysia, Serdang, Selangor, Malaysia


A. Karnwal, A. R. Mohammad Said Al-Tawaha (eds.), *Food Microbial Sustainability*, https://doi.org/10.1007/978-981-99-4784-3_2

## 2.1 Plant Microbiome Structure and Benefits for Sustainable Agriculture

Microbiomes play a significant part in sustainable agriculture. In recent years, there has been an increasing interest in gaining a better understanding of plant microbiomes and strengthening them in order to increase plant health and production. In sustainable agriculture, the promotion of the beneficial impacts of plant microbiomes can be accomplished through the use of a number of different ways. Included in this category is the application of biocontrol agents, biofertilizers, and plant probiotics. Biofertilizers are microbial inoculants that improve plant growth and production by fixing nitrogen and solubilizing phosphorus. Biofertilizers are also known as microbial nutrient adjuvants (Santos & Olivares, 2021). Biocontrol agents are microorganisms that protect plants from infections by either creating substances that inhibit the growth of pathogens or developing a systemic resistance in the plant. Plant probiotics are consortiums of microorganisms that stimulate plant development and defend plants against the adverse effects of environmental stressors (Santos & Olivares, 2021). Trivedi et al. (2021) explore the significance of microbiomes and their potential for contributing to the development of more environmentally responsible agricultural practices. They bring to light the importance of understanding the complicated interactions that occur between plants and bacteria, which might be different from one plant species to another, as well as depending on the environment and other aspects. Trivedi et al. (2021) contend that a deeper comprehension of these interactions can pave the way for the creation of targeted microbiome-based strategies for the growth and defense of plants. As a conclusion, the study of plant microbiomes and its applications in environmentally responsible agriculture are becoming increasingly important as a result of the positive effects that plant microbiomes have on the growth and productivity of plants.

## 2.2 Biofortification with Microorganisms

Prasanna et al. (2016) describe the notion of biofortification with microorganisms, which utilizes beneficial microbes to increase the nutrient content of crops. As a method to combat malnutrition in underdeveloped nations, this method is gaining popularity. Prasanna et al. (2016) emphasize the significance of choosing microorganisms that are good at solubilizing and mobilizing nutrients such as phosphorus, iron, and zinc, and are also capable of colonizing plant roots. Using microbial inoculants can boost crop yields and improve crop nutritional quality (Prasanna et al., 2016). Prasanna et al. (2016) emphasize that there are obstacles to the broad adoption of microorganism biofortification. The need for additional research to discover the best microbes for various crops and settings is a difficulty. Another obstacle is the approach's cost-effectiveness, as microbial inoculants can be costly to create and distribute. In addition, it is necessary to guarantee the safety of microbial inoculants and address regulatory issues. In spite of these obstacles, the biofortification with microbes has the potential to significantly contribute to global food security and nutrition. Future research should focus on producing cost-effective and sustainable microbial inoculants, as well as establishing appropriate delivery routes for these inoculants, according to the authors. The authors also underline the significance of educating farmers and other stakeholders about the advantages of microbial biofortification as a tool for enhancing crop yields and nutritional quality, as well as strengthening their capacity to do so.

## 2.3 Uses of Microbes for Sustainable Food Production

### *2.3.1 Fermentation*

Fermentation is an ancient method of food preservation that has been utilized for generations. Fermented foods have become increasingly popular in recent years as a result of their possible health benefits. Şanlier et al. (2019) discovered various health benefits connected with fermented food consumption. These advantages include better gut health, higher nutritional bioavailability, and better immunological function. Fermented foods also have anti-inflammatory and anti-carcinogenic qualities. One of the most important advantages of fermented foods is their capacity to boost intestinal health. Fermentation improves digestion, food absorption, and immunological function by increasing the amount and diversity of beneficial bacteria in the gut. Fermented foods have also been demonstrated to help with lactose intolerance and irritable bowel syndrome symptoms. Fermented foods also include a lot of bioavailable elements such vitamins, minerals, and amino acids. The fermentation process converts complicated nutrients into simpler ones that the body can absorb. Another advantage of fermented foods is that they improve immunological function. Beneficial bacteria found in fermented foods can assist to boost immunity and protect against illnesses. Fermented foods have also been demonstrated to lower the risk of several diseases, including cancer, cardiovascular disease, and diabetes.

### *2.3.2 Bread and Alcoholic Drinks*

Lactic acid bacteria (LAB) are a varied collection of microorganisms that are typically found in fermented foods and the human gastrointestinal tract. Several metabolites produced by these bacteria have been proven to have health-promoting benefits. Tang et al. (2023) have provided an in-depth analysis of the metabolites produced by LAB and their role in modifying gut microbiota. These metabolites have a variety of activities in the gut, from preventing harmful bacteria to stimulating good bacteria. Knowing the biosynthesis and activities of these metabolites can inform the creation of probiotics and prebiotics for enhancing gut health. To completely comprehend the mechanisms underlying these effects and to discover the appropriate amounts and sources of LAB metabolites for various individuals, additional research is required.

During the process of alcoholic fermentation, yeasts convert carbohydrates into ethanol, carbon dioxide, and other metabolic byproducts. These metabolic byproducts contribute to the fermented foodstuffs' chemical composition as well as its sensory qualities. Fermentation of alcoholic ingredients is the fundamental process behind the production of alcoholic beverages like wine and beer. The regulation of the fermentation process is typically regarded as an essential step in arriving at a conclusion regarding the product's overall level of quality. In this context, monitoring fermentation is an increasing necessity, which asks for technologies that are quick, low-cost, and nondestructive, giving real-time or online information in order to guarantee an efficient level of control throughout the entire process.

### *2.3.3 Cheese and Yogurt*

Microorganisms from the surrounding environment, such as bacteria, yeasts, molds, viruses, and bacteriophages, have the potential to make their way into the dairy supply chain at a number of different points and to have an impact on the chain as a whole. Microorganisms are the most important agents in the essential stages of cheese and dairy product production, which are responsible for maximizing the overall quality and safety, flavor, appearance, and typicality of the product. Milk fermentation and various biochemical events during manufacture and maturation are caused by microorganisms, giving cheese its distinctive textures and flavors. Microorganisms' metabolism and diverse enzymatic activities contribute significantly to the development of organoleptic qualities during the transformation of milk, during the manufacturing and maturation of cheeses. By acting as a barrier and producing several low molecular weight antibacterial chemicals, microorganisms also contribute to cheese's microbiological safety. A milk product called yogurt is fermented by bacterial culture. Yogurt can be made using any type of milk, although cow's milk is the one that is most frequently used (Dave & Shah, 1997). Lactose, the sugar contained in milk, is fermented by bacteria including Streptococcus salivarius, Streptococcus thermophiles, and Lactobacillus bulgaricus to produce lactic acid. LAB, which stands for "lactic acid bacteria," is the name given to this particular bacterial group (Oyeleke, 2009). Lactic acid is created as a waste product as the bacteria break down the lactose. Lactic acid causes the casein protein in milk to curdle, forming a solid mass known as curd. Yogurt gets its distinctive flavor and gelatinous consistency from the conversion of lactose into lactic acid (Sandoval-Castilla et al., 2004).

One benefit of the increased acidity in yogurt brought on by lactic acid generation is that it inhibits the development of potentially hazardous bacteria. Both pasteurized and unpasteurized milk can be used to make yogurt, which can then be fully fermented using a combination of two or more bacterial cultures. Yogurt can be flavored, sweetened, or have fruit added to it after it has fermented (Chollet et al., 2013).

### 2.3.4 *Biofertilizers*

Several studies have found that using biological fertilizers not only improves soil quality but also increases crop output. Biological fertilizers are less expensive than chemical fertilizers, making them a more cost-effective solution. Biofertilizers are especially useful in organic farming because they can transform insoluble phosphorus into forms that plants can use. They are critical in conserving soil fertility and ensuring the long-term viability of farming methods. Over reliance on chemical fertilizers is not a long-term strategy because it is costly in terms of local resources and foreign cash, including the creation and upkeep of fertilizer factories. Also, the environmental impact of chemical fertilizers should be addressed while deciding whether to utilize them. Biofertilizers, as an alternative, are a feasible supply of nutrients that can meet crop requirements. Beneficial bacteria such as Azotobacter, Azospirillum, Rhizobium, and Mycorrhizae are found in biofertilizers and play an important role in crop productivity. Plants can better tolerate adverse environmental circumstances by using biofertilizers. Overall, using biofertilizers is a more environmentally friendly and cost-effective alternative to chemical fertilizers, with the added benefit of enhancing soil quality and crop productivity. Many investigations have been conducted with the goal of optimizing the generation of biofertilizer through the cooperation of microorganisms and the relationships they form. On the basis of their characteristics and functions, they can be classified into the following categories.

#### 2.3.4.1 Rhizobium

Rhizobia and legumes have a complicated connection in which they exchange nod-gene-inducing signal molecules as well as lipo-chito-oligosaccharide nodulation components to form nodules. Through this process, both the rhizobia and their legume hosts are capable of synthesizing and releasing various phytohormones, such as IAA, lumichrome, riboflavin, lipo-chito-oligosaccharide Nod factors, rhizobitoxine, gibberellins, jasmonates, brassinosteroids, ethylene, cytokinins, and the enzyme 1-aminocyclopropane-1-carboxylate (ACC) deaminase, that can directly or indirectly stimulate plant growth. These features may aid plants in adapting to edapho-climatic difficulties such as nutrient scarcity, but understanding their mechanisms is required to optimize their benefits. Some N2-fixing rhizobia have been discovered to stimulate plant growth in the rhizosphere by solubilizing soil-bound phosphorus via the manufacture of gluconic acid under the direction of pyrroloquinoline quinone (PQQ) genes. Some rhizobia are known to synthesize and release siderophores for plant iron nutrition, to chelate heavy metals in contaminated soils, and to act as disease biocontrol agents. The ACC deaminase generated by rhizobia can also lessen the detrimental effects of ethylene on plant development, hence increasing growth. Plant mineral nutrition, insect pest and disease resistance, and other biological outcomes can all be influenced by symbiotic rhizobia, but understanding the processes involved is critical for maximizing the benefits of these compounds. Rather than investigating individual rhizobial or plant metabolites, it is critical to comprehend their synergistic impacts and how they can aid plant development and production.

#### 2.3.4.2 Arbuscular Mycorrhizal Fungi

Stresses from the environment can hurt plant growth and yield. Human actions, like using too much fertilizer and pesticides, and climate change have made these problems worse and hurt the environment. Because of this, we need environmentally friendly and sustainable ways to grow crops, like using arbuscular mycorrhizal fungi (AMF) biofertilizers. When AMF inoculation is used, it not only helps plants grow and produce more, but it also protects them from things like high temperatures, salt, drought, and heavy metals. AMF may also stop metabolic pathways from shutting down and make the host plant more resistant to bad conditions. As natural root symbionts, AMF give their host plants essential inorganic nutrients that help them grow and produce more. Using the biofertilizer AMF can help plants adapt to changing environmental conditions and improve their overall quality and productivity. Therefore, it is very important to do a full study of how AMF affects plant growth at different stages of development. This review aims to give up-to-date information about AMF, such as their benefits, how they can be used, and how important it is that AMF and nutrients work together to help crops grow better.

#### 2.3.4.3 Azotobacter

Concerns have been raised about the usage of chemical fertilizers to maintain the health of soil and plants because of the harmful impact that these fertilizers have on both the environment and human health. The usage of biofertilizers has been proposed as a more environmentally friendly alternative to conventional chemical fertilizers. A microbe known as Azotobacter, which has the ability to fix atmospheric nitrogen and stimulate the growth of plants, has demonstrated promise as a candidate for use in biofertilizer. The gram-negative bacterium known as Azotobacter can frequently be discovered in soil. It can change the nitrogen in the air into a form that plants can use, and it can also produce compounds that promote growth, including as vitamins, amino acids, and enzymes, which are beneficial to the expansion and maturation of plants. Sumbul et al. (2020) explored the possible advantages of utilizing Azotobacter as a biofertilizer to improve the health of soil and plants. The incorporation of Azotobacter into soil can improve its fertility, structure, and ability to retain water, which eventually results in a higher crop production. In addition, the usage of Azotobacter as a biofertilizer might lessen a farm's reliance on chemical fertilizers, which are known to have unfavorable outcomes for the surrounding ecosystem. The research also shed insight on the difficulties and restrictions associated with utilizing Azotobacter as a biofertilizer. They include the fact that its efficacy shifts depending on the soil and environment circumstances, that it engages in competition with other microbes, and that it may be responsible for the transfer of plant infections. In conclusion, the research carried out by Sumbul et al. (2020) offers a detailed analysis of the possible applications of Azotobacter as a biofertilizer for the management of the soil's and plants' overall health. It places a strong emphasis on the significance of sustainable agriculture as well as the possible advantages of using bi-fertilizers. On the other hand, additional study is required so that the effectiveness of Azotobacter can be maximized in a variety of circumstances. In general, the findings of this analysis shed light on the potential of biofertilizers as a more environmentally friendly approach to the management of soil and plant health.

#### 2.3.4.4 Cyanobacteria

Research conducted by Mishra and Pabbi (2004) looked into the viability of cyanobacteria as a biofertilizer for the cultivation of rice. Cyanobacteria are photosynthesis-based microbes that have the ability to remove nitrogen from the air and convert it into a form that plants can use. They are also responsible for the production of cytokinins, auxins, and gibberellins, which are chemicals that stimulate plant growth. The research highlighted the significance of rice as a staple food crop as well as the difficulties that are associated with its cultivation. These difficulties include the high demand for nitrogen and phosphate fertilizers, the lack of available water, and the environmental degradation that is caused by the use of chemical fertilizers. Mishra and Pabbi (2004) presented a comprehensive analysis of the many positive aspects associated with the growing of rice utilizing cyanobacteria as a form of biological fertilizer. It is possible for the introduction of cyanobacteria to raise the fertility of the soil, boost plant development and crop yield, and increase the amount of nutrients that are taken up by plants. It is possible to lessen dependency on chemical fertilizers and mitigate the negative impact that they have on the environment through the use of biofertilizers such as cyanobacteria. In conclusion, the research that was conducted by Mishra and Pabbi reveals important information regarding the feasible application of cyanobacteria as a biofertilizer in the cultivation of rice. The necessity of environmentally responsible farming practices and the possible advantages of utilizing biofertilizers are highlighted in this study. More study is required to determine how to maximize the efficiency of cyanobacteria in a variety of different environmental settings. In general, the findings of the study highlight the potential of biofertilizers as a more environmentally responsible approach to agriculture and as a means of ensuring enough food supplies.

## 2.4 Photosynthetic Microorganisms for Food Applications

In recent years, a significant amount of study has been conducted on the production and exploitation of photosynthetic microorganisms since these organisms present a prospective source of food that is both environmentally friendly and nutritionally sound. Biotechnology research on photosynthetic microorganisms for food and food-related applications is gaining traction due to a variety of causes such as increased consumer acceptance, sustainability, desire for eco-friendly food sources, and global economic concerns. Furthermore, non-toxic types of photosynthetic bacteria not only provide a sustainable source of nutrients but may also be helpful to human health. Recent research has shown that microalgae can be used as a source of protein and fatty acids. Lee et al. (2022) discovered that using the bacterial inoculant dramatically boosted tomato plant yield and quality, including size, color, and flavor. Furthermore, the injection altered the structure of the soil's bacterial community, favoring the growth of helpful microbes while suppressing the growth of detrimental ones. On the other hand, Lee et al. (2022) discovered that the bacterial inoculant increased tomato plant photosynthetic activity, as shown by higher chlorophyll content and photosynthetic rates. Microbial inoculants have the potential to boost agricultural productivity and sustainability while lowering dependency on chemical inputs by increasing the growth of beneficial microbes and enhancing plant photosynthesis (Lee et al. (2022).

## 2.5 Conclusion

Agriculture can greatly benefit from the huge number and variety of microbiomes found on and within plants. Microorganisms and their metabolites have developed into a dependable and effective tool for managing plant stress, improving nutrient uptake and output, and warding off pests as a result of advances in biological technology. The effectiveness and dependability of microbial technology, however, must be significantly enhanced. As a result, there is a push to develop eco-friendly biofertilizers and biopesticides that work well across a variety of soil types and crop types. Crop breeding programs should consider plant–microbe interactions for the creation of "microbe-optimized plants." Microbiome engineering can also be utilized to identify potential communities of beneficial microbes for plant growth. Only by integrating crop breeding efforts with microbiome engineering will food security issues be addressed and agricultural outputs maximized.

## References

Chollet, M., Gille, D., Schmid, A., Walther, B., & Piccinali, P. (2013). Acceptance of sugar reduction in flavored yogurt. *Journal of Dairy Science, 96*, 5501–5511.

Dave, R. I., & Shah, N. P. (1997). Viability of yoghurt and probiotic bacteria in yoghurts made from commercial starter cultures. *International Dairy Journal, 7*, 31–41.

Lee, S.-K., Chiang, M.-S., Hseu, Z.-Y., Kuo, C.-H., & Liu, C.-T. (2022). A photosynthetic bacterial inoculant exerts beneficial effects on the yield and quality of tomato and affects bacterial community structure in an organic field. *Frontiers in Microbiology, 13*, 959080. https://doi.org/10.3389/fmicb.2022.959080

Mishra, U., & Pabbi, S. (2004). Cyanobacteria: A potential biofertilizer for rice. *Resonance, 9*, 6–10.

Oyeleke, S. B. (2009). Microbial assessment of some commercially prepared yoghurt retailed in Minna, Niger state. *African Journal of Microbiology Research, 3*, 245–248.

Prasanna, R., Nain, L., Rana, A., & Shivay, Y. S. (2016). Biofortification with microorganisms: Present status and future challenges. In *Biofortification of food crops* (pp. 249–262). https://doi.org/10.1007/978-81-322-2716-8_19

Sandoval-Castilla, O., Lobato-Calleros, C., Aguirre-Mandujano, E., & Vernon-Carter, E. J. (2004). Microstructure and texture of yogurt as influenced by fat replacers. *International Dairy Journal, 14*, 151–159.

Şanlier, N., Gökcen, B. B., & Sezgin, A. C. (2019). Health benefits of fermented foods. *Critical Reviews in Food Science and Nutrition, 59*(3), 506–527.

Santos, L. F., & Olivares, F. L. (2021). Plant microbiome structure and benefits for sustainable agriculture. *Current Plant Biology, 26*, 100198. https://doi.org/10.1016/j.cpb.2021.100198

Sumbul, A., Ansari, R. A., Rizvi, R., & Mahmood, I. (2020). Azotobacter: A potential bio-fertilizer for soil and plant health management. *Saudi Journal of Biological Sciences, 27*(12), 3634–3640. Epub 2020 Aug 8. PMID: 33304174; PMCID: PMC7714982. https://doi.org/10.1016/j.sjbs.2020.08.004

Tang, H., Huang, W., & Yao, Y. F. (2023). The metabolites of lactic acid bacteria: Classification, biosynthesis and modulation of gut microbiota. *Microbial Cell, 10*(3), 49–62.

Trivedi, P., Mattupalli, C., Eversole, K., & Leach, J. E. (2021). Enabling sustainable agriculture through understanding and enhancement of microbiomes. Transley review. *The New Phytologist, 230*, 2129–2147. https://doi.org/10.1111/nph.17319

# Chapter 3
# Role of Microbes in Sustainable Food Preservation

**Hale İnci Öztürk and Sebnem Ozturkoglu-Budak**

**Abstract** Biopreservation is identified as the use of beneficial microorganisms or their naturally occurring metabolites to protect the food safety and to extend the shelf-life of food products via inhibiting pathogenic and spoilage microbial activity. Majorly, lactic acid bacteria (LAB) have been defined in terms of the mentioned duties as bioprotective cultures and/or biopreservation agents. However, other than LAB some other microbes also have a crucial role in the sustainability of food preservation. Biopreservatives are important to prevent the use of synthetic additives such as antimicrobial and antioxidant compounds, which have potential risks to consumer health. In recent years, corresponding to the awareness of consumers, there is a tendency toward natural foods that do not contain synthetic chemicals. Therefore, the discovery of potential new antimicrobial or antioxidant agents has become important in food industry. This chapter introduces potential microbial species as bioprotective cultures as well as their metabolites via their functionalities in terms of obtaining sustainable food preservation. Regarding this, subjects on the importance of food preservation, the role of microbial cultures in food preservation, the metabolites produced by cultures, the possibility of using culture metabolites and paraprobiotics in food preservation, and the application methods of bioprotective agents to foods are presented and discussed in this chapter.

**Keywords** Food safety · Lactic acid bacteria · Biopreservation · Protective culture · Culture metabolites · Antimicrobial agents · Probiotics · Paraprobiotics

---

H. İ. Öztürk
Konya Food and Agriculture University, Konya, Turkey

S. Ozturkoglu-Budak (✉)
Ankara University, Ankara, Turkey
e-mail: budak@ankara.edu.tr



A. Karnwal, A. R. Mohammad Said Al-Tawaha (eds.), *Food Microbial Sustainability*, https://doi.org/10.1007/978-981-99-4784-3_3

## 3.1 Introduction

Safe food is characterized as food that has physicochemical and microbiological properties suitable for consumption, does not constitute toxicological properties, is not harmful to human health, and provides beneficial nutritional components to the consumer. Therefore, food science and technology deals with sustainable food quality and food safety issues together. In this context, the focus points are to protect or improve the nutritional, sensory, and microbiological quality of a food product during its production or storage period. In addition, it has become a necessity to increase the shelf-life of foods and to prevent quality losses in this storage life. These situations have led to the emergence of the concept of food preservation, which aims to prevent or minimize physical, chemical, and microbial changes that may occur during production and storage, and to ensure food safety. To ensure sustainable food preservation, controlling the water content in foods, applying chemicals, developing different packaging techniques, thermal treatments, innovative and emerging non-thermal practices, and using biopreservatives are among the food preservation techniques (Horita et al., 2018; Michailidis & Krokida, 2015; Rifna et al., 2019).

Today, various synthetic food additives are used in the production of industrial foods in addition to applications such as heat treatment to extend the shelf-life and improve product quality. Among these synthetic food additives, antimicrobials, antioxidants, sweeteners, flavors, flavor enhancers, emulsifiers, and stabilizers are widely used in industrial production (Carocho et al., 2014). Unfortunately, the use of these synthetic chemicals above the limits specified in standards or the potential to accumulate in the human body over time jeopardizes food safety and human health (Kamal & Fawzia, 2018). Moreover, with the increasing consumer awareness, industrial products containing chemical additives have been avoided in recent years. In this regard, studies in recent years have focused on the development of natural additives from plant or microbial sources and their evaluation in various food products to reduce or eliminate the use of chemical additives (Aziz & Karboune, 2018; Bouarab Chibane et al., 2019; Khubber et al., 2022; Varsha & Nampoothiri, 2016). Additives from microbial origin come to the fore, since microorganisms can be developed in a shorter time and more easily compared to plants. Although a great number of studies using live cultures directly in foods have continued (Agriopoulou et al., 2020; Rodgers, 2008), the use of postbiotics, which are cellular metabolites, and paraprobiotics, which are dead bacterial cells, have gained popularity in recent studies (Barros et al., 2020; İncili et al., 2021, 2022; Sawada et al., 2016). Both concepts can be used for both probiotic and non-probiotic microorganisms. The fact that paraprobiotics are more stable and safer than the live microorganisms from which they are obtained makes them advantageous.

Paraprobiotics are obtained by inactivating microorganisms with thermal or emerging non-thermal technologies (de Almada et al., 2016). Various studies have shown that paraprobiotics have antioxidant, antimicrobial, and even immune system-strengthening effects (Aydın et al., 2021; Geraldo et al., 2020; Li et al., 2021a; Taverniti & Guglielmetti, 2011). In other respects, metabolites produced by microorganisms such as cell-free supernatants, cell wall components such as exopolysaccharides, vitamins, phenolic derived components, organic acids, bioactive peptides, short-chain fatty acids, and aromatic amino acids are considered as postbiotics (Marhuenda-Muñoz et al., 2019; Mayorgas et al., 2021). In addition to their potential antioxidant and antimicrobial properties (Mahdhi et al., 2017; Sornsenee et al., 2021), it has been determined that these postbiotics have also very important effects (such as anti-inflammatory, anticarcinogenic, antiallergenic, antidiabetic, and immunomodulatory activities) on human health (Bourebaba et al., 2022; Dahech et al., 2011; Riaz Rajoka et al., 2019). In short, postbiotics and paraprobiotics have the potential to be alternative natural and innovative food additives. Therefore, the evaluation of these components in the food industry will not only provide an alternative to chemical additives by extending the shelf-life of foods but also provide consumers with additional health advantages beyond meeting their nutritional needs due to the functional properties of these natural biopreservatives. Herewith, a step will also be taken to protect food safety.

This chapter reveals the protective roles of microorganisms and their effects on sustainable food preservation. Although there are emerging applications, the biological preservation of foods with living microorganisms, microbial metabolites, or killed microbial cells has been rapidly adopted both in the field of food science and technology and in health and nutrition. From this point of view, this chapter aims to provide the basic and important information necessary to develop various biopreservation strategies involving innovative applications. To that end, this chapter presents knowledge on the biological preservation of foods, the different usage patterns of microorganisms as biopreservatives, the protective roles of microbial cells or metabolites, procedures for applying bioprotective agents to foods, and developed analytical methods for the production and application of biological agents and for their control in the final product.

## 3.2 Food Biopreservation

Legislations around the world restrict the level of use of currently approved preservatives and their use in different foods (Smith, 2003). In addition, with the increasing consumer awareness, the demand for preservative-free, high-quality, long shelf-life, and minimally processed foods has also increased. All these situations have led to the need for innovative approaches to food preservation. Numerous studies have been conducted for many years on obtaining and evaluating natural plant-derived preservatives to restrict or eliminate the use of synthetic additives and still continue to be done (Nazari et al., 2019; Smith-Palmer et al., 2001; Thielmann et al., 2017; Tkaczewska et al., 2022). On the other hand, recently, the concept of biological preservation of foods as a new approach has come to the fore to increase the shelf-life of foods, protect their sensory properties, and control food safety. Stiles (1996) defined biological preservation as both increasing the storage time of foods and ensuring food safety with the natural microbiota and their antimicrobial metabolites in foods. In a sense, it was initially considered that the microbiota of foods plays a biological role in food preservation. Accordingly, it can be said that the current concept of biological preservation is inspired by the ancestral food fermentation used for food preservation. It is well known that fermented foods have a long shelf-life since organic acids, hydrogen peroxide, volatile compounds, and bacteriocins produced by microorganisms during fermentation inhibit the growth of spoilage and pathogenic microorganisms (Gálvez et al., 2010). However, due to reasons such as consumer preferences and the desired final characteristics of the product, it is not possible to preserve every food by fermentation. For this, adding the aforementioned antimicrobial components that can be produced during fermentation directly to the product during or after the process will also be a protective approach.

Today, the concept of biopreservation includes the application of preservative cultures with a long history of safe use, or a wide variety of their metabolites, to foods in different ways to preserve the quality characteristics of the product, increase the shelf-life, and ensure food safety. The use of preservative cultures can help to obtain the desired technological and organoleptic properties in the final product (Viana de Souza & Silva Dias, 2017) as well as being an alternative to chemical additives. LAB are the main focus in the biological preservation of foods. Indeed, their fermentation abilities, their recognition as safe, and their presence as a natural microbiota in foods have made LAB the most studied bacterial group (García et al., 2010). Moreover, protective strains of *Carnobacterium maltaromaticum*, *Carnobacterium divergens,* and *Hafnia alvei* species are commercially available and used in the food industry as an alternative to LAB (*Lactococcus* and *Lactobacillus* species) (Borges et al., 2022). However, there is a limited number of commercial products in terms of microbial metabolites, thus it will be important to take an attempt in this area for the food industry. Currently, bacteriocin nisin, which is approved for use in most countries, is commercially available as a microbial metabolite for food preservation (Shin et al., 2016). From the view of food safety and food preservation, there is a need for future studies to reveal the Generally Recognized as Safe (GRAS) status of different microbial metabolites as food biopreservatives, to develop production technologies, to determine their activity spectrum, and to make them suitable for commercial production.

As mentioned before, limiting the use of chemical substances in foods and increasing human awareness are among the main factors that pave the way for the emergence of biological protection methods. Increasing the shelf-life of foods with these applications is also important from an economic point of view. In addition, exposure to high levels of chemical additives through food has been reported to cause various disorders such as allergies, hyperactivity, asthma, liver and kidney dysfunction, behavioral disorders, reproductive disorders, nervous system disorders, toxic effects, and cancer (Kamal & Fawzia, 2018). Therefore, biological preservation of food is seen as a potential application in terms of eliminating these effects and increasing food safety. As a result, the purpose of biopreservation of foods is summarized in Fig. 3.1.

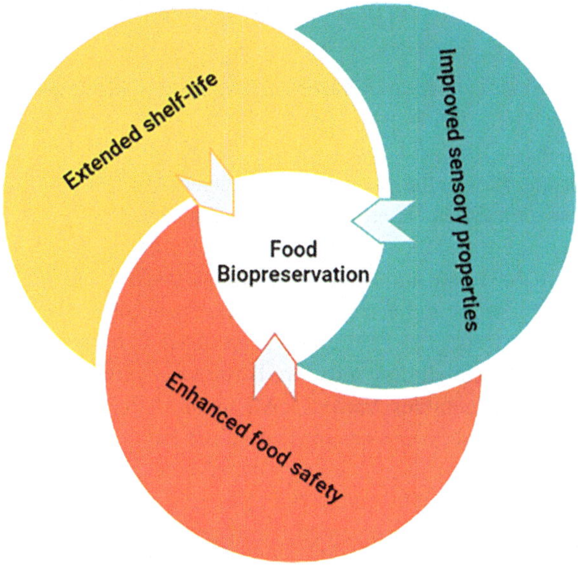


**Fig. 3.1** The goals of food biopreservation

## 3.3 Protective Cultures

Numerous microorganisms can have potential use as protective cultures. Protective cultures generally consist of bacteria with GRAS status, and they reduce or prevent the growth of spoilage and pathogenic microorganisms. Although lactic acid bacteria have a big concern (Varsha & Nampoothiri, 2016), some other species can be used for this regard (Ran et al., 2022). In this context, the other species that have been studied extensively have been different strains of *Propionibacterium* (*P.*) *freudenreichii* (Ran et al., 2022; Tangyu et al., 2022). Especially in these studies, it has been shown that the use of protective strains of this species together with lactic cultures increases their protective effects. Therefore, LAB plays an important role in the biopreservation of foods due to its enhancing effect on the activities of other protective species and their direct usage as protective cultures in food processing. In this chapter, we aimed to provide information about the LAB containing the commonly used species for preventive approaches. In addition, some species other than LAB are given in the probiotic section. Since the protective activity is targeted regardless of the direct and indirect effects of microorganisms, we also included paraprobiotics, which refer to killed bacterial cells, under the title of protective cultures. We believe that the paraprobiotic concept is a very promising application for the future.

### 3.3.1 Lactic Acid Bacteria as Food Biopreservatives

Lactic acid bacteria are a group of bacteria consisting of Gram-positive, non-spore-forming, cocci or rod-shaped *Lactobacillus*, *Streptococcus*, *Lactococcus*, *Pediococcus*, *Leuconostoc*, *Enterococcus* genera (Khalid, 2011). This group of bacteria has important roles in the preservation of fermented foods and the development of their sensory properties. They are used as bioprotective cultures to extend the shelf-life and ensure the safety of many foods (Castellano et al., 2008). These protective effects are due to the antimicrobial components they produce such as organic acids (mainly lactic acid and acetic acid), bacteriocins, fatty acids, enzymes, and hydrogen peroxide (Reis et al., 2012). LAB are used as biopreservatives in a wide range of food products, including meat products, dairy products, marinades, bakery products, fruits, and vegetables. Various LAB species/strains used as protective cultures in different product groups are presented in Table 3.1. In addition, it has been suggested in some studies that higher protection can be achieved by using combined LAB strains instead of a single culture (Essia Ngang et al., 2015).

Various LAB cultures are commercially available (Varsha & Nampoothiri, 2016). FreshQ (Chr.Hansen), Holdbac (Dupont Danisco), and Lyofast are used for the fermented milk products, while SafePro ImPorous (Chr.Hansen) and Bactoferm (Chr. Hansen) are used for the meat products. Recent studies have focused on evaluating the protective effects of different LAB strains or new isolates (Ibrahim et al., 2021; Salomskiene et al., 2019). Ran et al. (2022) reported that 15 different LAB strains inhibited the growth of *Aspergillus niger*, *Aspergillus flavus*, and *Penicillium crustosum* by producing acetic acid and lactic acid at a large rate. The growth of *Clostridium* spp. is kept under control by adding nitrate in fermented meat products. Di Gioia et al. (2016) followed the growth of *Clostridium* spp. in pork mince using two *Lactobacillus* (*Lb.*) strains as a preservative culture instead of nitrate. They have suggested that lactic cultures consisting of *Lb. plantarum* PCS20 and *Lb. delbrueckii* DSM 20074 can be used as protective cultures, thus reducing the use of nitrate, which can pose a health risk, in products. In another study, it was reported that bacteriocin-producing LAB could be used as a solution to control the number of listeria in fresh cheeses during cold storage conditions (Coelho et al., 2014). It has also been claimed that the use of bacteriocin-producing lactococcal strains in raw sausages is a protective strategy against *Listeria monocytogenes* (Benkerroum et al., 2003).

Dairy products provide a unique growth environment for many opportunistic spoilage and pathogenic microorganisms. In industrial productions, *Staphylococcus aureus*, *Listeria monocytogenes*, *Salmonella* spp., and *Escherichia coli* are among the most detected pathogens in dairy products, mostly due to the use of raw milk, contaminations, or improper processing of milk (Cancino-Padilla et al., 2017). It has been reported that pathogen growth in dairy products can be controlled by the application of LAB consisting of *Lb. plantarum*, *Lb. helveticus*, *Lb. reuteri*, *Lb. rhamnosus*, *Lc. lactis*, and *Enterococcus faecium* species alone or in combination (Al-Gamal et al., 2019). Among these LABs, the most potential effect was detected

**Table 3.1** Protective lactic cultures for different food fields and their potential implications

| Species/strain | Product | Function | References |
|---|---|---|---|
| *Lb. plantarum* L9 | Bakery product | Antifungal | Ran et al. (2022) |
| *Lb. plantarum* PCS20 *Lb. delbrueckii* DSM 20074 | Fermented pork meat | Anticlostridial | Di Gioia et al. (2016) |
| *Lactococcus* (*Lc.*) *lactis* L3A21M1 | Fresh cheese | Antilisterial | Coelho et al. (2014) |
| *Lb. helveticus* CNRZ 32 | White cheeses | Antibacterial | Al-Gamal et al. (2019) |
| *Pediococcus* (*P.*) spp. A19 *P.* spp. A21, *Lb. plantarum* B4496, *Lb. brevis* 207, *Lb. sanfranciscensis* BB12 (5/13) | Fermented vegetables | Antifungal | Essia Ngang et al. (2015) |
| *Lb. plantarum* (commercial Holdbac YM-XPM, YM-XPK) | Fermented dairy products | Antifungal | Leyva Salas et al. (2017) |
| *Lc. lactis* subsp. *lactis* | Poultry products | Antimicrobial against *Staphylococcus aureus* | Akbar and Anal (2014) |
| *Streptomyces natalensis* (Commercial Natamax) | Fruit juices, wine, dry-ripened food, dairy, bakery products | Antifungal | Leyva Salas et al. (2017) |
| *Lb. sakei* | Meat products | Antimicrobial activity against *Staphylococcus aureus*, *listeria monocytogenes*, and *salmonella* typhimurium | Zagorec and Champomier-Vergès (2017) |
| *Lb. strains* CIT3 and V7B3 | Fruit products | Antimicrobial | Siroli et al. (2015) |
| *Lc. lactis* M | Raw sausages | Antilisterial | Benkerroum et al. (2003) |

especially in *Lb. helveticus*, with ≥95% reduction in the number of undesirable microorganisms.

The ability of the secretion of antimicrobial compounds of LAB strains offers the potential to extend the shelf-life of tomato paste and reduce the microbial load (Awojobi et al., 2016). Moreover, *Pediococcus* and *Lactobacillus* species isolated from fermented cocoa can be used in the biological control of ochratoxin A-producing molds in cocoa production (Essia Ngang et al., 2015). As for minimally processed foods, *Lb. plantarum* CIT3 and V7B3 strains have been suggested to increase shelf-life without causing any sensory defects (Silva et al., 2018). Besides, the application of these bioprotective cultures with other natural preservatives such as thyme EO resulted in a 1-week longer shelf-life. Therefore, strategies of using different strains in combination and taking advantage of the synergistic effects of other natural antimicrobials can be applied to increase the bioprotective effects of LAB.

### 3.3.2 Probiotics as Food Biopreservatives

Probiotics are the beneficial bacteria present in fermented food products. Dairy commodities like yogurt are enriched with these bacteria. The major class of probiotic bacteria is LAB, which includes its further species and metabolites. Their use as food biopreservative is increased due to their GRAS status and minimum interaction with the gut microbiome. Bacterial metabolites include mainly antimicrobial/bioactive peptides and bacteriocins. Shehata et al. (2019) stated that the metabolites from a novel *Lactobacillus* spp. RM1 strain have exceptional antifungal properties. The GC-MS analysis of the CFS of RM1 strain showed components that exhibits vast antimicrobial and antioxidant properties. The treatment (tenfold concentrated CFU) on wheat grains showed complete inhibition of *Aspergillus parasiticus.* The strain also inhibits aflatoxin $B_1$ and ochratoxin A when used in the concentration of 15 mg/mL.

Altieri et al. (2005) stated the role of potential probiotic *Bifidobacterium* (*B.*) *bifidum* as biopreservative. It was observed that dipping fillets of fish plaices in a mixture thymol essential oil and *B. bifidum* significantly reduces the spoilage capacity under cold storage. They also evaluated the effects of different packing conditions on storage time. The obtained results revealed the synergistic effects of applications of modified atmosphere packaging and vacuum packaging on the activity of the biopreservative culture, both during storage at 4 °C and 12 °C. Therefore, the use of hurdle technology, in which different protection strategies are combined, is seen as an effective application to increase the effectiveness of biopreservatives.

It is obvious that a high or a certain number of living probiotic cells must be present in the product to benefit from their protective effects. Related to this, microencapsulation technology has been a popular application to obtain high bacterial viable cell counts in the final product. Motalebi Moghanjougi et al. (2020) have described microencapsulation of probiotic bacteria as an innovative approach to enhance the shelf-life of white-brined cheeses. Bioactive cellulose film with encapsulated probiotics, *Lb. acidophilus* and *B. animalis*, provided antifungal properties to the brined cheeses. Moreover, the shelf-life studies showed sodium alginate microencapsulated *Lb acidophilus* drastically decreased the *Aspergillus niger* colonies on the cheese surface. In addition, it has been demonstrated that the use of encapsulated probiotic microorganisms in foods with low pH is a successful application due to the damage to the viability of probiotics in foods with high acidity (Pourjafar et al., 2020). Accordingly, to benefit from their protective properties, the addition of probiotics into the food matrix with proper pre-applications depending on the intrinsic factors of the relevant food is crucial for sustainable food preservation.

Apart from *Lactobacillus* species, some other bacterial and yeasts strains are also regarded as potential biopreservatives. Different microbial probiotic species and their functionalities determined in studies are given in Table 3.2. Radiati et al. (2022) studied the antimicrobial and antioxidant activities of probiotic *Saccharomyces cerevisiae* in kefir made from goat milk. The usage of this probiotic yeast in kefir not only distinctly inhibited pathogenic *Escherichia* (*E.*) *coli*, *Klebsiella* spp., and *Salmonella* Typhimurium, but also contributed toward high antioxidant activity and the production of enzyme ß-galactosidase. de Almeida Godoy et al. (2022) have reported that the combined incorporation of probiotic *Saccharomyces boulardii* 17 and *Lb. paracasei* DTA83 metabolites increases the shelf-life of raw chicken sausages or semi-processed chicken products, provided that the cold chain is maintained. In this application, it has been claimed that the antimicrobial properties are concentration dependent. Some non-saccharomyces yeasts were also identified to have potential anti-salmonella properties along with antibiofilm characteristics. Ceugniez et al. (2017) showed that *Kluyveromyces*

**Table 3.2** Some probiotic species and their determined functions in different food products

| Probiotic species | Product | Function | References |
|---|---|---|---|
| *Lb. plantarum* S61 | Juice and purees | Activity against *Penicillium digitatum* and *Candida pelliculosa* | Abouloifa et al. (2021) |
| LAB strains (*Lb. casei* MYSRD 108 and *Lb. plantarum* MYSRD 71) and their cell free supernatants | Laboratory pathogens | Antimicrobial activity against *Salmonella* spp. and *Salmonella* typhimurium. Reduction in biofilm formation of described pathogens | Divyashree et al. (2021) |
| Lyophilized LAB strains (*Lb. acidophilus* LA5, *Lb. rhamnosus*, and *Lb. casei* 01) | Marination liquid for meat | High antioxidant activity, antimicrobial activity (against 11 pathogens including *E. coli* and *Listeria monocytogenes*) | Gargi and Sengun (2021) |
| Postbiotic mixture of *Saccharomyces boulardii* 17 and *Lb. paracasei* DTA83 | Chicken products (Sausages and ready-to-cook products) | Production of biocides, complete inhibition of contaminants by adding 3% concentration, extension of shelf-life of products | de Almeida Godoy et al. (2022) |
| *Saccharomyces cerevisiae* and *Saccharomyces boulardii* | Fermented juices of black carrots, radish, and red beet | Highly percentage of bioactive and phenolic compounds (significant quantity of vanillic acid in black carrot), increased bio-accessibility of antioxidants | Değirmencioğlu et al. (2016) |
| *Bacillus coagulans* BDU3 | Fermented fish product | Low molecular weight bacteriocin (presence of novel bacteriocin), temperature and stability against various enzymes, inhibitory activity against *Bacillus cereus* and *Enterococcus* spp | Abdhul et al. (2015) |

*marxianus* and *Kluyveromyces lactis* presented high inhibitory activity against various *Salmonella* strains.

Probiotic *Bacillus coagulans* strains are tolerant to extreme environmental conditions due to their spore-forming properties (Cao et al., 2020). These features make them the most preferred probiotic species for foods produced or prepared with difficult processes such as high temperature treatments, drying, tempering, pressing, etc. *Bacillus coagulans* BDU3 strain producing bacteriocin was reported to have antimicrobial properties on a broad spectrum of foodborne pathogens (Abdhul et al., 2015). Recently, many products (teas, chocolate bars, snacks, etc.) containing various strains of *Bacillus coagulans* have started to take place on the market shelves. There is an industrial trend in the use of this probiotic species in various product formulations. The use of strains of *Bacillus coagulans* in food products produced with harsh processes seems more advantageous compared to probiotic LAB.

## 3.4 Paraprobiotics as Food Biopreservatives

The term paraprobiotic was first proposed by Taverniti and Guglielmetti (2011). Paraprobiotics are referred to as inactivated or ghost cells, providing benefits to consumers when taken in sufficient quantities. They can be dead cells of both probiotic and non-probiotic microorganisms (de Almada et al., 2016). The general properties of paraprobiotics are presented in Fig. 3.2. Because paraprobiotics are

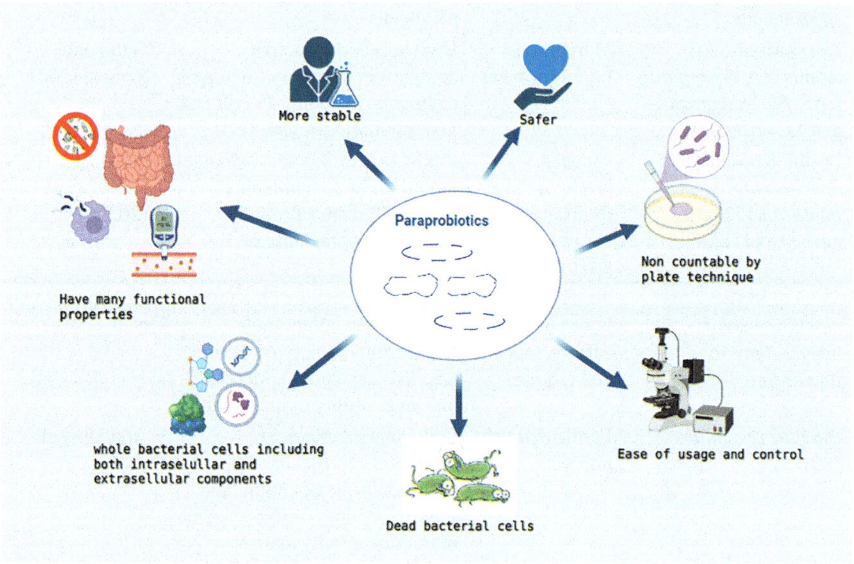


**Fig. 3.2** Main characteristics of paraprobiotics

not viable, they cannot be grown or counted by culture-dependent methods. Paraprobiotics are obtained when microorganisms completely lose their viability depending on changing cell structures such as DNA fragments, membrane structures and enzyme mechanisms by subjecting them to various stress factors. Technologies used to obtain paraprobiotics focus on reducing the loss of viability in the microorganism completely or substantially. In this context, thermal and innovative non-thermal technologies are used to destroy cell viability (Cuevas-González et al., 2020). Although many studies are generally done on inactivation by heat treatment, non-thermal innovative technologies such as high pressure, radiation, ultrasound, exposure to oxidative stress, pulsed electric field, and atmospheric pressure plasma are used in the inactivation of cells (de Almada et al., 2016; Salminen et al., 2021).

Although the effects of probiotics on health have been demonstrated by many studies, there are some concerns due to the fact that probiotics are living cells (Didari et al., 2014). Recently, it has been revealed that the use of probiotics in high-risk groups (older adults, hospitalized patients, cancer patients) poses a safety problem. In particular, bacteremia, sepsis, and endocarditis problems have been reported to be experienced (Boumis et al., 2018; Koyama et al., 2019; Sendil et al., 2020). In addition, it has been also reported that problems such as strain-specific behavior, unknown molecular mechanisms, and horizontal antibiotic gene transfer may occur (Suez et al., 2019). These problems led to the emergence of the paraprobiotic concept. Paraprobiotics are inactivated microorganisms and therefore the hazards foreseen for living probiotic cells are out of the question. Paraprobiotics are more stable and safer for food applications than the live microorganisms from which they are derived.

Although various studies have revealed many functional properties of paraprobiotics such as modulating the immune system, inhibiting pathogens and regulating the intestinal microbiota, improving diarrhea, alleviating lactose tolerance, and stopping the development of some types of cancer (de Almada et al., 2016), there is no study available on its use in foods as a bioprotective agent. Ishikawa et al. (2010) demonstrated on various animal models that paraprobiotics show antimicrobial activity as a result of reducing the adherence of pathogens to intestinal epithelial cells. Paraprobiotics can act directly and indirectly. It has been suggested that paraprobiotics may show antioxidant and antimicrobial properties depending on their interactions with various molecules or receptors (Cuevas-González et al., 2020; de Almada et al., 2016). With in situ experiments of paraprobiotics, it is very important to define their antimicrobial properties in food and to reveal their potential mechanisms in terms of food preservation.

## 3.5 Culture Metabolites as Bioprotective Agent

Culture metabolites, called postbiotics, are among the bioprotective agents with the potential for application in foods. Postbiotics are metabolites that are secreted from live bacteria cells or released after the lysis of bacteria. Postbiotics with a wide variety and different functions are produced by microorganisms. Postbiotics are not only metabolites produced by probiotic bacteria, but also metabolites produced by LAB can be grouped under the title of postbiotic. It has been reported that the effects of postbiotics vary between species or strains due to various mechanisms (Thorakkattu et al., 2022). These metabolic products produced by microorganisms can be secreted out of the cell or released after cell lysis. In the selection of postbiotics that can be used as biopreservatives, the criteria are to have GRAS status, to be stable during the production, storage, and distribution of foods, not to adversely affect sensory properties, to prevent the growth of spoilage and pathogenic microorganisms in foods, and not to cause the formation of any substance that may harm humans (Ben Said et al., 2019). Metabolites produced by microorganisms and important in food preservation are described in this section under different groups.

### *3.5.1 Antimicrobial Agents*

In the scope of biopreservation, LAB have a crucial role due to their diverse metabolite productions with antimicrobial activity. These compounds include low molecular weight substances including diacetyl, fatty acids, reuterin, reutericyclin, antifungal compounds such as phenyl lactate, propionate, hydroxyphenyl lactate as well as hydrogen peroxide, carbon dioxide, lactic acid, acetic acid, and bacteriocins (Castellano et al., 2017). In addition, the fact that LAB creates a competitive situation for nutrients in the environment can also be included in their antimicrobial activity. Due to their antimicrobial properties, LAB emerges as bioprotective agents.

Organic acids are one of the main products of LAB after fermentation. The main acids produced by LAB are lactic acid, acetic acid, propionic acid, and formic acid, as well as some other acids depending on the strain (Punia Bangar et al., 2022). These acids, after penetrating through the membrane of target microorganisms, lower the cytoplasmic pH and stop cell metabolic activities depending on the decreasing pH. It has been reported that 0.5% (v/v) lactic acid concentration inhibits the growth of pathogens such as *Escherichia coli*, *Listeria monocytogenes*, and *Salmonella* spp. (Wang et al., 2015). Acetic acid has been shown to have good antibacterial properties against Gram-negative microorganisms, especially *Pseudomonas aeruginosa*, even at very low concentrations of 0.166% (Fraise et al., 2013). Formic acid is used in poultry feed to limit the growth of *Salmonella* spp. (Ricke et al., 2020). Salomskiene et al. (2019) have reported that the organic acid production capabilities of LAB differ. According to their results, *Lb. helveticus* R showed the highest antimicrobial activity by producing the highest amount of lactic acid while

*Lc. lactis* 140/2 achieved this activity with various organic acids produced as complexes. LAB also produces short-chain fatty acids (SCFA) as a final product after the fermentation of dietary carbohydrates and dietary fibers. SCFAs such as isovaleric, isobutyric, and butyric have been suggested to have antimicrobial properties (Huang et al., 2011).

Some LAB strains synthesize antimicrobial peptides known as bacteriocins in their ribosomes. Most bacteriocins act on foodborne pathogens and spoilage bacteria, the main ones being nisin, diploxin, acidophilin, plantaricin, bulgaricin, helveticin, lactacin, and pediocin (Zimina et al., 2020). Bacteriocins are generally ineffective against Gram-negative bacteria as they cannot penetrate the peptidoglycan layer due to their cell wall. Nisin, a bacteriocin produced by *Lc. lactis*, is the only bacteriocin approved for use in more than 50 countries by the U.S. Food and Drug Administration (Juturu & Wu, 2018). Numerous studies have been conducted on bacteriocins produced by LAB in the last 25 years (De Vuyst & Leroy, 2007; Mokoena, 2017), however, their applications in the industry did not achieve equal success. The effect of food composition on the efficacy of bacteriocins is among the reasons that may explain the limited industrial applications of bacteriocins.

It has been reported that glycoside hydrolase β-1,3-glucanase, which is commonly found in plants, fungi, and bacteria, has antifungal properties in fermentation technology (Wu et al., 2018). This enzyme acts by hydrolyzing β-glucans, which are sugars found in the cell walls of microorganisms. It has been suggested that β-glucanase can be used as a preservative against yeasts that cause spoilage in wine (Enrique et al., 2010). Considering its mechanism of action, β-glucanase can be used as a control strategy for spoilage yeasts and molds rather than bacteria.

The production of hydrogen peroxide ($H_2O_2$) by the LAB can also be useful in food preservation. It has been reported that LAB, which synthesizes $H_2O_2$, inhibits the growth of pathogenic and psychrophilic microorganisms at refrigerated temperatures (Reis et al., 2012). LAB produce $H_2O_2$ by oxidizing lactic acid. The $H_2O_2$ production capabilities of the LAB may vary from species to species. Adesokan et al. (2010) reported that *Leuconostoc* (*Leu.*) *mesenteroides* produced higher amounts of $H_2O_2$ than *Lb. brevis* and *Lb. plantarum*. They found that $H_2O_2$ from *Leu. mesenteroides* showed high antimicrobial activity against *Pseudomonas aeruginosa*.

Lipopeptides can also be evaluated among antimicrobial components. Bacterial lipopeptides contain a hydrophobic fatty acid chain combined with a hydrophilic polypeptide (7–10 amino acid residues). Lipopeptides produced by *Bacillus* species have been shown to have antibacterial activity against *Listeria monocytogenes* and *Bacillus cereus* as well as its antifungal properties (Perez et al., 2017). Lipopeptides synthesized by *Bacillus* species interact with the microbial cell membrane and form a pore, thereby destabilizing the membrane structure and exerting an antimicrobial effect (Zhang et al., 2022). Various volatile compounds produced by bacteria also act as antimicrobials. However, since these volatile components are discussed under a separate sub-title, antimicrobial volatile components are not explained in this section.

### 3.5.2 Antioxidant Agents

A wide variety of reactive oxygen species can occur in the human body and food systems due to oxidation. Oxidative damage plays an important pathological role in human diseases. In addition, oxidative reactions occurring in foods cause significant losses in sensory properties such as taste, odor, and color of the product (Franklin et al., 2017). Therefore, in terms of both human health and increased shelf-life, oxidative reactions can be prevented by taking antioxidant supplements or adding antioxidants to foods. Various synthetic and natural antioxidants have been reported (Pokorný, 2007; Viana da Silva et al., 2021). Antioxidants derived from natural sources are preferred because of the safety concerns and long-term adverse effects of synthetic antioxidants on health. In this context, in addition to herbal natural antioxidants, microbial antioxidants are also encountered. Some microbial cell metabolites can act as antioxidants. Their antioxidant properties are related to the defense mechanism of these postbiotics against destructive reaction oxygen species.

Catalase, peroxide dismutase, and glutathione peroxidase are considered as antioxidant microbial enzymes (Aghebati-Maleki et al., 2021). These enzymes suppress reactive oxygen species. In case of targeting the use of these enzymes as antioxidants, the pH values of the foods should be taken into account to ensure enzymatic activity. Sanders et al. (1995) demonstrated that *Lc. lactis* showed antioxidative superoxide dismutase enzyme activity as a stress response. Moreover, it has been demonstrated by various studies that exopolysaccharides (EPS) produced by bacteria also have antioxidant properties. Bomfim et al. (2020) associated the antioxidant property of an EPS produced by *Lb. plantarum* CNPC003 with its ability to bind iron ions. Liu et al. (2010) suggested that EPS acts as an electron donor and reacts with free radicals, transforming them into more stable products. Mahdhi et al. (2017) have reported that EPS produced by *Lb. plantarum* inhibits linoleic acid peroxidation.

Peptides, proteins, glutathione, pyrrole, and cyclic compounds produced by microbial strains can also be considered among antioxidant molecules (Chang et al., 2021). Bioactive peptides are specific peptide sequences with various functional properties, including antioxidant properties, depending on their amino acid sequence and three-dimensional structure (Romero-Luna et al., 2022). It is well known that bioactive peptides are present in postbiotic fractions after fermentation (especially in dairy products) with lactic cultures. LAB, which contains cell envelope-associated proteinases, causes the breakdown of caseins during milk fermentation to form antioxidant peptides (Brown et al., 2017). Aguilar-Toalá et al. (2019) reported that the proteins and glutathione in the postbiotic of *Lb. casei* CRL-431 may be responsible for its antioxidant activity. Chang et al. (2021) examined the component profile of the *Lb. plantarum* postbiotic by GC-MS and reported that the existing antioxidant activity might be due to pyrrole and cyclic compounds.

Ou et al. (2009) investigated the antioxidant activity of intact cells and intracellular extracts of *B. longum* and *Lb. bulgaricus*. The obtaining results showed that both treatments showed high DPPH radical binding activity (70.4–75.1%), inhibited liposome peroxidation by 25–31%, and significantly reduced malondialdehyde formation. Ou et al. (2006) determined that *Streptococcus* (*S.*) *thermophilus* ATCC 19258 and *Lb. bulgaricus* ATCC 11842 strains showed antioxidant effects against hydrogen peroxides that cause toxicity. Amaretti et al. (2013) investigated the antioxidant properties of different LAB with various methods and determined that the antioxidant properties were strain specific. Aguilar-Toalá et al. (2019), who examined antioxidant properties of the intracellular content and of the cell wall of different LAB, reported that the intracellular composition had a higher antioxidant content than the cell wall fraction. The variations in the observed activity are due to the change of active ingredients. Accordingly, the selection of the appropriate component, production method, and strain will be essential to benefit from effective and high level of antioxidant activity for food preservation.

### *3.5.3 Volatile Compounds Synthesized by Starter LAB (SLAB) and Non-Starter LAB (NSLAB)*

During the fermentation, LAB (starter and/or non-starter) degrade the food components especially proteins, peptides, fats, and lactose present in milk and hence, the breakdown of these macromolecules not only give textural properties to products but also cause expression of aromatic and volatile compounds such as aldehydes, ketones, aromatic hydrocarbons, esters, and alcohols that are specific to the culture added and the type of products (Gallegos et al., 2017). These volatiles also take role in food biopreservation in terms of their inhibiting effect on some spoilage, contaminating, and/or unwanted microorganisms in food structure. This output is a very important point in terms of improving food quality and increasing the shelf-life. Therefore, the volatile compounds they produce play an important role in biopreservation due to their antimicrobial activities.

Homofermentation occurred by the homofermentative LAB consists of the metabolism of lactose to almost pure lactic acid. However, during heterofermentation, ethyl alcohol, carbon dioxide, hydrogen peroxide, diacetyl, acetoin, acetic aldehyde, etc. are formed apart from lactic acid. Diacetyl (2,3-butanedione) produced as a metabolic by-product of heterofermentative LAB was reported to inhibit both Gram-positive and Gram-negative bacteria such as *Bacillus cereus*, *Staphylococcus aureus*, *Escherichia coli*, *Salmonella anatum*, *Listeria monocytogenes*, *Yersinia*, and *Aeromonas* spp. (Singh, 2018). However, due to the buttery aroma of this compound and high concentration requirement for food preservation, its usage as a food preservative has been limited (Šušković et al., 2010). Similarly, acetaldehyde, produced by heterofermentative LAB, has also been demonstrated to have an antibacterial effect against various pathogenic bacteria (Singh, 2018). Yogurt is a fermented product produced by the activity of two bacterial species of *S. thermophilus* and *Lb. bulgaricus*. Dan et al. (2017) discussed the contribution of both bacteria, separately and in combination, on the volatile profile of yogurt.

According to above study, a total of 53 volatile compounds were produced by *S. thermophilus*, 43 by *Lb. bulgaricus* and 32 were produced by a mixture of both. Among them the high ratios were found for acetaldehyde.

In the study of Sgarbi et al. (2013), the characteristics of two NSLAB cultures as *Lb. casei* and *Lb. rhamnosus* were discussed. The SPME-GC analysis of lysed cell medium revealed the presence of 6,6 dimethyl-undecane (an alkane) in high percentage in head space. Additionally, a variety of ketones were observed which included 4-methyl-2-heptanone, 2-heptanone. The most probable reason for the presence of these volatile compounds is the fatty acid metabolism of NSLAB cultures, in which the fatty acids such as butyric acid also showed an inhibition effect on some microorganisms such as yeasts and molds (Huang et al., 2011).

In terms of LAB, all LAB species show certain impact on the commodity, but some are deemed to produce specific compounds. This ability is also affected by the presence of other non-starter bacteria and/or yeasts. Álvarez-Martín et al. (2008) determined the highest concentrations of compounds produced by various lactic species in the presence of yeasts in UHT skimmed milk. According to the observations, *Lc. lactis* was responsible for producing high levels of diacetyl, *Leu. citreum* produced most ethanol, while *Lc. lactis* produced highest acetaldehyde compared to others. In another study by Guarrasi et al. (2017), the presence of volatile compounds was determined in semi-hard cheese called Caciocsvallo palermitano. For such cheese type, the contribution of started cultures along with nine non-starter cultures was studied. Collectively 43 volatile compounds were detected with the help of GC-MS analysis, including ketones, aldehydes, esters, alcohols which have inhibiting effect on unwanted microorganisms. *P. pentosaceus*, as an NSLAB, contributed toward the highest percentage of ketones while LAB cultures like *Lb. rhamnosus* and *Lb. casei* contributed toward production of 2,3 butanedione or diacetyl. Among the alcohols, 2-butanol was the most produced alcohol. In terms of esters, while they were present in lower concentrations, but they contributed more in terms of flavor development.

During the ripening stages of dairy products such as cheese, the inhibitory effect of LAB/NSLAB keep increasing through the different metabolites having antimicrobial role. A study performed by Li et al. (2021b) showed the distinct change and development of volatile compounds in traditional cheese called as Kazakh cheese. This flavor production was observed within different time frames of 20-, 40-, and 60-days of storage, to study the activity of three NSLAB species including *Lb. rhamnosus, S. epidermidis*, and *P. acidilactici*. After 20 days of ripening, main volatile components present in the cheese were esters including ethyl caprylate, alkanes including n-pentadecane, and alcohols. After 40 days of cheese aging, aldehydes including nonanal, alcohols including benzyl alcohol, and acids including capric acid and caprylic acid were among the most dominating volatile compounds. Additionally, 60 days maturation inclined the volatile components of cheese to as much as 46. Hence, it was evident that above mentioned NSLAB species modify the volatile profile of Kazakh cheese. In another similar study by Battelli et al. (2019) the effect of non-starter culture including *Lb. casei* and *Enterococcus* (*E.*) *lactis* on the goat cheese was evaluated after ripening time of 30 and 60 days. During the

30 days, it was observed that the most dominant volatile compounds were acetic acid, 2-methylpropanoic acid, and 3-methylbutanoic acid. As the ripening time was extended to 60 days, ketones were dominated in cheese made with *E. lactis* culture inoculation while volatile compounds like esters, alcohols like butan-2-one, dimethyl disulfide, propanoic acid and phenols were dominated in cheese having inoculated *Lb. casei* culture.

### *3.5.4 Novel Compounds*

Innovation in technological aspects of food sciences radicalizes the plethora of chemical products that are used to extent the shelf-life of food commodities. Perishable food products especially dairy is prone to spoilage hence, the probability of spoilage is increased in these products. Preservatives are chemical or natural compounds that preserve the quality of food products by acting as antimicrobials, antifungal, and antioxidant compounds. The use of chemical preservatives in high concentrations may lead to serious health hazards in humans as pathogens sometimes develop resistance to chemicals components. Hence, the increased awareness in population has led manufacturers to divert toward using natural preservatives for food preservation. These are known as food biopreservatives. Various compounds are obtained from beneficial plants and their essential oils (Dhankhar et al., 2021), probiotic bacteria, and their metabolites etc.

Peptides are naturally occurring compounds the have shown their activity against spoilage bacteria. Among them bacteriocins have gained special importance due to their antibacterial properties. Liu et al. (2016) identified a novel bacteriocin named as Plantaricin Q7 from *Lactobacillus plantarum* Q7. Plantaricin Q7 showed broad range antimicrobial activity against various Gram-positive as well as, Gram-negative pathogen bacteria including *E. Coli, Pseudomonas* spp. and some *Shigella* spp. Another similar study by Ismael et al. (2023), identified another novel bacteriocin as Plantaricin Bio-LP1 with unique peptide sequence. This bacteriocin works best at temperature of 4 °C and at pH 6.0 hence, could be used as a potential biopreservative agent in foods. Studies on the identification of new bacteriocins and the determination of their effect spectra are still ongoing. Some compounds studied as potential food biopreservatives in recent years were given in Table 3.3 with their determined functional activities.

Among the biocontrol methods, bacteriophage applications are also among the innovative methods. Therefore, bacteriophages can also be recommended as natural food preservatives. *Listeria monocytogenes* targeted phage cocktails have been approved for food applications (Bai et al., 2016). Recent studies have focused on the characterization of various phages for different pathogenic species and their application in food (Dewanggana et al., 2022; Waturangi et al., 2021). Bacteriophages have potential use in the food industry as biocontrol agents. However, phage applications have some limitations in food applications due to the possibility of the targeted pathogenic bacteria to develop phage resistance. Therefore, the use of cocktail

**Table 3.3** Some novel biopreservative agents used in food products

| Compounds | Source | Activity | References |
|---|---|---|---|
| AMP (antimicrobial peptides) As-CATH4, As-CATH5, Hc-CATH | Marine and reptile | Antimicrobial activity against various spoilage bacteria (broad spectrum), antibiofilm properties, shelf-life enhancement | Zhang et al. (2021) |
| Nanoliposomes with phenolic compounds | Pistachio green hulls | Absence of *S. aureus* and sharp decline in Enterobacteriaceae | Rafiee et al. (2018) |
| Spermine analogous compounds | *Lb. harbinensis* K.V9.3.1Np from milk | Antifungal activity against yeasts and molds | Mosbah et al. (2018) |
| Bacteriophages | Beef intestine, beef lung, chicken intestine, chicken skin | Antimicrobial activity against enteropathogenic *E. coli* | Waturangi et al. (2021) |

phages is the first solution considered. High acidity in foods also inactivates phages, therefore the use of encapsulated phages can be offered as a solution (Waturangi et al., 2021). As other solution methods, development of the low pH resistant phages or increasing application dose of bacteriophage can be given (Gildea et al., 2022).

Gene interactions are thought to have important effects on metabolites produced by microorganisms. Therefore, this may lead to the formation of new bioprotective agents. It has been demonstrated that gene interactions between microorganisms lead to the formation of new antimicrobial compounds or produce a higher rate of an existing antimicrobial compound (Tyc et al., 2017). Regarding this, the existing microbiota of foods has an important effect on providing the desired protective properties, especially in the case of using live protective cultures from biopreservation methods. This can be a strengthening effect as well as an alleviating effect. Protective cultures should be selected with this in mind. In future studies, the relationship of protective cultures in different food product groups should be examined with transcriptomic and metabolomic studies. Especially for products produced using starter cultures, the relationship between starter cultures and protective cultures should be clarified.

## 3.6 Applications of Bioprotective Agents (Cultures or Culture Metabolites) in Food Systems

Techniques for applying bioprotective agents to foods are shown in Fig. 3.3. The application of bioprotective agents to foods can be evaluated in two categories as in situ and ex situ. In situ application is involved in the direct use of living cells, creating a competitive environment, or producing protective metabolites in the product. For this purpose, commercially available protective culture preparations containing different species are directly used in the formulations of fermented products (mostly

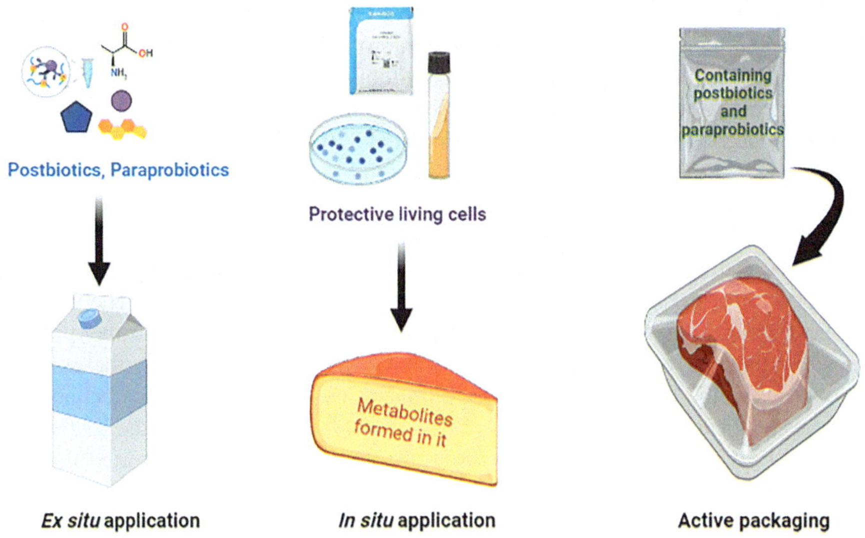


**Fig. 3.3** Application techniques of bioprotective agents to foods

dairy and meat products) (Varsha & Nampoothiri, 2016). In addition to adding preservative cultures to food products during manufacture, they can also be added to final products in a variety of ways, such as by spraying, dipping or surface coating (Ben Said et al., 2019). Spanu et al. (2017) reported that the application of *Carnobacterium* spp. protective culture to Ricotta cheese by spraying reduced the growth of *Pseudomonas* spp. In another study examining yogurt, cheese, and sour cream models, antifungal properties were observed in products produced by using protective LAB cultures directly in their production (Leyva Salas et al., 2018). They also reported that the antifungal cultures used in this study did not affect the post-acidification of the products. Siroli et al. (2015) determined that the application of protective *Lb. plantarum* CIT3 and *Lb. plantarum* V7B3 cultures by immersion method increased the shelf-life of minimally processed foods.

Ex situ technique, on the other hand, involves adding the purified agent to the food as a food preservative. However, in situ applications show cost advantages compared to ex situ ones (R. H. Perez et al., 2015). Indeed, ex situ techniques come to the fore in the use of postbiotics and paraprobiotics as bioprotective agents. Postbiotics and paraprobiotics can be applied to foods similarly to treatments (dipping, spraying, direct addition, etc.) of living protective cells. Arrioja-Bretón et al. (2020) added cell-free supernatants of *Lb. plantarum* to the meat marination liquid. They reported that this application improved the sensory properties of fresh beef and increased its shelf-life. Moradi et al. (2019) supplemented cell-free supernatants of various LAB species directly into milk and ground meat at a rate of 10–45 mg/mL-g and determined the highest antimicrobial effect against *Listeria monocytogenes* in postbiotics of *Lb. salivarus*. İncili et al. (2021) applied the supernatants of *P. acidilactici* to chicken breast fillets by dipping method (3 min). They reported that the shelf-life of the product was increased up to 12 days in this application. In other respects, investigations on paraprobiotic applications in foods are not frequently encountered in the literature. Although it has been reported by many researchers that it can be promising application, studies mostly focused on the functional properties of paraprobiotics (Kumar et al., 2021; Lee et al., 2022). There is a need to reveal the roles of paraprobiotics in food preservation with in situ applications. For this purpose, similar to postbiotic supplementation ways, paraprobiotics can also be integrated into foods and their effects can be evaluated.

It can also benefit from active packaging systems in the use of postbiotics and paraprobiotics in the biopreservation of foods (Hosseini et al., 2021). The stability of postbiotics in the food matrix is substantial for the antimicrobial or other properties that their provide. Therefore, the instability of postbiotics due to the intrinsic properties of the food matrix can be eliminated with active packaging technology. Studies on the use of bacteriocins, peptides, and organic acids among the postbiotics in food packaging have been carried out. Contessa et al. (2021) reported that the microbiological stability of cream cheese was enhanced by packaging with a bioplastic film containing *Lb. sakei* bacteriocin and chitosan/agar-agar. This active ingredient used in packaging decreased the number of coagulase positive staphylococci by 2.62 log CFU/g. Nithya et al. (2013) studied the development and application of active film for food packaging by purification of antimicrobial peptide produced by *Bacillus licheniformis* Me1. In this study, in which low-density polyethylene (LDPE) and cellulose materials were used as packaging materials, the number of *Listeria monocytogenes* Scott A decreased by 2.00 log CFU/g in cheese samples. In addition, these researchers claimed that the film material also affected the antimicrobial activity. The use of this antimicrobial peptide with LDPE had a bacteriostatic effect while its use with cellulose provide a bactericidal effect. Ressutte et al. (2022) evaluated the effect of coating with citric acid-containing chitosan film on the shelf-life of ripened cheeses. This active packaging application improved the sensory properties of ripened cheeses and there was a significant decrease in the number of aerobic mesophilic bacteria. However, there is a need for studies evaluating paraprobiotics in active packaging. In addition to the active packaging, edible coating seems also to be advantageous in the application of bioprotective agents. Although most of the current studies in this area focus on the use of living cells in edible films (Pech-Canul et al., 2020; Shigematsu et al., 2018), applying postbiotics and paraprobiotics in this way will be a supportive approach to food preservation.

## 3.7 Analytical Methods for Controlling of Biopreservative Agents

Live cultures, their metabolites, non-living cultures, and even bacteriophages can be used as biocontrol agents. It is important in terms of food technology to know the production, application, and control methods to benefit from their effects. These conditions may be different for each biological agent. However, the availability of bioprotective agents in the desired concentration and in the form that will show the desired activity is the priority of food preservation and safety.

Although preservative cultures are sold commercially under various trademarks (Varsha & Nampoothiri, 2016), current studies have focused on investigating the protective effects of isolated lactic cultures as the natural microbiota of foods (Hossain et al., 2020; Janashia et al., 2018; Sirichokchatchawan et al., 2018). In these studies, the potential antimicrobial or bioprotective effects of the isolates were investigated after confirmation of lactic acid bacteria isolated by culture-dependent methods with 16S rRNA analyses. As a result of these studies, there are isolates suggested as potential protective cultures. Either the direct use of commercial protective cultures or the use of isolated cultures with confirmed protective properties is included in the bioprotection approach. Commercially available preservative cultures are designed for bulk production. However, isolates need to go through certain preliminary stages such as being activated, concentrated, and lyophilized for their application in the food industry. Therefore, future studies need to focus on making the isolates suitable for industrial production.

To benefit from the protective effect, the presence of certain living bacterial cells in the final product is essential (Berninger et al., 2018). In this way, a competitive environment is provided for spoilage or pathogenic microorganisms. Moreover, the number of viable cells is an important issue, especially for bioprotective cultures used to produce an antimicrobial metabolite in situ (Silva et al., 2018). The number of viable protective cells in the final product can be controlled by the plate count or spectrophotometric methods after production or during the storage period (Hazan et al., 2012). In particular, the plate technique is officially used to assess cell viability in foods (Barros et al., 2020). These techniques are considered as culture-dependent techniques, and microbial enumerations can also be performed with non-culture-dependent methods. Culture independent enumeration methods include real-time (RT) quantitative polymerase chain reaction (PCR) and matrix-assisted laser desorption/ionization time-of-flight (MALDI-TOF) mass spectrometry, which are generally performed based on the presence of nucleic acids (Vinderola et al., 2019).

Among bioprotective agents, postbiotics, which are microbial metabolites, and paraprobiotics, which are non-viable cultures, have a more potential application area because they are more stable and safe as compared to living cells (Barros et al., 2020). Furthermore, it is an advantage for these agents that the viability factor is not important after the production of foods. Therefore, technologies for obtaining paraprobiotics and postbiotics focus entirely on the loss of cell viability. Various technologies used for this purpose are presented in Fig. 3.4. The mixtures obtained with these technologies may contain living cells, inactivated intact cells, and metabolites, depending on the applied processes and conditions. Hence, additional steps such as microfiltration are needed to separate postbiotics (Cuevas-González et al., 2020). On the other hand, when it is aimed to obtain postbiotics by using live cells directly after certain incubation conditions., metabolites in the supernatant are recovered by centrifugation or filtration of the medium. In a study examining the effectiveness of postbiotics produced on a laboratory or industrial scale, it was reported that the production scale did not change their protective effects (Dunand et al., 2019). The more important point here is to design the production process to maintain their stability for long periods of time. At this point, it is a necessity to ensure that production is scalable by choosing or developing cost-effective methods.

Numerous analytical techniques are available for the determination of changes in the cell structures of paraprobiotics and the identification of metabolites. The choice of appropriate techniques for their characterization depends on the intended purpose and both the qualitative and quantitative properties of the substance under study (Cuevas-González et al., 2020). For postbiotics, techniques that allow the identification of the focused metabolites and the identification of new metabolites should be used, while it would be more appropriate to use microscopic techniques in which cell morphologies can be examined in the characterization of paraprobiotics. In general, characterization of paraprobiotics and postbiotics can be performed by analytical methods such as gas and liquid chromatography or flow cytometry combined with mass spectrometry (de Almada et al., 2016; Shenderov et al., 2020). Intracellular and extracellular metabolomic analyzes are important tools in the identification of postbiotics. In these methods, chromatographic methods are generally preferred for the characterization of postbiotics. Azami et al. (2022) determined the chemical composition of the supernatants of *Lb. casei* PTCC 1608 by gas

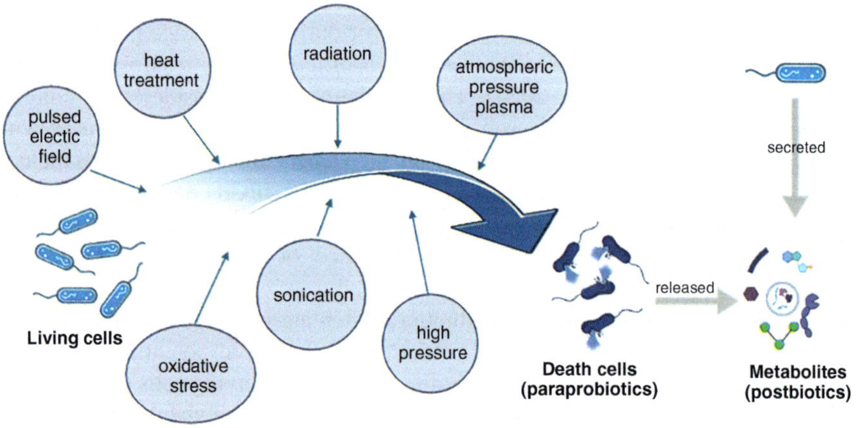


**Fig. 3.4** Current technologies used to kill bacteria in the production of paraprobiotics and postbiotics

chromatography-mass spectrometry (GC-MS) and liquid chromatography-mass spectrometry (LC-MS). They identified 18 postbiotics with GC-MS, and also identified new postbiotics consisting of succinic acid and propionic acid by LC-MS. Kim et al. (2011) analyzed the cell wall components of *Lb plantarum* K8 with MALDI-TOF mass spectrometry and determined that lipoteichoic acid can be responsible for detected anti-inflammatory properties.

Since the non-living cell structure needs to be preserved in the concept of paraprobiotics, analyzing the physiological state of bacterial cells with flow cytometry and determining bacterial viability with fluorescent microscopy using different dyes are important approaches in the characterization of paraprobiotics (Cuevas-González et al., 2020). In addition, paraprobiotics can be examined with immunofluorescence, immunoelectron, and scanning electron microscopy (Montanaro et al., 2015). Depending on the processes applied, the way bacteria die can also be analyzed with these techniques. Because bacteria can be inactivated by various mechanisms without membrane degradation. Ananta and Knorr (2009) have reported that high-pressure application irreversibly damages the membrane structure while temperature application does not always cause membrane deterioration. Chu-Chyn Ou et al. (2011) determined that the cell surfaces of lactic acid bacteria killed by heat treatment by field emission scanning electron microscopy (FE-SEM) were rougher than those of living cells. They even observed that this change increased considerably as the applied temperature increased. Accordingly, the protective effects can be examined by evaluating the effect of different treatments on paraprobiotics with the mentioned analytical techniques. In this way, it will be possible to choose the most suitable inactivation method to benefit from the desired feature.

Phages also have the potential to be used as bioprotective agents. However, providing phage stability depending on intrinsic and extrinsic properties is an important concern. Extrinsic factors such as visible and UV rays and intrinsic factors such as water activity, pH, and osmotic shock affect phage stability (Ramos-Vivas et al., 2021). For this reason, various strategies should be used in food applications of phages as bioagents. With respect to this, studies on the use of encapsulated phages by freeze-drying technology have been initiated (Petsong et al., 2021). Characterization of phages is mostly carried out using transmission electron microscopy (TEM) and PCR techniques (Akhwale et al., 2019). Specific phages or phage cocktails can be applied directly to foods (Guenther et al., 2012). The phage counts applied or found in foods are controlled by double-layer agar plate techniques. Therefore, based on the results obtained from these analyzes, application strategies can be developed for the relevant phage. Currently, there is still a need for innovative analytical techniques to develop phages that are more resistant to production technologies for food applications.

## 3.8 Concluding Remarks

This chapter highlights the importance of microorganisms in food preservation. In this context, microorganisms appear as bioprotective agents. Numerous scientific papers specifically confirm the positive effects of LAB on food preservation and safety. In the preservation of foods, the use of cellular metabolites or non-living cells has become the focus of recent studies, as well as directly benefiting from living cells. Hereby, the risks associated with the viability factor are also minimized. Postbiotics and paraprobiotics have the potential to be alternative natural additives. Therefore, the evaluation of these components in the food industry will not only provide an alternative to chemical additives and extend the shelf-life of foods but also provide consumers with additional health advantages beyond meeting their nutritional needs due to their functional properties. Currently, nisin bacteriocin is the only commercially available postbiotic. Commercialization of different postbiotics and paraprobiotics besides live cultures as preservative agents for the food industry is promising. In addition to designing large-scale production processes for the production of paraprobiotics and postbiotics, the discovery of new potential preservative cultures, their postbiotics, and their paraprobiotic forms and the identification of new techniques for their characterization are of great importance.

**Acknowledgments** All the artworks were constructed using BioRender (2022 version) online program (https://biorender.com/).

## References

Abdhul, K., Ganesh, M., Shanmughapriya, S., Vanithamani, S., Kanagavel, M., Anbarasu, K., & Natarajaseenivasan, K. (2015). Bacteriocinogenic potential of a probiotic strain *Bacillus coagulans* [BDU3] from Ngari. *International Journal of Biological Macromolecules, 79*, 800–806.

Abouloifa, H., Gaamouche, S., Rokni, Y., Hasnaoui, I., Bellaouchi, R., Ghabbour, N., Karboune, S., Brasca, M., D'Hallewin, G., Ben Salah, R., Saalaoui, E., & Asehraou, A. (2021). Antifungal activity of probiotic *lactobacillus* strains isolated from natural fermented green olives and their application as food bio-preservative. *Biological Control, 152*, 104450.

Adesokan, I., Ekanola, Y., & Okanlawon, B. (2010). Influence of cultural conditions on hydrogen peroxide production by lactic acid bacteria isolated from some Nigerian traditional fermented foods. *African Journal of Microbiology Research, 4*(19), 1991–1996.

Aghebati-Maleki, L., Hasannezhad, P., Abbasi, A., & Khani, N. (2021). Antibacterial, antiviral, antioxidant, and anticancer activities of postbiotics: A review of mechanisms and therapeutic perspectives. *Biointerface Research in Applied Chemistry, 12*, 2629–2645.

Agriopoulou, S., Stamatelopoulou, E., Sachadyn-Król, M., & Varzakas, T. (2020). Lactic acid bacteria as antibacterial agents to extend the shelf life of fresh and minimally processed fruits and vegetables: Quality and safety aspects. *Microorganisms, 8*(6), 952.

Aguilar-Toalá, J. E., Astiazarán-García, H., Estrada-Montoya, M. C., Garcia, H. S., Vallejo-Cordoba, B., González-Córdova, A. F., & Hernández-Mendoza, A. (2019). Modulatory effect of the intracellular content of *lactobacillus casei* CRL 431 against the aflatoxin b1-induced oxidative stress in rats. *Probiotics and Antimicrobial Proteins, 11*(2), 470–477.

Akbar, A., & Anal, A. K. (2014). Occurrence of *Staphylococcus aureus* and evaluation of anti-staphylococcal activity of *Lactococcus lactis* subsp. *lactis* in ready-to-eat poultry meat. *Annales de Microbiologie, 64*(1), 131–138.

Akhwale, J. K., Rohde, M., Rohde, C., Bunk, B., Spröer, C., Boga, H. I., Klenk, H. P., & Wittmann, J. (2019). Isolation, characterization and analysis of bacteriophages from the haloalkaline lake Elmenteita, Kenya. *PLoS One, 14*(4), e0215734.

Al-Gamal, M. S., Ibrahim, G. A., Sharaf, O. M., Radwan, A. A., Dabiza, N. M., Youssef, A. M., & El-ssayad, M. F. (2019). The protective potential of selected lactic acid bacteria against the most common contaminants in various types of cheese in Egypt. *Heliyon, 5*(3), e01362.

Altieri, C., Speranza, B., Del Nobile, M. A., & Sinigaglia, M. (2005). Suitability of bifidobacteria and thymol as biopreservatives in extending the shelf life of fresh packed plaice fillets. *Journal of Applied Microbiology, 99*(6), 1294–1302.

Álvarez-Martín, P., Flórez, A. B., Hernández-Barranco, A., & Mayo, B. (2008). Interaction between dairy yeasts and lactic acid bacteria strains during milk fermentation. *Food Control, 19*(1), 62–70.

Amaretti, A., di Nunzio, M., Pompei, A., Raimondi, S., Rossi, M., & Bordoni, A. (2013). Antioxidant properties of potentially probiotic bacteria: In vitro and in vivo activities. *Applied Microbiology and Biotechnology, 97*(2), 809–817.

Ananta, E., & Knorr, D. (2009). Comparison of inactivation pathways of thermal or high pressure inactivated *lactobacillus rhamnosus* ATCC 53103 by flow cytometry analysis. *Food Microbiology, 26*(5), 542–546.

Arrioja-Bretón, D., Mani-López, E., Palou, E., & López-Malo, A. (2020). Antimicrobial activity and storage stability of cell-free supernatants from lactic acid bacteria and their applications with fresh beef. *Food Control, 115*, 107286.

Awojobi, K., Adeyemo, S., & Sanusi, O. (2016). Biosynthesis of antimicrobial compounds by lactic acid bacteria and its use as biopreservative in pineapple juice. *Frontiers in Science, 6*(1), 17–24.

Aydın, B., Çiydem, T., Kaya, E., & Açık, L. (2021). Evaluation of the antioxidant effects of postbiotics and paraprobiotics in lactic acid bacteria isolated from traditional fermented sausages. *European Journal of Science and Technology, 28*, 849–852.

Azami, S., Arefian, E., & Kashef, N. (2022). Postbiotics of *lactobacillus casei* target virulence and biofilm formation of *Pseudomonas aeruginosa* by modulating quorum sensing. *Archives of Microbiology, 204*(2), 157.

Aziz, M., & Karboune, S. (2018). Natural antimicrobial/antioxidant agents in meat and poultry products as well as fruits and vegetables: A review. *Critical Reviews in Food Science and Nutrition, 58*(3), 486–511.

Bai, J., Kim, Y.-T., Ryu, S., & Lee, J.-H. (2016). Biocontrol and rapid detection of food-borne pathogens using bacteriophages and endolysins. *Frontiers in Microbiology, 7*, 474.

Barros, C. P., Guimarães, J. T., Esmerino, E. A., Duarte, M. C. K. H., Silva, M. C., Silva, R., Ferreira, B. M., Sant'Ana, A. S., Freitas, M. Q., & Cruz, A. G. (2020). Paraprobiotics and postbiotics: Concepts and potential applications in dairy products. *Current Opinion in Food Science, 32*, 1–8.

Battelli, G., Scano, P., Albano, C., Cagliani, L. R., Brasca, M., & Consonni, R. (2019). Modifications of the volatile and nonvolatile metabolome of goat cheese due to adjunct of non-starter lactic acid bacteria. *LWT—Food Science and Technology, 116*, 108576.

Ben Said, L., Gaudreau, H., Dallaire, L., Tessier, M., & Fliss, I. (2019). Bioprotective culture: A new generation of food additives for the preservation of food quality and safety. *Industrial Biotechnology, 15*(3), 138–147.

Benkerroum, N., Daoudi, A., & Kamal, M. (2003). Behaviour of *Listeria monocytogenes* in raw sausages (merguez) in presence of a bacteriocin-producing lactococcal strain as a protective culture. *Meat Science, 63*(4), 479–484.

Berninger, T., González López, Ó., Bejarano, A., Preininger, C., & Sessitsch, A. (2018). Maintenance and assessment of cell viability in formulation of non-sporulating bacterial inoculants. *Microbial Biotechnology, 11*(2), 277–301.
Bomfim, V. B., Pereira Lopes Neto, J. H., Leite, K. S., de Andrade Vieira, É., Iacomini, M., Silva, C. M., Olbrich dos Santos, K. M., & Cardarelli, H. R. (2020). Partial characterization and antioxidant activity of exopolysaccharides produced by *Lactobacillus plantarum* CNPC003. *LWT—Food Science and Technology, 127*, 109349.
Borges, F., Briandet, R., Callon, C., Champomier-Vergès, M. C., Christieans, S., Chuzeville, S., Denis, C., Desmasures, N., Desmonts, M. H., Feurer, C., Leroi, F., Leroy, S., Mounier, J., Passerini, D., Pilet, M. F., Schlusselhuber, M., Stahl, V., Strub, C., Talon, R., & Zagorec, M. (2022). Contribution of omics to biopreservation: Toward food microbiome engineering. *Frontiers in Microbiology, 13*, 951182.
Bouarab Chibane, L., Degraeve, P., Ferhout, H., Bouajila, J., & Oulahal, N. (2019). Plant antimicrobial polyphenols as potential natural food preservatives. *Journal of the Science of Food and Agriculture, 99*(4), 1457–1474.
Boumis, E., Capone, A., Galati, V., Venditti, C., & Petrosillo, N. (2018). Probiotics and infective endocarditis in patients with hereditary hemorrhagic telangiectasia: A clinical case and a review of the literature. *BMC Infectious Diseases, 18*(1), 65.
Bourebaba, Y., Marycz, K., Mularczyk, M., & Bourebaba, L. (2022). Postbiotics as potential new therapeutic agents for metabolic disorders management. *Biomedicine & Pharmacotherapy, 153*, 113138.
Brown, L., Pingitore, E. V., Mozzi, F., Saavedra, L., Villegas, M., & Hebert, E. (2017). Lactic acid bacteria as cell factories for the generation of bioactive peptides. *Protein and Peptide Letters, 24*(2), 146–155.
Cancino-Padilla, N., Fellenberg, M. A., Franco, W., Ibáñez, R. A., & Pérez, E. V. B. (2017). Foodborne bacteria in dairy products: Detection by molecular techniques. *Ciencia e Investigación Agraria, 44*(3), 215–229.
Cao, J., Yu, Z., Liu, W., Zhao, J., Zhang, H., Zhai, Q., & Chen, W. (2020). Probiotic characteristics of *Bacillus coagulans* and associated implications for human health and diseases. *Journal of Functional Foods, 64*, 103643.
Carocho, M., Barreiro, M. F., Morales, P., & Ferreira, I. C. F. R. (2014). Adding molecules to food, pros and cons: A review on synthetic and natural food additives. *Comprehensive Reviews in Food Science and Food Safety, 13*(4), 377–399.
Castellano, P., Belfiore, C., Fadda, S., & Vignolo, G. (2008). A review of bacteriocinogenic lactic acid bacteria used as bioprotective cultures in fresh meat produced in Argentina. *Meat Science, 79*(3), 483–499.
Castellano, P., Pérez Ibarreche, M., Blanco Massani, M., Fontana, C., & Vignolo, G. M. (2017). Strategies for pathogen biocontrol using lactic acid bacteria and their metabolites: A focus on meat ecosystems and industrial environments. *Microorganisms, 5*(3), 38.
Ceugniez, A., Coucheney, F., Jacques, P., Daube, G., Delcenserie, V., & Drider, D. (2017). Anti-*salmonella* activity and probiotic trends of *Kluyveromyces marxianus* S-2-05 and *Kluyveromyces lactis* S-3-05 isolated from a French cheese, Tomme d'Orchies. *Research in Microbiology, 168*(6), 575–582.
Chang, H. M., Foo, H. L., Loh, T. C., Lim, E. T. C., & Abdul Mutalib, N. E. (2021). Comparative studies of inhibitory and antioxidant activities, and organic acids compositions of postbiotics produced by probiotic *Lactiplantibacillus plantarum* strains isolated from Malaysian foods. *Frontiers in Veterinary Science, 7*, 602280.
Coelho, M. C., Silva, C. C. G., Ribeiro, S. C., Dapkevicius, M. L. N. E., & Rosa, H. J. D. (2014). Control of listeria monocytogenes in fresh cheese using protective lactic acid bacteria. *International Journal of Food Microbiology, 191*, 53–59.
Contessa, C. R., da Rosa, G. S., & Moraes, C. C. (2021). New active packaging based on biopolymeric mixture added with bacteriocin as active compound. *International Journal of Molecular Sciences, 22*(19), 10628.

Cuevas-González, P. F., Liceaga, A. M., & Aguilar-Toalá, J. E. (2020). Postbiotics and paraprobiotics: From concepts to applications. *Food Research International, 136*, 109502.

Dahech, I., Belghith, K. S., Hamden, K., Feki, A., Belghith, H., & Mejdoub, H. (2011). Antidiabetic activity of Levan polysaccharide in alloxan-induced diabetic rats. *International Journal of Biological Macromolecules, 49*(4), 742–746.

Dan, T., Wang, D., Jin, R. L., Zhang, H. P., Zhou, T. T., & Sun, T. S. (2017). Characterization of volatile compounds in fermented milk using solid-phase microextraction methods coupled with gas chromatography-mass spectrometry. *Journal of Dairy Science, 100*(4), 2488–2500.

de Almada, C. N., Almada, C. N., Martinez, R. C. R., & Sant'Ana, A. S. (2016). Paraprobiotics: Evidences on their ability to modify biological responses, inactivation methods and perspectives on their application in foods. *Trends in Food Science and Technology, 58*, 96–114.

de Almeida Godoy, C. L., Costa, L. M., Guerra, C. A., de Oliveira, V. S., de Paula, B. P., Lemos Junior, W. J. F., da Silva Duarte, V., Luchese, R. H., Bautitz, I. R., & Guerra, A. F. (2022). Potentially postbiotic-containing preservative to extend the use-by date of raw chicken sausages and semifinished chicken products. *Sustainability, 14*(5), 2646.

De Vuyst, L., & Leroy, F. (2007). Bacteriocins from lactic acid bacteria: Production, purification, and food applications. *Microbial Physiology, 13*(4), 194–199.

Değirmencioğlu, N., Gurbuz, O., & Şahan, Y. (2016). The monitoring, via an in vitro digestion system, of the bioactive content of vegetable juice fermented with *Saccharomyces cerevisiae* and *Saccharomyces boulardii*. *Journal of Food Processing & Preservation, 40*(4), 798–811.

Dewanggana, M. N., Evangeline, C., Ketty, M. D., Waturangi, D. E., & Yogiara, M. (2022). Isolation, characterization, molecular analysis and application of bacteriophage DW-EC to control enterotoxigenic *Escherichia coli* on various foods. *Scientific Reports, 12*(1), 495.

Dhankhar, S., Argade, A., Thakur, N., Bishnoi, S., & Ahlawat, S. (2021). Application of nanoemulsion technology for development of novel functional foods with essential oils encapsulation: A review. *Pharma Innovation journal, 10*(3), 454–458.

Di Gioia, D., Mazzola, G., Nikodinoska, I., Aloisio, I., Langerholc, T., Rossi, M., Raimondi, S., Melero, B., & Rovira, J. (2016). Lactic acid bacteria as protective cultures in fermented pork meat to prevent *Clostridium* spp. growth. *International Journal of Food Microbiology, 235*, 53–59.

Didari, T., Solki, S., Mozaffari, S., Nikfar, S., & Abdollahi, M. (2014). A systematic review of the safety of probiotics. *Expert Opinion on Drug Safety, 13*(2), 227–239.

Divyashree, S., Anjali, P. G., Somashekaraiah, R., & Sreenivasa, M. Y. (2021). Probiotic properties of *Lactobacillus casei*—MYSRD 108 and lactobacillus plantarum-MYSRD 71 with potential antimicrobial activity against *Salmonella paratyphi*. *Biotechnology Reports, 32*, e00672.

Dunand, E., Burns, P., Binetti, A., Bergamini, C., Peralta, G. H., Forzani, L., Reinheimer, J., & Vinderola, G. (2019). Postbiotics produced at laboratory and industrial level as potential functional food ingredients with the capacity to protect mice against *Salmonella* infection. *Journal of Applied Microbiology, 127*(1), 219–229.

Enrique, M., Ibáñez, A., Marcos, J. F., Yuste, M., Martínez, M., Vallés, S., & Manzanares, P. (2010). Beta-Glucanases as a tool for the control of wine spoilage yeasts. *Journal of Food Science, 75*(1), M41–M45.

Essia Ngang, J.-J., Yadang, G., Sado Kamdem, S. L., Kouebou, C. P., Youte Fanche, S. A., Tsochi Kougan, D. L., Tsoungui, A., & Etoa, F.-X. (2015). Antifungal properties of selected lactic acid bacteria and application in the biological control of ochratoxin a producing fungi during cocoa fermentation. *Biocontrol Science and Technology, 25*(3), 245–259.

Fraise, A. P., Wilkinson, M. A. C., Bradley, C. R., Oppenheim, B., & Moiemen, N. (2013). The antibacterial activity and stability of acetic acid. *The Journal of Hospital Infection, 84*(4), 329–331.

Franklin, L. M., Chapman, D. M., King, E. S., Mau, M., Huang, G., & Mitchell, A. E. (2017). Chemical and sensory characterization of oxidative changes in roasted almonds undergoing accelerated shelf life. *Journal of Agricultural and Food Chemistry, 65*(12), 2549–2563.

Gallegos, J., Arce, C., Jordano, R., Arce, L., & Medina, L. M. (2017). Target identification of volatile metabolites to allow the differentiation of lactic acid bacteria by gas chromatography-ion mobility spectrometry. *Food Chemistry, 220*, 362–370.

Gálvez, A., Abriouel, H., Benomar, N., & Lucas, R. (2010). Microbial antagonists to food-borne pathogens and biocontrol. *Current Opinion in Biotechnology, 21*(2), 142–148.

García, P., Rodríguez, L., Rodríguez, A., & Martínez, B. (2010). Food biopreservation: Promising strategies using bacteriocins, bacteriophages and endolysins. *Trends in Food Science and Technology, 21*(8), 373–382.

Gargi, A., & Sengun, I. Y. (2021). Marination liquids enriched with probiotics and their inactivation effects against food-borne pathogens inoculated on meat. *Meat Science, 182*, 108624.

Geraldo, B. M. C., Batalha, M. N., Milhan, N. V. M., Rossoni, R. D., Scorzoni, L., & Anbinder, A. L. (2020). Heat-killed *Lactobacillus reuteri* and cell-free culture supernatant have similar effects to viable probiotics during interaction with *Porphyromonas gingivalis*. *Journal of Periodontal Research, 55*(2), 215–220.

Gildea, L., Ayariga, J. A., Robertson, B. K., & Villafane, R. (2022). P22 phage shows promising antibacterial activity under pathophysiological conditions. *Archives of Microbiology & Immunology, 6*(1), 81–100.

Guarrasi, V., Sannino, C., Moschetti, M., Bonanno, A., Di Grigoli, A., & Settanni, L. (2017). The individual contribution of starter and non-starter lactic acid bacteria to the volatile organic compound composition of Caciocavallo Palermitano cheese. *International Journal of Food Microbiology, 259*, 35–42.

Guenther, S., Herzig, O., Fieseler, L., Klumpp, J., & Loessner, M. J. (2012). Biocontrol of *Salmonella typhimurium* in RTE foods with the virulent bacteriophage FO1-E2. *International Journal of Food Microbiology, 154*(1), 66–72.

Hazan, R., Que, Y.-A., Maura, D., & Rahme, L. G. (2012). A method for high throughput determination of viable bacteria cell counts in 96-well plates. *BMC Microbiology, 12*(1), 259.

Horita, C. N., Baptista, R. C., Caturla, M. Y. R., Lorenzo, J. M., Barba, F. J., & Sant'Ana, A. S. (2018). Combining reformulation, active packaging and non-thermal post-packaging decontamination technologies to increase the microbiological quality and safety of cooked ready-to-eat meat products. *Trends in Food Science and Technology, 72*, 45–61.

Hossain, M. I., Mizan, M. F. R., Ashrafudoulla, M., Nahar, S., Joo, H.-J., Jahid, I. K., Park, S. H., Kim, K.-S., & Ha, S.-D. (2020). Inhibitory effects of probiotic potential lactic acid bacteria isolated from kimchi against *listeria monocytogenes* biofilm on lettuce, stainless-steel surfaces, and MBEC™ biofilm device. *LWT—Food Science and Technology, 118*, 108864.

Hosseini, S. A., Abbasi, A., Sabahi, S., & Khani, N. (2021). Application of postbiotics produced by lactic acid bacteria in the development of active food packaging. *Biointerface Research in Applied Chemistry, 12*, 6164–6183.

Huang, C. B., Alimova, Y., Myers, T. M., & Ebersole, J. L. (2011). Short- and medium-chain fatty acids exhibit antimicrobial activity for oral microorganisms. *Archives of Oral Biology, 56*(7), 650–654.

Ibrahim, S. A., Ayivi, R. D., Zimmerman, T., Siddiqui, S. A., Altemimi, A. B., Fidan, H., Esatbeyoglu, T., & Bakhshayesh, R. V. (2021). Lactic acid bacteria as antimicrobial agents: Food safety and microbial food spoilage prevention. *Food, 10*(12), 3131.

İncili, G. K., Karatepe, P., Akgöl, M., Kaya, B., Kanmaz, H., & Hayaloğlu, A. A. (2021). Characterization of *Pediococcus acidilactici* postbiotic and impact of postbiotic-fortified chitosan coating on the microbial and chemical quality of chicken breast fillets. *International Journal of Biological Macromolecules, 184*, 429–437.

İncili, G. K., Karatepe, P., Akgöl, M., Güngören, A., Koluman, A., İlhak, O. İ., Kanmaz, H., Kaya, B., & Hayaloğlu, A. A. (2022). Characterization of lactic acid bacteria postbiotics, evaluation in-vitro antibacterial effect, microbial and chemical quality on chicken drumsticks. *Food Microbiology, 104*, 104001.

Ishikawa, H., Kutsukake, E., Fukui, T., Sato, I., Shirai, T., Kurihara, T., Okada, N., Danbara, H., Toba, M., Kohda, N., Maeda, Y., & Matsumoto, T. (2010). Oral administration of heat-killed

*Lactobacillus plantarum* strain b240 protected mice against *Salmonella enterica* serovar typhimurium. *Bioscience, Biotechnology, and Biochemistry, 74*(7), 1338–1342.

Ismael, M., Wang, T., Yue, F., Cui, Y., Yantin, Q., Qayyum, N., & Lü, X. (2023). A comparison of mining methods to extract novel bacteriocins from *Lactiplantibacillus plantarum* NWAFU-BIO-BS29. *Analytical Biochemistry, 661*, 114938.

Janashia, I., Choiset, Y., Jozefiak, D., Déniel, F., Coton, E., Moosavi-Movahedi, A. A., Chanishvili, N., & Haertlé, T. (2018). Beneficial protective role of endogenous lactic acid bacteria against mycotic contamination of honeybee beebread. *Probiotics and Antimicrobial Proteins, 10*(4), 638–646.

Juturu, V., & Wu, J. C. (2018). Microbial production of bacteriocins: Latest research development and applications. *Biotechnology Advances, 36*(8), 2187–2200.

Kamal, A. A., & Fawzia, S. A.-S. (2018). Toxicological and safety assessment of tartrazine as a synthetic food additive on health biomarkers: A review. *African Journal of Biotechnology, 17*(6), 139–149.

Khalid, K. (2011). An overview of lactic acid bacteria. *International Journal of Biosciences, 1*(3), 1–13.

Khubber, S., Marti-Quijal, F. J., Tomasevic, I., Remize, F., & Barba, F. J. (2022). Lactic acid fermentation as a useful strategy to recover antimicrobial and antioxidant compounds from food and by-products. *Current Opinion in Food Science, 43*, 189–198.

Kim, H. G., Lee, S. Y., Kim, N. R., Lee, H. Y., Ko, M. Y., Jung, B. J., Kim, C. M., Lee, J. M., Park, J. H., Han, S. H., & Chung, D. K. (2011). *Lactobacillus plantarum* lipoteichoic acid down-regulated *Shigella flexneri* peptidoglycan-induced inflammation. *Molecular Immunology, 48*(4), 382–391.

Koyama, S., Fujita, H., Shimosato, T., Kamijo, A., Ishiyama, Y., Yamamoto, E., Ishii, Y., Hattori, Y., Hagihara, M., Yamazaki, E., Tomita, N., Nakajima, H., & On behalf of the Yokohama Cooperative Study Group for, H. (2019). Septicemia from *Lactobacillus rhamnosus* GG, from a probiotic enriched yogurt, in a patient with autologous stem cell transplantation. *Probiotics and Antimicrobial Proteins, 11*(1), 295–298.

Kumar, H., Schütz, F., Bhardwaj, K., Sharma, R., Nepovimova, E., Dhanjal, D. S., Verma, R., Kumar, D., Kuča, K., & Cruz-Martins, N. (2021). Recent advances in the concept of paraprobiotics: Nutraceutical/functional properties for promoting children health. *Critical Reviews in Food Science and Nutrition*, 1–16.

Lee, J.-H., Woo, K.-J., Kim, M.-A., Hong, J., Kim, J., Kim, S.-H., Han, K.-I., Iwasa, M., & Kim, T.-J. (2022). Heat-killed *Enterococcus faecalis* prevents adipogenesis and high fat diet-induced obesity by inhibition of lipid accumulation through inhibiting C/EBP-alpha and PPAR-gamma in the insulin signaling pathway. *Nutrients, 14*(6), 1308.

Leyva Salas, M., Mounier, J., Valence, F., Coton, M., Thierry, A., & Coton, E. (2017). Antifungal microbial agents for food biopreservation—A review. *Microorganisms, 5*(3), 37.

Leyva Salas, M., Thierry, A., Lemaître, M., Garric, G., Harel-Oger, M., Chatel, M., Lê, S., Mounier, J., Valence, F., & Coton, E. (2018). Antifungal activity of lactic acid bacteria combinations in dairy mimicking models and their potential as bioprotective cultures in pilot scale applications. *Frontiers in Microbiology, 9*, 1787.

Li, H., Abraham, A., Gazzola, D., Hu, Y., Beamer, G., Flanagan, K., Soto, E., Rus, F., Mirza, Z., Draper, A., Vakalapudi, S., Stockman, C., Bain, P., Urban, J. F., Ostroff, G. R., & Aroian, R. V. (2021a). Recombinant paraprobiotics as a new paradigm for treating gastrointestinal nematode parasites of humans. *Antimicrobial Agents and Chemotherapy, 65*(3), e01469–e01420.

Li, S., Li, Y., Du, Z., Li, B., Liu, Y., Gao, Y., Zhang, Y., Zhang, K., Wang, Q., Lu, S., Dong, J., Ji, H., & Li, Y. (2021b). Impact of NSLAB on Kazakh cheese flavor. *Food Research International, 144*, 110315.

Liu, J., Luo, J., Ye, H., Sun, Y., Lu, Z., & Zeng, X. (2010). *In vitro* and *in vivo* antioxidant activity of exopolysaccharides from endophytic bacterium *Paenibacillus polymyxa* EJS-3. *Carbohydrate Polymers, 82*(4), 1278–1283.

Liu, H., Zhang, L., Yi, H., Han, X., & Chi, C. (2016). Identification and characterization of plantaricin Q7, a novel plantaricin produced by *Lactobacillus plantarum* Q7. *LWT—Food Science and Technology, 71*, 386–390.
Mahdhi, A., Leban, N., Chakroun, I., Chaouch, M. A., Hafsa, J., Fdhila, K., Mahdouani, K., & Majdoub, H. (2017). Extracellular polysaccharide derived from potential probiotic strain with antioxidant and antibacterial activities as a prebiotic agent to control pathogenic bacterial biofilm formation. *Microbial Pathogenesis, 109*, 214–220.
Marhuenda-Muñoz, M., Laveriano-Santos, E. P., Tresserra-Rimbau, A., Lamuela-Raventós, R. M., Martínez-Huélamo, M., & Vallverdú-Queralt, A. (2019). Microbial phenolic metabolites: Which molecules actually have an effect on human health? *Nutrients, 11*(11), 2725.
Mayorgas, A., Dotti, I., & Salas, A. (2021). Microbial metabolites, postbiotics, and intestinal epithelial function. *Molecular Nutrition & Food Research, 65*(5), 2000188.
Michailidis, P. A., & Krokida, M. K. (2015). In S. Bhattacharya (Ed.), *Drying and dehydration processes in food preservation and processing* (pp. 1–32). Wiley Blackwell.
Mokoena, M. P. (2017). Lactic acid bacteria and their bacteriocins: Classification, biosynthesis and applications against uropathogens: A mini-review. *Molecules, 22*(8), 1255.
Montanaro, J., Inic-Kanada, A., Ladurner, A., Stein, E., Belij, S., Bintner, N., Schlacher, S., Schuerer, N., Mayr, U. B., Lubitz, W., Leisch, N., & Barisani-Asenbauer, T. (2015). *Escherichia coli* Nissle 1917 bacterial ghosts retain crucial surface properties and express chlamydial antigen: An imaging study of a delivery system for the ocular surface. *Drug Design, Development and Therapy, 9*, 3741–3754.
Moradi, M., Mardani, K., & Tajik, H. (2019). Characterization and application of postbiotics of *Lactobacillus* spp. on *Listeria monocytogenes in vitro* and in food models. *LWT—Food Science and Technology, 111*, 457–464.
Mosbah, A., Delavenne, E., Souissi, Y., Mahjoubi, M., Jéhan, P., Le Yondre, N., Cherif, A., Bondon, A., Mounier, J., Baudy-Floch, M., & Le Blay, G. (2018). Novel antifungal compounds, spermine-like and short cyclic polylactates, produced by *Lactobacillus harbinensis* K.V9.3.1Np in yogurt. *Frontiers in Microbiology, 9*, 1–10.
Motalebi Moghanjougi, Z., Rezazadeh Bari, M., Alizadeh Khaledabad, M., Almasi, H., & Amiri, S. (2020). Bio-preservation of white brined cheese (feta) by using probiotic bacteria immobilized in bacterial cellulose: Optimization by response surface method and characterization. *LWT—Food Science and Technology, 117*, 108603.
Nazari, M., Ghanbarzadeh, B., Samadi Kafil, H., Zeinali, M., & Hamishehkar, H. (2019). Garlic essential oil nanophytosomes as a natural food preservative: Its application in yogurt as food model. *Colloid and Interface Science Communications, 30*, 100176.
Nithya, V., Murthy, P. S. K., & Halami, P. M. (2013). Development and application of active films for food packaging using antibacterial peptide of *Bacillus licheniformis* Me1. *Journal of Applied Microbiology, 115*(2), 475–483.
Ou, C.-C., Ko, J.-L., & Lin, M.-Y. (2006). Antioxidative effects of intracellular extracts of yogurt bacteria on lipid peroxidation and intestine 407 cells. *Journal of Food and Drug Analysis, 14*(3), 304–310.
Ou, C.-C., Lu, T.-M., Tsai, J.-J., Yen, J.-H., Chen, H.-W., & Lin, M.-Y. (2009). Antioxidative effect of lactic acid bacteria: Intact cells vsintracellular extracts. *Journal of Food and Drug Analysis, 17*(3), 209–216.
Ou, C.-C., Lin, S.-L., Tsai, J.-J., & Lin, M.-Y. (2011). Heat-killed lactic acid bacteria enhance immunomodulatory potential by skewing the immune response toward Th1 polarization. *Journal of Food Science, 76*(5), M260–M267.
Pech-Canul, A. C., Ortega, D., García-Triana, A., González-Silva, N., & Solis-Oviedo, R. L. (2020). A brief review of edible coating materials for the microencapsulation of probiotics. *Coatings, 10*(3), 197.
Perez, R. H., Perez, M. T. M., & Elegado, F. B. (2015). Bacteriocins from lactic acid bacteria: A review of biosynthesis, mode of action, fermentative production, uses, and prospects. *International Journal of Philippine Science and Technology, 08*(2), 61–67.

Perez, K. J., Viana, J. S., Lopes, F. C., Pereira, J. Q., dos Santos, D. M., Oliveira, J. S., Velho, R. V., Crispim, S. M., Nicoli, J. R., Brandelli, A., & Nardi, R. M. D. (2017). *Bacillus* spp. isolated from puba as a source of biosurfactants and antimicrobial lipopeptides. *Frontiers in Microbiology, 8*, 61.

Petsong, K., Benjakul, S., & Vongkamjan, K. (2021). Optimization of wall material for phage encapsulation via freeze-drying and antimicrobial efficacy of microencapsulated phage against *salmonella*. *Journal of Food Science and Technology, 58*(5), 1937–1946.

Pokorný, J. (2007). Are natural antioxidants better—And safer—Than synthetic antioxidants? *European Journal of Lipid Science and Technology, 109*(6), 629–642.

Pourjafar, H., Noori, N., Gandomi, H., Basti, A. A., & Ansari, F. (2020). Viability of microencapsulated and non-microencapsulated lactobacilli in a commercial beverage. *Biotechnology Reports, 25*, e00432.

Punia Bangar, S., Suri, S., Trif, M., & Ozogul, F. (2022). Organic acids production from lactic acid bacteria: A preservation approach. *Food Bioscience, 46*, 101615.

Radiati, L. E., Hati, D. L., Fardiaz, D., & Sari, L. R. H. (2022). Effect of *Saccharomyces cerevisiae on* probiotic properties of goat milk kefir. *IOP Conference Series: Earth and Environmental Science, 1041*(1), 012028.

Rafiee, Z., Barzegar, M., Sahari, M. A., & Maherani, B. (2018). Nanoliposomes containing pistachio green hull's phenolic compounds as natural bio-preservatives for mayonnaise. *European Journal of Lipid Science and Technology, 120*(9), 1800086.

Ramos-Vivas, J., Elexpuru-Zabaleta, M., Samano, M. L., Barrera, A. P., Forbes-Hernández, T. Y., Giampieri, F., & Battino, M. (2021). Phages and enzybiotics in food biopreservation. *Molecules, 26*(17), 5138.

Ran, Q., Yang, F., Geng, M., Qin, L., Chang, Z., Gao, H., Jiang, D., Zou, C., & Jia, C. (2022). A mixed culture of *Propionibacterium freudenreichii* and *Lactiplantibacillus plantarum* as antifungal biopreservatives in bakery product. *Food Bioscience, 47*, 101456.

Reis, J. A., Paula, A. T., Casarotti, S. N., & Penna, A. L. B. (2012). Lactic acid bacteria antimicrobial compounds: Characteristics and applications. *Food Engineering Reviews, 4*(2), 124–140.

Ressutte, J. B., da Silva Saranti, T. F., de Moura, M. R., dos Santos Pozza, M. S., da Silva Scapim, M. R., Stafussa, A. P., & Madrona, G. S. (2022). Citric acid incorporated in a chitosan film as an active packaging material to improve the quality and duration of matured cheese shelf life. *The Journal of Dairy Research, 89*(2), 201–207.

Riaz Rajoka, M. S., Zhao, H., Mehwish, H. M., Li, N., Lu, Y., Lian, Z., Shao, D., Jin, M., Li, Q., Zhao, L., & Shi, J. (2019). Anti-tumor potential of cell free culture supernatant of *Lactobacillus rhamnosus* strains isolated from human breast milk. *Food Research International, 123*, 286–297.

Ricke, S. C., Dittoe, D. K., & Richardson, K. E. (2020). Formic acid as an antimicrobial for poultry production: A review. *Frontiers in Veterinary Science, 7*, 563.

Rifna, E. J., Singh, S. K., Chakraborty, S., & Dwivedi, M. (2019). Effect of thermal and non-thermal techniques for microbial safety in food powder: Recent advances. *Food Research International, 126*, 108654.

Rodgers, S. (2008). Novel applications of live bacteria in food services: Probiotics and protective cultures. *Trends in Food Science and Technology, 19*(4), 188–197.

Romero-Luna, H. E., Hernández-Mendoza, A., González-Córdova, A. F., & Peredo-Lovillo, A. (2022). Bioactive peptides produced by engineered probiotics and other food-grade bacteria: A review. *Food Chemistry: X, 13*, 100196.

Salminen, S., Collado, M. C., Endo, A., Hill, C., Lebeer, S., Quigley, E. M. M., Sanders, M. E., Shamir, R., Swann, J. R., Szajewska, H., & Vinderola, G. (2021). The International Scientific Association of Probiotics and Prebiotics (ISAPP) consensus statement on the definition and scope of postbiotics. *Nature Reviews. Gastroenterology & Hepatology, 18*(9), 649–667.

Salomskiene, J., Jonkuviene, D., Macioniene, I., Abraitiene, A., Zeime, J., Repeckiene, J., & Vaiciulyte-Funk, L. (2019). Differences in the occurence and efficiency of antimicrobial

compounds produced by lactic acid bacteria. *European Food Research and Technology, 245*(3), 569–579.
Sanders, J. W., Leenhouts, K. J., Haandrikman, A. J., Venema, G., & Kok, J. (1995). Stress response in *Lactococcus lactis*: Cloning, expression analysis, and mutation of the lactococcal superoxide dismutase gene. *Journal of Bacteriology, 177*(18), 5254–5260.
Sawada, D., Sugawara, T., Ishida, Y., Aihara, K., Aoki, Y., Takehara, I., Takano, K., & Fujiwara, S. (2016). Effect of continuous ingestion of a beverage prepared with *Lactobacillus gasseri* CP2305 inactivated by heat treatment on the regulation of intestinal function. *Food Research International, 79*, 33–39.
Sendil, S., Shrimanker, I., Mansoora, Q., Goldman, J., & Nookala, V. K. (2020). *Lactobacillus rhamnosus* bacteremia in an immunocompromised renal transplant patient. *Cureus, 12*(2), e6887.
Sgarbi, E., Lazzi, C., Tabanelli, G., Gatti, M., Neviani, E., & Gardini, F. (2013). Nonstarter lactic acid bacteria volatilomes produced using cheese components. *Journal of Dairy Science, 96*(7), 4223–4234.
Shehata, M. G., Badr, A. N., El Sohaimy, S. A., Asker, D., & Awad, T. S. (2019). Characterization of antifungal metabolites produced by novel lactic acid bacterium and their potential application as food biopreservatives. *Annals of Agricultural Science, 64*(1), 71–78.
Shenderov, B. A., Sinitsa, A. V., Zakharchenko, M. M., & Lang, C. (2020). Cellular metabiotics and metabolite metabiotics. In B. A. Shenderov, A. V. Sinitsa, M. M. Zakharchenko, & C. Lang (Eds.), *Metabiotics* (pp. 63–75). Springer International.
Shigematsu, E., Dorta, C., Rodrigues, F. J., Cedran, M. F., Giannoni, J. A., Oshiiwa, M., & Mauro, M. A. (2018). Edible coating with probiotic as a quality factor for minimally processed carrots. *Journal of Food Science and Technology, 55*(9), 3712–3720.
Shin, J. M., Gwak, J. W., Kamarajan, P., Fenno, J. C., Rickard, A. H., & Kapila, Y. L. (2016). Biomedical applications of nisin. *Journal of Applied Microbiology, 120*(6), 1449–1465.
Silva, C. C. G., Silva, S. P. M., & Ribeiro, S. C. (2018). Application of bacteriocins and protective cultures in dairy food preservation. *Frontiers in Microbiology, 9*, 594.
Singh, V. P. (2018). Recent approaches in food bio-preservation—a review. *Open Veterinary Journal, 8*(1), 104–111.
Sirichokchatchawan, W., Temeeyasen, G., Nilubol, D., & Prapasarakul, N. (2018). Protective effects of cell-free supernatant and live lactic acid bacteria isolated from Thai pigs against a pandemic strain of porcine epidemic diarrhea virus. *Probiotics and Antimicrobial Proteins, 10*(2), 383–390.
Siroli, L., Patrignani, F., Serrazanetti, D. I., Tabanelli, G., Montanari, C., Gardini, F., & Lanciotti, R. (2015). Lactic acid bacteria and natural antimicrobials to improve the safety and shelf-life of minimally processed sliced apples and lamb's lettuce. *Food Microbiology, 47*, 74–84.
Smith, J. (2003). Legislative aspects. In N. J. Russell & G. W. Gould (Eds.), *Food preservatives* (pp. 321–347). Springer US.
Smith-Palmer, A., Stewart, J., & Fyfe, L. (2001). The potential application of plant essential oils as natural food preservatives in soft cheese. *Food Microbiology, 18*(4), 463–470.
Sornsenee, P., Chatatikun, M., Mitsuwan, W., Kongpol, K., Kooltheat, N., Sohbenalee, S., Pruksaphanrat, S., Mudpan, A., & Romyasamit, C. (2021). Lyophilized cell-free supernatants of *Lactobacillus* isolates exhibited antibiofilm, antioxidant, and reduces nitric oxide activity in lipopolysaccharide-stimulated RAW 264.7 cells. *PeerJ, 9*, e12586.
Spanu, C., Scarano, C., Piras, F., Spanu, V., Pala, C., Casti, D., Lamon, S., Cossu, F., Ibba, M., Nieddu, G., & De Santis, E. P. L. (2017). Testing commercial biopreservative against spoilage microorganisms in MAP packed ricotta fresca cheese. *Food Microbiology, 66*, 72–76.
Stiles, M. E. (1996). Biopreservation by lactic acid bacteria. *Antonie van Leeuwenhoek, 70*(2–4), 331–345.
Suez, J., Zmora, N., Segal, E., & Elinav, E. (2019). The pros, cons, and many unknowns of probiotics. *Nature Medicine, 25*(5), 716–729.
Šušković, J., Kos, B., Beganović, J., Leboš Pavunc, A., Habjanič, K., & Matošić, S. (2010). Antimicrobial activity–the most important property of probiotic and starter lactic acid bacteria. *Food Technology and Biotechnology, 48*(3), 296–307.

Tangyu, M., Fritz, M., Ye, L., Aragão Börner, R., Morin-Rivron, D., Campos-Giménez, E., Bolten, C. J., Bogicevic, B., & Wittmann, C. (2022). Co-cultures of *Propionibacterium freudenreichii* and *Bacillus amyloliquefaciens* cooperatively upgrade sunflower seed milk to high levels of vitamin B12 and multiple co-benefits. *Microbial Cell Factories, 21*(1), 48.

Taverniti, V., & Guglielmetti, S. (2011). The immunomodulatory properties of probiotic microorganisms beyond their viability (ghost probiotics: Proposal of paraprobiotic concept). *Genes & Nutrition, 6*(3), 261–274.

Thielmann, J., Kohnen, S., & Hauser, C. (2017). Antimicrobial activity of *Olea europaea* Linné extracts and their applicability as natural food preservative agents. *International Journal of Food Microbiology, 251*, 48–66.

Thorakkattu, P., Khanashyam, A. C., Shah, K., Babu, K. S., Mundanat, A. S., Deliephan, A., Deokar, G. S., Santivarangkna, C., & Nirmal, N. P. (2022). Postbiotics: Current trends in food and pharmaceutical industry. *Food, 11*(19).

Tkaczewska, J., Zając, M., Jamróz, E., & Derbew, H. (2022). Utilising waste from soybean processing as raw materials for the production of preparations with antioxidant properties, serving as natural food preservatives—a pilot study. *LWT—Food Science and Technology, 160*, 113282.

Tyc, O., de Jager, V. C. L., van den Berg, M., Gerards, S., Janssens, T. K. S., Zaagman, N., Kai, M., Svatos, A., Zweers, H., Hordijk, C., Besselink, H., de Boer, W., & Garbeva, P. (2017). Exploring bacterial interspecific interactions for discovery of novel antimicrobial compounds. *Microbial Biotechnology, 10*(4), 910–925.

Varsha, K. K., & Nampoothiri, K. M. (2016). Appraisal of lactic acid bacteria as protective cultures. *Food Control, 69*, 61–64.

Viana da Silva, M., Santos, M. R. C., Alves Silva, I. R., Macedo Viana, E. B., Dos Anjos, D. A., Santos, I. A., Barbosa de Lima, N. G., Wobeto, C., Jorge, N., & Lannes, S. C. D. S. (2021). Synthetic and natural antioxidants used in the oxidative stability of edible oils: An overview. *Food Review International*, 1–24.

Viana de Souza, J., & Silva Dias, F. (2017). Protective, technological, and functional properties of select autochthonous lactic acid bacteria from goat dairy products. *Current Opinion in Food Science, 13*, 1–9.

Vinderola, G., Reinheimer, J., & Salminen, S. (2019). The enumeration of probiotic issues: From unavailable standardised culture media to a recommended procedure? *International Dairy Journal, 96*, 58–65.

Wang, C., Chang, T., Yang, H., & Cui, M. (2015). Antibacterial mechanism of lactic acid on physiological and morphological properties of *salmonella* Enteritidis, *Escherichia coli* and *Listeria monocytogenes*. *Food Control, 47*, 231–236.

Waturangi, D. E., Kasriady, C. P., Guntama, G., Sahulata, A. M., Lestari, D., & Magdalena, S. (2021). Application of bacteriophage as food preservative to control enteropathogenic *Escherichia coli* (EPEC). *BMC Research Notes, 14*(1), 336.

Wu, Q., Dou, X., Wang, Q., Guan, Z., Cai, Y., & Liao, X. (2018). Isolation of β-1,3-glucanase-producing microorganisms from Poria cocos cultivation soil via molecular biology. *Molecules, 23*(7), 1555.

Zagorec, M., & Champomier-Vergès, M.-C. (2017). *Lactobacillus sakei:* A starter for sausage fermentation, a protective culture for meat products. *Microorganisms, 5*(3), 56.

Zhang, F., Zhang, M., Chen, Y., Ouyang, J., Wang, Y., Yang, H., Luo, X., Zhang, D., Lu, Y., Yu, H., & Wang, Y. (2021). Antimicrobial, anti-biofilm properties of three naturally occurring antimicrobial peptides against spoilage bacteria, and their synergistic effect with chemical preservatives in food storage. *Food Control, 123*, 107729.

Zhang, B., Xu, L., Ding, J., Wang, M., Ge, R., Zhao, H., Zhang, B., & Fan, J. (2022). Natural antimicrobial lipopeptides secreted by *Bacillus* spp. and their application in food preservation, a critical review. *Trends in Food Science and Technology, 127*, 26–37.

Zimina, M., Babich, O., Prosekov, A., Sukhikh, S., Ivanova, S., Shevchenko, M., & Noskova, S. (2020). Overview of global trends in classification, methods of preparation and application of bacteriocins. *Antibiotics, 9*(9), 553.

# Chapter 4
# Food Fermentation: Role of Microorganism in Food Production

**Harshika Joshi, Gaurav Pant, Manu Pant, and Gaurav Kumar**

**Abstract** Microorganism plays a crucial role in manufacturing food products we consume daily, including traditional oriental foods, single-cell proteins, alcoholic beverages, dairy products, and many more. This chapter is an overall discussion about the role of microorganisms in food production from pre-historic times to till today. This chapter deals with fermentation and how microbes such as bacteria and fungi help obtain products at the industrial scale. In general, fermentation involves generating safe and healthy food that is organoleptically valuable. Besides, it gives us a glimpse of how manipulating organisms and products can yield a determined outcome with many implementations in the real world. In addition, the manufacturing of new-age probiotics and their usage in many fields that apply to human and animal health benefits are also described. The probiotic proposal is alluring as it is rebuilding the natural condition, thus repairing a deficiency rather than adding foreign chemicals to the body having toxic outcomes. Microorganisms are also used in the production of protein biomass called single-cell protein (SCP) that helps deliver beneficial food nutrients containing excellent properties, antiviral, anti-inflammatory, hypoglycemic, anti-cancerous, and many more. Lactic acid bacteria (LAB) are specified strains that carry probiotic properties and thus transmit peculiar sensory properties to food products contributing to their texture and aroma. LAB, along with probiotics, help elevate and enhance the human immune system, thereby increasing its defiance against many disease problems. Microorganisms also play an essential role in biopreservation, extending shelf-life, eco-friendly replacement of chemical methods, and production of higher digestible nutrients (nutraceuticals).

---

H. Joshi · G. Pant (✉)
Department of Microbiology, Graphic Era Deemed to be University, Dehradun, Uttarakhand, India

M. Pant
Department of Biotechnology, Graphic Era Deemed to be University, Dehradun, Uttarakhand, India

G. Kumar (✉)
Department of Microbiology, School of Bioengineering and Biosciences, Lovely Professional University, Punjab, India



A. Karnwal, A. R. Mohammad Said Al-Tawaha (eds.), *Food Microbial Sustainability*, https://doi.org/10.1007/978-981-99-4784-3_4

**Keywords** Fermentation · Single-cell proteins · Biopreservation · Prebiotics and probiotics · Biogenic

## 4.1 Introduction

Since primitive times, we have seen that microorganisms have been involved in preparing food (locally available) as a traditional practice (Materia et al., 2021). This knowledge has been passed on to future generations with no perception of the budding role of microorganisms in the process. The reasons that govern the development of these spontaneously fermented foods are decisive environment, raw materials accessibility, socio-cultural ethos, and ethical choices. These adaptation determined the selection of microorganisms in the preparation of traditional fermented foods to the substrate or by regulating fermentation conditions and are generally present either in the surroundings, in the dishes, or on the ingredients (Mannaa et al., 2021). A variety of fermented food is soy sauce (shoyu), black rice vinegar (kurosu), kimchi, papad, bhalle, makgeolli, dahi, idli, dosa, etc.

## 4.2 The Industrial Microbial Products

The branch of biotechnology that applies microbial science to produce industrial products in mass quantities by using microbial factories is called industrial microbiology. The manipulation of an organism is done to yield a specific product with many utilizations to the real world, like the building of antibiotics, amino acids, vitamins, solvents, alcohol, enzymes, dairy products, and the food industry (Singh et al., 2016). Basically, industrial microbiology adds up to obtain a product or service of economic value using microbes. Breaking down sizeable organic sugar molecules into simpler compounds with the help of the microorganism is called "fermentation" (Sharma et al., 2020). Fermentation is any process where a quality product of economic importance is mediated by or produced by involving microorganisms (Dimidi et al., 2019). To allow food fermentation an apt substratum, relevant microorganism (s) and correct environmental conditions are required, such as the right amount of moisture, temperature, pH, etc. These fermented foods carry constructive outcomes that are exercised by the fusion of living microorganisms and biologically active elements released into food because of the fermentation process (Mathur et al., 2020). There are two broad categories based on the type of fermenter microorganism: (a) Spontaneous fermentation—The type of fermentation where the naturally available microorganism is present in pure food or processing surrounding, e.g., kimchi, dahi (Sharma et al., 2022) and (b) Starter culture bacteria—To obtain desired products, a specific type of microorganism is directly inoculated in the food materials for predictable changes. It is extensively used in the food industry with beneficial properties like nutrition and health benefits (Sharma et al., 2022). Microorganism plays an essential role in industrial production. It should have properties including (a) being able to generate desired product at a large scale in less time, (b) proficient in growth, (c) less price, (d) growing in the liquid medium, (e) non-pathogenic, and (f) induced genetic manipulation.

## 4.3 Fermented Food

### *4.3.1 Oriental Fermented Food*

Mixed cultured fermentation is noticeably more common in the far east than in the western world. These include countries like Japan, India, Taiwan, China, Korea, Pakistan, Thailand, and Indonesia (Ray et al., 2014). In oriental fermentation, a few genera and species of microorganisms are selected, which in the case of fungi is restricted to *Rhizopus*, *Monascus*, *Aspergillus*, and *Neurospora*. The essential ingredients used in traditional fermented foods and beverages are cereals (rice, wheat, sorghum, and corn) (Sharma et al., 2022). The principal advantages of fermented food over processed food are adding flavors and preventing spoilage. Some of the oriental fermented foods are:

(a) Oriental soybean food: For Asian people, soybean has been an important source of proteins, fats, and flavors for thousands of years. Besides red fermented tofu and fermented black soybean, China and Japan derived the first soybean-fermented food. The main fermented soybean foods are shoyu (soya sauce), miso (fermented soy paste), sufu (Chinese soybean cheese), and natto (fermented whole soybeans) (Dimidi et al., 2019). The preparation of Japanese food Natto and Korean Food Cheonggukjang requires a two-step procedure, i.e., cooking followed by fermentation. The microbial strain *Bacillus subtilis* arises from ambient surroundings or inoculation and is used in the fermentation of cooked soybean food for 48 h. Fungal strains like *Aspergillus* and *Rhizopus* are used in the fermentation of some other cooked soybean food, e.g., tempeh, soy sauce, and doenjang resulting in large-scale disintegration of soy elements and the formation of novel bioactive compounds. (Jang et al., 2021).
(b) Korean fermented functional food: The symbolic representation of Korean culture includes kimchi, the name of a traditional fermented vegetable. Chinese cabbage (*Brassica pekinensis),* radish, garlic, green onion, ginger, red pepper, carrot, salt, etc., are the critical ingredients of kimchi (Jung et al., 2022). Due to its nutritive and health-boosting properties, like anti-microbial, anti-oxidative, anti-cancer, immune-stimulatory, and weight-controlling, kimchi is well-liked and commercialized. Heterofermentative LAB is preferred over homofermentative LAB as heterofermentative LAB produces mannitol, ethanol, and $CO_2$ as by-products and produces less acid than the homofermentative LAB, which

gives rise to finer quality of kimchi production with a pleasant taste and fresh flavor. Therefore, heterofermentative LAB is good starter culture (Chang, 2018).

(c) Japan's fermented functional food: The food stores in Japan sell a variety of fermented food products like natto, kurosu (black rice vinegar), miso (soybean-barley paste), tempeh, shoyu (soy sauce), and also food containing probiotic bacteria (Jung et al., 2022). Of many exported Japanese fermented foods, the most popular one is Miso, a soybean-fermented paste with a tangy and spicy taste often used by chefs and cooks as a flavoring or in desserts. It is most trending in Western cuisine (Allwood et al., 2021). Preparation of miso requires two consecutive steps. The first step involves the infusion of the substrate with *Aspergillus oryzae* (a mold) and is left for fermentation to yield a by-product called "koji." In the second step, Koji is fermented with bacteria and yeast with the subsequent addition of soybean mash and salt (Shurtleff & Aoyagi, 2018). During the formation of koji and miso, the microorganism involved played a vital role in imparting its distinctive taste, consistency, and nutritious form (Hui et al., 2017).

(d) Indonesia's fermented functional food: In Indonesia, soybeans, cassava, and glutinous rice are involved in preparing many types of native fermented food, i.e., tempe, kecap (soy sauce), tauco (fermented whole soybean), cassava, and glutinous rice (tape ketan) (Jung et al., 2022). Tempe, a functional food originating in Central Java, Indonesia, is the only classic soy-form fermented food not developed in Japan or China. Its primary preparation methods include washing, soaking in water, dehulling, boiling, inoculating with the tempe starter, packaging, and incubating, thereby changing soybean's biochemical and chemical nature (Romulo and Surya 2021).

### 4.3.2 *Indian Fermented Food*

The concept of "ethno-microbiology" is to acknowledge the manufacturing of valuable food products derived by natural fermentation and organoleptically and culturally more acceptable (Tamang & Lama, 2022). In multi-ethnic communities, about 1000 exotic, common, uncommon, and artisan fermented foods are consumed in different geographical regions in India. Some of the Indian fermented foods are shown in Fig. 4.1. The most important aspect of traditional fermented food is its nutritional content which has been explored repeatedly and scientifically documented. For example, Haria, a fermented beverage made of rice, shows the phenomenon of nutraceutical production at the time of its preparation which is healthy and nutritious (Sarma & Gupta, 2022). Nowadays, one of the most potent trends among consumers is their inclination toward natural food items, which is witnessing high growth in different products. For example, a dairy milk drink mixed with a large turmeric amount is called turmeric latte (Chaudhary et al., 2018).

Consumption of fermented foods, i.e., cultured dairy products or kimchi, is becoming habitual nowadays due to omega-3 fatty acids and probiotics that provide

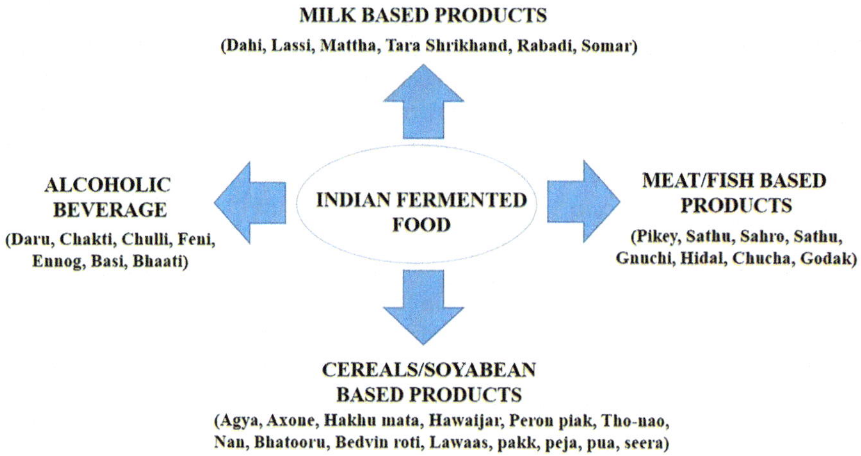


**Fig. 4.1** Indian fermented foods

**Table 4.1** Indian fermented food with their local names and details of microbes

| Substrate | Fermented foods items | Microbes | References |
|---|---|---|---|
| Milk and its derivatives | Dahi, lassi, mattha, tara shrikhand, rabadi, somar | Yeasts and LAB | Tamang (2021) |
| A mixture of rice or legumes | Uthappam, idli, dhokla, dosa | Yeasts and LAB | Joshi et al. (2019) |
| Cereals | Nan, bhatooru, bedvin roti, lawaas, pakk, peja, pua, seera | Yeasts and LAB | Tamang (2021) |
| Meat | Pikey, sathu, pila, sahro, sathu, satchu, lukter, khyopeh, kheuri | LAB and CNS | Joshi et al. (2019) |
| Fish | Gnuchi, hidal, hentak, chucha, bordia, ayaiba, godak | LAB and CNS | Thapa (2016) |
| Legumes | Amriti, bhallae, bijori, borhe, khaman, papad, rakhiya bari, vadai, bari | LAB and yeasts | Tamang (2021) |
| Soybean | Agya, axone, hakhu mata, hawaijar, peron piak, tho-nao | Bacillus spp. and LAB | Joshi et al. (2019) Tamang (2015) |
| Vegetables | Anishi, cutocie, gundruk, goyang, kahudi, kharoli, ehalpi | LAB | Tamang (2021) |
| Bamboo tender shoot | Bastanga, ekung, eup, hiring, mesu, soibum, soidon, thunbin | LAB | Behera and Balaji (2021) |
| Alcoholic beverage | Daru, chakti, chulli, feni, ennog, basi, bhaati | Filamentous molds and LAB | Tamang (2021) Joshi et al. (2019) |

several health benefits like nutrient absorption or improved digestion. Yogurt, soy yogurt, dahi, etc., are good sources of probiotics that contain bifidobacteria and LAB as major microbial domains and hence consumed (Chaudhary et al., 2018). LAB and bifidobacterial are the major microbial domains used as probiotics and are often consumed through yogurt, soy yogurt, and dahi (Mathur et al., 2020). LAB includes homo-lactic and hetero-lactic acid organisms, e.g., *Lactobacillus, Streptococcus, Lactococcus, and Bifidobacterium.* Details of microbes used in preparing some Indian fermented food with their local names are mentioned in Table 4.1.

Idli is the most common type of Indian fermented food, which is eaten in almost every part of India and is produced locally, which is a steam-cooked cake that is small, soft, white in color, and spongy made from rice and a pulse (*Phaseolus mungo*) with a slightly sour flavor. It is a rich source of protein and vitamins, specifically B complex, and at the same time, lessened in phytic acid and calories. In developing countries, it is used as a dietary additive for treating kwashiorkor and protein malnutrition (Tamang, 2021). Dosa is analogous to idli batter, except in the preparation of batter in dosa, the black gram and rice are very finely ground and somewhat thinner than idli and is fried like a pancake (Tamang, 2021). Dahi is another typical Indian fermented milk food that includes dahi which is an outcome acquired by lactic acid fermentation by the action of LAB (both single and mixed strains) or by yeast fermentation on milk (Mathur et al., 2020). The strains found in plain dahi brands are *lactobacillus acidophilus* and *Streptococcus thermophilus* (Antony et al., 2020). Rabadi is a traditional fermented food made from cereal and buttermilk-based recipes in regions of western India. The fermentation process raises the minerals (Ca, P, Fe) at different significance levels in Rabadi of all types. It is prepared by mixing buttermilk with cereal flours (pearl millet, wheat flour, and refined wheat flour) with salt and water, followed by fermentation. Buttermilk is used as a starter culture for fermentation. The antinutritional factor gets reduced after dehulling due to increased enzymatic activity (Gupta & Nagar, 2014). Traditional fermented food of Manipur made from a mixture of aroid plants petioles (*Alocasia macrorrhiza*) and sun-dried fish powder (*Esomus danricus Hamilton*) is known as hentak. It is served as curry or seasoning, or sauce. Pregnant women also consume it in their final trimester or by patients having injury or sickness. Bacteria (*Lactobacillus lactis subsp., Enterococcus faecium, L. fructosus, L. amylophilus, L. coryniformis subsp. Torquens, L. plantarum, Bacillus subtilis, B. pumilus, and Micrococcus*) and yeasts (*Candida* sp. and *Saccharomycopsis*) can be obtained from hentak (Thapa, 2016). Hawaijar is an alkaline traditional fermented food of Manipur prepared from soybean. It can be eaten directly, as sauce or seasoning, or used in curry making when mixed with vegetables. Microorganisms like *Bacillus subtilis* (dominant functional bacterium), *Bacillus licheniformis, Bacillus cereus*, and other nonbacilli bacteria, e.g., *Staphylococcus aureus, Staphylococcus sciuri*, and *Alcaligenes sp.* are used in its fermentation (Tamang, 2015). Gundruk is a non-salted fermented, dried vegetable product of Nepal prepared by leaves of mustard, rayo, cauliflower, cabbage, and radish. One of the highly rated native products of Nepal can be offered as an appetizer, a soup, or a side dish along with the main dish.

Dominant microbes used in its fermentation are *Pediococcus pentasaceus*, *Lactobacillus cellubiosus*, and *Lactobacillus plantarum* (Anal, 2019). *Lactobacillus plantarum*, *L. casei*, *L. brevis*, and *Tetragenococcus halophilus* are the microorganisms present in the ekung. The shelf-life and properties like anti-fungal and anti-microbial can be conversed by the action mechanism of *Lactobacillus plantarum* in ekung. An ethnic fermented food product of Arunachal prepared from bamboo shoots. It can be cooked with vegetables, fish, or meat (Behera & Balaji, 2021).

## 4.4 Single-Cell Protein (SCP)

The world's challenge was the increasing population, which also generated the problem of providing necessary food and nutrients (Kameshwari et al., 2020). Therefore in 1996, new sources such as bacteria, Yeast, fungi, and algae, were used to build called single-cell protein, protein biomass (Gour et al., 2015). These microorganisms utilize the available carbon and nitrogen in these materials and turn them into proteins of high quality and can be grown using inexpensive substrates like wood, sawdust, corn cobs, and even on human and animal waste. It was Carol L. Wilson who coined the term single-cell protein. For the problem of protein deficiency faced by humanity, manufacturing SCP offers an unconventional but plausible solution. A dried mass of microorganisms used as a protein supplement for humans, and so for animal feed, is called single-cell protein.

Its characteristics include high protein content (60–82%), carbohydrates, nucleic acids, minerals, vitamins, and fats; the most vital are lysine and methionine. Lysine and methionine are regarded as the limiting factors for human growth and development; hence also called treasures of nutrients. These SCPs incorporate a variety of nutrients like proteins, fats, carbohydrates, ash ingredients, water, and many other elements such as phosphorus and potassium. Not only does single-cell protein carries nutritional benefits but also it can be produced throughout the year. Moreover, there are many advantages with SCP as in its preparation, and waste material is used as a substrate, it requires a small piece of land for its production and can be made in less time. Depending upon the microorganism used, the nutritional and food values of SCP depend. They are also added as a partial replacement for fishmeal. They also help in decreasing pollution by applying agro-industrial residues in bioprocesses. Therefore, it is becoming favorable in the market and helps waste recycling.

**Bacteria** Certain qualities make bacteria suitable for SCP production, including high-speed growth and short propagation time. Carbohydrates (starch or sugars), gaseous or liquid hydrocarbons (methane and petroleum fractions to petrochemicals methanol, ethanol) are raw materials for bacterial growth (Gour et al., 2015). *Bacillus megaterium*, *Lactobacillus species*, *Aeromonas hydrophila*, *Bacillus subtilis*, *Cellulomonas species*, etc., are various bacterial species used in the production of SCP.

**Fungi** Many fungal species are used as a source of protein-rich food (Bhalla et al. 2007), and many other filamentous species, i.e., *Chaetomium celluloliticum, Aspergillus fumigates, Aspergillus niger, Aspergillus oryzae* are also used as a source of single-cell protein. At a second international conference convened at MIT in 1973, it was revealed that these filamentous fungi and actinomycetes could produce protein from various substrates. During World War II, the cultures of Fusarium and Rhizopus were used as trials to grow in fermentation as a protein food source (Gour et al., 2015).

**Algae** Algae are a choice of microbes that are used as SCP. The thread-like microscopic cyanobacterium belonging to the Oscillatoriaceae family is called *Spirulina.* Linoleic acid [LA], linolenic acid [GLA], phycobiliproteins like fatty acids, minerals like iron, phycocyanin, and allophycocyanin-like amino acids being the most important; the highest values being leucine, valine, and isoleucine (Kameshwari et al., 2020). The cellulose cell walls are absent in the *Spirulina,* so it does not need any physical and chemical processes to become digestible. About 65–70% of crude proteins and almost all vitamins are present in *Spirulina*. Thus, it can be used as a significant source for protein deficiency. *Spirulina* can be utilized as a considerable means for protein deficiency, comprising eight necessary and eight non-necessary amino acids. One of the essential phytochemicals rich in protein is C-phycocyanin, also present in Spirulina (Kameshwari et al., 2020). So, this concludes that microalgae *Spirulina* owns large numbers of nutrients and vitamins and can be used as a regular nutrient source and medicine for diseases (Kameshwari et al., 2020). The health benefits of *Spirulina* are shown in Fig. 4.2.

*Spirulina* carries several nutrients, essential amino acids, vitamins, beta carotene, vitamin B, etc. It has 5100% more iron than spinach and nearly 180% more calcium than milk, making it an extraordinary nutrient and antioxidant agent (Kameshwari et al., 2020). It contains sulfated polysaccharides, which release its antiviral activity and repress the replication. The viral multiplication gets lessened by 50% by the effectual concentration of Spirulina's calcium (Kameshwari et al., 2020). The suitable anti-inflammatory property of *Spirulina* is due to its vast phycocyanin content and anti-oxidative properties. Anti-peroxidative property of *Spirulina fusiformis* protects against rheumatoid arthritis. (Kameshwari et al., 2020). Hyperglycerolemia and hypercholestolemia can be regulated by practicing *Spirulina* as a dietary additive (Kameshwari et al., 2020). *Spirulina* as a viable food helps maintain a positive intestinal environment, especially Bifidus and Lactobacillus, promoting an active population of these species and maintaining healthy surroundings with lesser potential disputes from pathogens, e.g., *Candida albicans* and *E. coli* (Kameshwari et al., 2020). *Spirulina* encourages elevating immunity and strengthening the body against viral infection. The cells of the innate immune system get determined by Spirulina to enhance the immune system and mucosal elements. *Spirulina* is efficacious in blocking carcinogens also. The consumption of Spirulina is a primary mucosal barrier against infection that conserves the functional intestinal epithelium. It is also helpful as a treatment against hay fever and

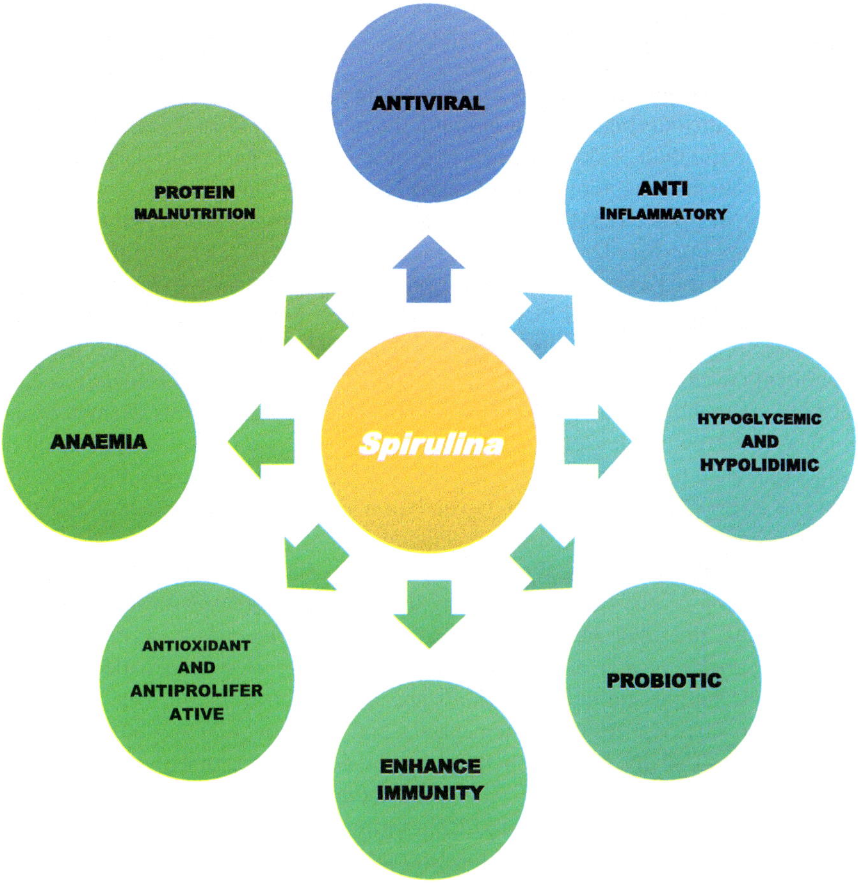


**Fig. 4.2** Health benefits of *Spirulina*

allergic rhinitis (Kameshwari et al., 2020). It has antiproliferative properties against cancer cells. Hepatocellular carcinoma cells (HEPG2) and Human colon carcinoma cells (HCT116) can be inhibited by the water extracted from *Spirulina platensis* found in invitro studies. In humans and mice, Spirulina extracts help to inhibit the growth of liver and lung cancer cells (Kameshwari et al., 2020). Constant *Spirulina* supplements prescribed to aged people suffering from anemia result in an increased concentration of mean capsular hemoglobin in the blood, eventually leading to increased CBC (Kameshwari et al., 2020).

## 4.5 Alcohol Beverages

Almost in every culture on every continent, one of the essential parts of the human diet is alcoholic beverages. It was not until the mid-1600s when the revelation came that the fermentation of alcoholic beverages was connected to a microorganism with the observations of Antonie van Leeuwenhoek in the Netherlands. Two hundred years later, in France (1850–1860), Louis Pasteur assuredly illustrated the conversion of worts into beer by its metabolic action and fruit juices into wines. In 1882, after the work of Pasteur, Hansen used pure cultures of yeasts to transform worts into beer and demonstrated that these are responsible for beer fermentation. This was how, for alcoholic fermentation, the "pure culture" technology began. The most commonly used microorganism in alcohol preparation is yeast. It is widely used in alcohol or ethanol production at the commercial level because of its characteristics, such as fast growth rates, efficient glucose repression, efficient ethanol production, and a tolerance for environmental stresses, like high ethanol concentration and low oxygen levels (Britannica, 2020). Given below in Table 4.2 are types of alcoholic beverages with different preparation techniques using microbes.

Fermentation is the critical process in order to procure any alcoholic beverage. By application of distillation, the strength of alcohol increases. In manufacturing beers, wines, and other alcoholic drinks, the critical fermenter or alcohol producer is yeast. *Saccharomyces cerevisiae* is the dominant yeast species called brewer's yeast. Both natural and cultured strains are used in fermentation. Due to its easy adjustment, sufficient fermentative capacity, and fast growth, it is used in alcoholic fermentation. Usually, most non-*Saccharomyces* yeasts cannot pull through the concentrations of $SO_2$, but *S. Cerevisiae* can endure it (Maicas, 2020). Whiskey is a kind of alcoholic beverage prepared from fermented grain mash like wheat, corn, barley, malted rye, etc. it tastes like roasted malted grain flavor with oak undertones.

ABV is 40–50% or higher. It is matured in oak casks for a certain period. Rum is another distilled beverage prepared from by-products obtained from sugarcanes like molasses or sugarcane juices that undergo distillation followed by fermentation and aging. Irrespective of the base, rum has a sweet flavor that can vary widely depending on its type. The alcohol content or ABV is 40% which can be increased to 75%. The essential factors that affect rum's quality, taste, color, quality, and viscosity are its fermenting material and method of fermentation that involves steps of sugar extraction, fermentation by yeast for 36–48 h, distillation, aging, blending, and bottling. The blending process involves mixing different ages of heavy and light rums, which involves different styles to generate a peculiar blend of tastes (Yildirum, 2021). Brandy is a distilled alcoholic beverage prepared from grapes and can also be made from other fruits like apples, peaches, apricots, and cherries. Brandy typically contains 40% ABV (alcohol by volume). It is aged in oak and often blended. It tastes like fruity burnt wine with a pleasant oakiness. Beer processing involves the conversion of starch grains into sugars. This sugar is extracted with water by fermentation with yeast strains to give light, carbonated beverages.

**Table 4.2** Types of alcoholic beverages

| Alcoholic beverage | Beer | Wine | Whisky | Rum | Brandy |
|---|---|---|---|---|---|
| Type | Brewed and fermented | Brewed and fermented | Distilled type | Distilled type | Distilled type |
| Prepared from | Malt (kiln-dried germinated barley), hops (*Humulus lupulus*) | Technically from grapes, but also made from apples, cranberries, plums | Mashed grains of wheat, corn, barley, malted rye | Sugarcane by-product called molasses or distillation of sugarcane juices | Grapes and can also be prepared from apples, apricots, peaches |
| Microorganism involved | Yeast<br>Top strain<br>*Saccharomyces cerevisiae*<br>Bottom strain<br>*Saccharomyces carlsbergensis* | LAB (*Lactobacillus plantarum*) responsible for malolactic fermentation followed by yeast for alcoholic fermentation | *L. brevis*, *L. fermentum*, *L. paracasei*, *Saccharomyces cerevisiae* | *Saccharomyces cerevisiae* | *Saccharomyces cerevisiae* |
| Flavor | Crisp, tart, malt, and sweet | Can be sweet, salty, sour, bitter depending upon wine region | Roasted malted grain flavor with oak undertones | Sweet, grassy, funky, earthy | Fruity burnt wine with a pleasant oakiness |
| Alcohol by volume (ABV) | 4.2–5% | 5–23%<br>Average: 12% | 40–50% or higher | 40% and can go up to 75% | 40% |

Beer production requires malting followed by milling, mashing, extract separation, hop addition, boiling, removal of hops, precipitation, cooling, aeration, fermentation, and yeast from young beer. The last steps involve aging, maturing, and packaging (Thomas W. Young, 2022). Beers contain starch as it goes through the process of enzymatic splitting during preparation (Britannica, 2020). Wine involves two steps in its production. The first step involves LAB for malolactic fermentation (MLF), which enhances wine's aroma, upgrades microbes' stability, and decreases wine's acidity. The lactic acid bacteria convert dicarboxylic malic acid into monocarboxylic malic acid and $CO_2$ with no free intermediate product (Virdis et al., 2021). The second step involves alcoholic fermentation by the yeast *Saccharomyces cerevisiae* to improve the consistency of the final product. Traditionally wines are prepared from grape juices, but they can also be made from apples, cranberries, and plums. The availability of carbon and nitrogen sources in grape juices gets rapidly consumed by *S. cerevisiae* along with high tolerance of ethanol, giving an advantage over other yeast. Nitrogen, the critical nutrient of wine fermentation, increases biomass production and accelerates sugar utilization, while its insufficiency causes slow and slack fermentation (Comitini et al., 2021).

## 4.6 Fermented Dairy Products

Microorganisms exclusively integrate Vitamin B12, which humans absorb from food; therefore, the excellent sources of this Vitamin B12 are foods of ruminant origin, so to meet the official daily B12 intake, dairy and meat products efforts in doing so. Since ancient times dairy products like milk and fermented products have been part of the human diet, which provides nutrients, as mentioned in Fig. 4.3.

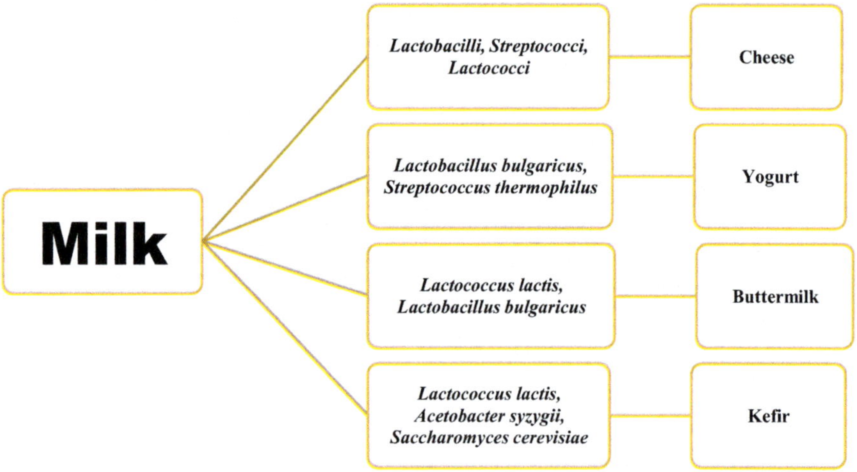


**Fig. 4.3** Fermented milk products and microorganisms involved

Behind the production of milk or its derivative products lies the diverse microbiota (*Lactobacillus, Lactococcus, and Leuconostoc),* which elevates the organoleptic and physiochemical characteristics of food. The shelf-life of the farmed milk can be extended by fermentation while maintaining its safety as it is a good source of hydration and nutrients (Reyes-Gavilán et al., 2015). Primarily, the fermentation of lactic acid is accomplished by lactic acid bacteria, where the acidification is carried out by starter lactic acid bacteria (LAB). In contrast, other LAB, molds, and yeast become overpowering during ripening and are responsible for the development of aroma and textures in these dairy products (Anal, 2019). Through the action of these microbial starter cultures, not only the flavor and texture of fermented milk but also play the chief role in manufacturing bioactive components that transmit antioxidant, anti-diabetic, anti-allergic, and anti-hypertensive properties to the raw material.

The leading players during lactic acid fermentation are LAB which convert lactose into lactic acid, which results in more acidity, making the growth condition of the microorganism other than LAB increasingly unfavorable. The milk proteins get precipitated by lactic acid bacteria and thus usually provides a thicker consistency than milk. *Lactobacillus*, *Streptococcus*, *Enterococcus*, and *Lactococcus are the LAB species present in milk* most often used (Quigley et al., 2013). Recently, the interest of consumers in healthy products has increased. To provide good health benefits, the high scope of probiotics led to expanding commercial and scientific interests in using microbial administration as a health-promoting approach. The kind of microorganisms presents inside these products depends on the quality of taste, flavor, and strong abilities of the product. Microorganisms are integral to yogurt, cream, sour, cheese, and other products. The consumption of yogurt, kefir, and other fermented foods has been considered healthy due to various nutritional benefits. *Streptococcus thermophilus* and *Lactobacillus delbrueckii*, *Bifidobacterium*, and *Lactobacillus* strains are the starter cultures that are added due to their probiotic properties. *Lactobacillus bulgaricus* and *Streptococcus thermophilus* are the two leading bacterial species responsible for the fermentation of Classic yogurt (Nagaoka, 2019). Many studies concluded that yogurt consumption and other fermented foods might upgrade intestinal and extraintestinal health, thus helping in the treatment of infectious diarrhea, increasing anti-inflammatory responses and immune responses, decreasing respiratory infections, etc. (Kok & Hutkins, 2018). Cheese is one of the famous and classic fermented dairy products, which is a good source of fat and milk protein with LAB as a microbial strain. In the manufacturing of cheese, the steps are somewhat similar to yogurt, but acidification is done, production of cheese is similar to yogurt.

However, after acidification, usually, with *Lactobacilli*, *Streptococci*, and *Lactococci*, the intermediate product is separated from the whey, and it is best suited to lactose-intolerant people due to low lactose content (Zheng et al., 2021). Depending on the type of cheese, the solids could go straight to packaging, or other bacteria or mold could be added for additional fermentation (e.g., *Penicillium* mold for blue cheese). In past decades, the changes in intestinal microbiota have greatly affected global health. It was considered that the creation of intestinal microbiota in childhood is the pillar in human disease development, such as obesity, allergies, etc., thereby stating that intestinal microbiota plays an essential role in maintaining a good healthy lifestyle. The collection of microorganisms present in the human intestinal tract, including helminths, bacteria, and fungi, is called intestinal microbiota. Most microorganisms in fermented foods are advantageous as they can act on the intestine and hold out against the digestive system. Fermented foods can change the balance of intestinal microbiota, enhance intestinal permeability regulation, help in vitamins and fatty chains formation, and mobilize digestive enzymes. Cancer, diabetes, cardiovascular diseases, and metabolic syndrome are the risks that get lessened with the consumption of these fermented foods as they carry antimicrobial, anti-oxidative, and anti-inflammatory properties in addition to peptides and bioactive compounds (Rosa et al., 2017). Kefir is a fermented drink prepared by a single culture of yeast and acetic and lactic acid bacteria. These cultures live in symbiotic association with each other. It is a fermented product with several health benefits, such as anti-inflammatory, anti-allergic, anti-bacterial, and anti-fungal properties. In addition, due to the presence of bioactive compounds in kefir and peptides, kefir has substantial capability to restrain the development and introduction of apoptosis in tumor cells. Kefir is anti-carcinogenic as it helps in the retardation of cancer and also helps in the induction of apoptosis in cancer cells. Many studies showed that kefir reacts against lung carcinoma, breast cancer, malignant T-lymphocytes, and different types of cancers (Azizi et al., 2021). Significant differences between kefir and yogurt are shown in Table 4.3.

Kefir carries probiotic properties along with physiological, prophylactic, and therapeutic properties that have several health benefits to humans. The reason for these health benefits is the production of bioactive compounds at the time of fermentation, along with a variety of diversified microbiota that acts harmoniously to impact these health benefits (Peluzio et al., 2021). It is acidic with low alcohol content. The carbonation of kefir grains with water or milk prepares it. Kefir, when prepared with milk, is more beneficial compared to non-dairy products. Regardless

**Table 4.3** Difference between kefir and yogurt

| Dairy products | Milk-kefir | Milk-yogurt |
|---|---|---|
| Starter cultures | Mesophilic culture | Mesophilic and thermophilic culture |
| Major microbial domain | *Streptococcus thermophilus, Lactobacillus delbrueckii, Bifidobacterium, Lactobacillus bulgaricus* | Lactic acid bacteria (*Lactococcus. lactis subsp. Lactis)*, acetic acid bacteria (*Acetobacter syzygii*), yeast (*Saccharomyces cerevisiae, S. unisporus, Candida kefyr, and Kluyveromyces marxianus* ssp.) |
| Bacteria location | Acts as probiotics; colonizes the intestinal tract | Called transient bacteria, balancing the healthy digestive tract |
| Flavor | Tart and sour | Mild to tangy |
| Characteristic property | Kefir carries probiotic properties along with physiological, prophylactic, and therapeutic properties that have several health benefits for humans | It has an excellent probiotic property which keeps the intestine healthy |

of its probiotic nature, it is unsuitable for lactose-intolerant vegan users who are allergic to dairy products. Hence, another method of its production includes the usage of non-dairy products as substrates (Sharifi et al., 2017). The insufficiency of the enzyme intestinal β-galactosidase causes lactose indigestion in many consumers, making them lactose-intolerant. However, this enzyme intestinal β-galactosidase is present naturally in kefir grains and causes a reduction of lactose content at the time of its fermentation, making the end-product (Kefir drink) or the outcome its consumption fit for the lactose-intolerant people (Peluzio et al., 2021).

## 4.7 Health Benefits and Other Biological Properties

The shelf-life of fermented foods obtained from milk, plants, or meat is usually extended than that of natural raw material as these raw food get easily spoiled because of their high nutritive value and high water content. Food fermentation utilizes LAB, yeast, and molds in its production. Therefore, these LAB occur naturally in fermented foods, playing an essential role in fermentation. LAB prepared from single or mixed cultures have several benefits to human health due to its potential to form promising compounds that help balance healthy gut microbiota, thus improving the digestive tract. They also can ameliorate immune responses against pathogenic bacteria. *Pediococcus, Bacillus, and Bifidobacterium Lactobacillus* are the microbial genera used as probiotics (Soemarie et al., 2021). LAB are Gram-positive, non-sporous, immotile bacteria that depend on carbohydrate fermentation for their active growth. The final product of lactic acid fermentation could be an abundant quantity of lactic acid (homofermentative) or in combination with ethanol, $CO_2$, and acetic acid (heterofermentative). During fermentation, these deliver a range of health-regulating elements and signaling molecules that interrelate with the host's intestinal microbiome creating an active and healthful microbiome. Thus, LAB-fermented foods describe secure, affordable, and authentic food production, enhancing human health in general (Castellone et al., 2021).

### 4.7.1 *Biopreservation*

The shelf-life of food items can be increased via pure or measured microbiota or anti-microbials. This procedure is called biopreservation. LAB and their metabolites are the chief interest organism used in this preservation technique. It is a preservation process that involves the exploitation of pathogenic microorganism and their metabolites, thereby maintaining the safety and quality of food. It is one of the most suitable preservation techniques effective in increasing food shelf-life with lesser nutritional losses and organoleptically safe. Microorganisms are the new branch of science used in biopreservation that renders food rotting and enhances the texture and flavor of food (Singh et al., 2022). The generally regarded as safe

(GRAS) status in the United States and qualified presumption of safety (QPS) status in Europe, the most favorable replacement for chemical preservatives is biopreservation through LAB in the dairy industry (Shi and Maktabdar 2022). The primary compound produced by LAB is bacteriocin, hydrogen peroxide, and organic acids. NISIN is FDA-approved and is the most common LAB Bacteriocin.

### *4.7.2 Extended Shelf-Life*

Evidence supports that fermentation extends the shelf-life of foods (Mannaa et al., 2021). A broad extent of reaction causes the rotting of food items, which can be significantly prevented through many preservation techniques that inhibit the growth of these microbes responsible for decomposing reactions. These procedures include conversing, freezing, chilling, drying, curing, adding preservatives, fermenting, and vacuum packing acidifying. Nowadays, the increasing appeal of the users to keep food fresh has led to the search for new ways of keeping food fresh along with health qualities. Fruits and vegetables are generally harvested when they are entirely ripened, but the problem is that they continue to rip when stored, resulting in short shelf-life and economic and commercial loss as they get wilted and undergo weight loss. This made the food industry search for new methods to protect the quality of foods by using microorganisms (Alsoufi & Aziz, 2017).

Biological product types:

(a) **Pullulan**: Pullulan ($C_6H_{10}O_5$) $H_2O$ is a neutral, water-soluble polysaccharide prepared by fermentation processing of starch by the fungus *Aureobasidium pullulans*. It is non-toxic to human as well as animals, is colorless, odorless, tasteless, and thin and has suitable adhering property so that it can be applied as coating and act as a barrier between $CO_2$ and $O_2$ that inhibit the growth of microorganism, ensuring food safety from microbes and protecting food from decaying. This pullulan is significant in the pharmaceutical, food, chemical, and agricultural industries (Alsoufi & Aziz, 2017).

(b) **Killer toxin**: Bakery yeast (*Saccharomyces cerevisiae*) can make certain types of proteins that inhibit the growth of harmful microorganisms responsible for food decaying by participating in the same growth medium as them. These proteins are acidic and depend upon surrounding factors such as ventilation, temperature, and pH. They can prevent the growth of microorganisms either through direct influence or by inhibiting bacterial toxins from reaching receptor cells. So yeast can inhibit pathogenic bacteria colonization in the GI tract through contesting for nutrients that decrease the effect of other microorganisms (Alsoufi & Aziz, 2017). Therefore, these biological products are being used to prevent and inhibit the growth of microbes responsible for food decomposition, extending the shelf-life of food and making it economically and commercially a more sustainable product for the market. Pasteurization, irradiation, and sterilization are the procedures that extend shelf-life by adding inactivating microbes. Ultrahigh pressure, electroporation, mono-thermosonication, and the addition of bacteriolytic enzymes are nowadays new trending preservation techniques.

(c) **Higher digestible nutrients**: Nutraceuticals have been launched as new food labels in the pharmaceutical and food markets over the past few years. In producing these nutraceuticals, particular *Lactococcus lactis* (LAB) are ideal cell factories. Their characteristic properties involved malignancy prevention, anti-aging property, and disease treatment and prevention (AlAli et al., 2021). There are significant sources of bioactive derivatives brought together with palatable sources containing the following properties like (a) antioxidants, (b) phytochemicals, (c) amino acids, (d) fatty acids, (e) probiotic properties, (f) low-calorie sugars, (g) B-vitamins.

### *4.7.3 Food Additives*

Many fungal and bacterial strains of microbes are used to prepare fermented foods like milk, fruits, vegetables, meats, etc., which imparts color, texture, odor, taste, and other benefits and qualities. For example, the ingestion of *lactobacilli* through cultured milk is easily digested by the lactose-intolerant individual because Lactobacilli engulf this lactose in the gastrointestinal tract. The microorganisms selected to establish and improve food production depends on the underlying metabolic pathways applicable to manufacturing food additives. In the case of naturally overproducing microorganisms, testing of the demanded product is done, but in the case of non-naturally produced microorganisms, it is rational to use the microbial host, which could be genetically handled (Kallscheuer, 2018). Common conditions needed for microbial strain production include (a) Quick growth, (b) low-cost in cultivation, (c) non-pathogenic, (d) high biomass concentration concerning defined growth media. Therefore, these food additives should be able to generate food products that have GRAS ("generally recognized as safe") status (Kallscheuer, 2018). Classification of microbially accessible food additives according to the metabolic pathways being involved in their biosynthesis.

**Carboxylic Acids** Carboxylic acids are primarily used as acid regulators and preservatives. They are mainly listed in the E number Catalog that can be obtained through the glycolysis or TCA cycle called the tricarboxylic acid cycle. These include fumaric acid, succinic acid, citric acid, lactic acid, malic acid, etc. (Kallscheuer, 2018).

**Amino Acids** EU has approved three proteinogenic amino acids as flavor modifiers, namely L-glutamic acid (E620), glycine (E640), and L-cysteine (E920). The reason behind the "*umami*" taste of foodstuff is the L-glutamic acid and, in particular, its monosodium salt (E621). L-Cysteine is also used as a flour treatment agent to enhance baking functionality. These three amino acids are hydrolyzed from proteins and can also be produced from engineered microorganisms (Kallscheuer, 2018).

**Terpenoids** The aliphatic compounds obtained from units of isoprene (2-methyl-1,3-butadiene) are mostly terpenoids (also referred to as isoprenoids). They are used as antioxidants and color enhancers (Kallscheuer, 2018).

### *4.7.4 Probiotics*

According to the food and Agriculture Organization, probiotics are defined as "Live microorganisms which, when administered in adequate amounts, confer a health benefit on the host" (Dahiya & Nigam, 2022). Lilly and Stillwell in 1963 first used the word "probiotics." They are substances that encourage the growth of another organism. Probiotics may be a desirable option to re-establish microbial equilibrium and prevent diseases. Besides a biogenic effect, fermented food also has health benefits due to a probiotic activity, which is the interactions of host with the ingested live microorganisms that must be supervised in sufficient amounts to converse a beneficial role on the host. It was acknowledged that in humans developing metabolic diseases like inflammatory bowel diseases, obesity, and diabetes, the gut microbiota may play a part in the growth of these diseases.

Interestingly, the usage of it has been shown to improve the diagnosis and control of those diseases (Singh et al., 2022). Microorganisms have been preferred for years as probiotics that can enhance health benefits. These probiotics generally belong to the LAB and bifidobacteriaceae group, and in the case of others, i.e., Yeast is *Enterococcus spp.* and *Saccharomyces boulardii.* Probiotics are discussed over a long period due to their inherent aspiring nature and probability of carrying resistance genes of anti-microbials (Castellone et al., 2021). The health benefits assigned to probiotic microorganisms are complicated. Features of probiotics include the synthesis of beneficial compounds, resistance toward pathogenic bacteria, and induction and regulation of immune responses. Probiotics concerning the intestine should exhibit the following characteristics: (a) should be acid-resistant, (b) must have the capability of capturing intestinal wall, (c) should battle for nutrients, (d) should be able to survive in the severe and discriminating surrounding of GI tract, (e) should show the disaggregating effect on biliary salts. Probiotics are essential in curing and avoiding gastrointestinal diseases, so their consumption is encouraged. For example, garlic is often used as a supportive therapy for high blood pressure and cholesterol (AlAli et al., 2021). A prebiotic is a non-digestible food ingredient that confers a benefit on the host by selectively stimulating one bacterium or group of bacteria in the colon with probiotic properties (Dahiya & Nigam, 2022). Therefore, they are indigestible ingredients that encourage the development of rich microorganisms that would affect gut microbiota activity. Prebiotics are used in functional foods to improve gastric health, and their examples are fructo-oligosaccharides and insulin. Both the prebiotic and probiotic are together called synbiotics. Various microorganisms are used in probiotic preparation, including bacteria (*Lactobacillus*, *Streptococcus*, *Bifidobacterium*, *Enterococcus*, etc.) and fungi (*Saccharomyces*, etc.).

## 4.8 Conclusion

Revolutions in technologies and scientific areas made by western countries revolved the fermentation activity from a household to a controlled procedure apt for industrial scale production suited for mass marketplace. Fermentation involves generating safe and healthy food organoleptically valuable, hence offering the tremendous unexploited potential for value-added products. Therefore fermentation improves the nutritional value of food by boosting micronutrient bioavailability and degrading its antinutritional components and also through the biosynthesis of amino acids, proteins, vitamins, and fiber digestibility. Fermented foods and beverages are beneficial to health due to active microorganisms and are marketed globally as functional, therapeutic, nutraceutical, bio-foods, and health foods. These fermented foods thus benefit the consumers compared to simple food in terms of antioxidant, organoleptic qualities, probiotic properties, production of peptides, strengthening immune system, prevention from malignancy, and anti-microbial activity. Fermentation not only provides health advantages but also increases the shelf-life of food, providing microbial safety of food, reducing the toxicity of substrate, and making food more digestible. Advancement of research in the field of lactic acid bacteria helps provide better strains for food fermentation, benefiting both producer and consumer. The purpose of this chapter is to review the conception and possible constructive valuable properties of microorganisms in food fermentation.

## References

AlAli, M., Alqubaisy, M., Aljaafari, M. N., AlAli, A. O., Baqais, L., Molouki, A., Abushelaibi, A., Lai, K. S., & Lim, S. H. E. (2021). Nutraceuticals: Transformation of conventional foods into health promoters/disease preventers and safety considerations. *Molecules, 26*, 2540. https://doi.org/10.3390/molecules26092540

Allwood, J. G., Wakeling, L. T., & Bean, D. C. (2021). Fermentation and the microbial community of Japanese Koji and Miso. *Journal of Food Science, 86*, 2194–2207. https://doi.org/10.1111/1750-3841.15773

Alsoufi, M. A., & Aziz, R. A. (2017). Extending shelf life of fruits by using some microorganisms biological products. *International Journal of Molecular Biology: Open Access, 2*(5), 141–144. https://doi.org/10.15406/ijmboa.2017.02.00032

Anal, A. K. (2019). Quality ingredients and safety concerns for traditional fermented foods and beverages from Asia: A review. *Fermentation, 5*(1), 8. https://doi.org/10.3390/fermentation5010008

Antony, U., Ilango, S., & Chelliah, R. (2020). *Ethnic fermented foods and beverages of Tamil Nadu*. Springer. https://doi.org/10.1007/978-981-15-1486-9

Azizi, N. F., Kumar, M. R., Yeap, S. K., Abdullah, J. O., Khalid, M., Omar, A. R., Osman, M. A., Mortadza, S. A. S., & Alitheen, N. B. (2021). Kefir and its biological activities. *Food, 10*(6), 1210. https://doi.org/10.3390/foods10061210

Bhalla, T. C., Sharma, N. N., & Sharma, M. (2007). Production of metabolites, industrial enzymes, aminoacids, organic acids, antibiotics, vitamins and single cell proteins. *J. Environ., 6*, 34–78. http://nsdl.niscair.res.in/jspui/handle/123456789/129

Behera, P., & Balaji, S. (2021). Health benefits of fermented bamboo shoots: The twenty-first century green gold of Northeast India. *Applied Biochemistry and Biotechnology, 193*, 1800–1812. https://doi.org/10.1007/s12010-021-03506-y

Britannica, T. E. (2020). *Sake – Alcoholic beverages*. Retrieved August 5, 2020, from https://www.britannica.com/topic/sake

Castellone, V., Bancalari, E., Rubert, J., Gatti, M., Neviani, E., & Bottari, B. (2021). Eating fermented: Health benefits of lab-fermented foods. *Food, 10*(11), 2639. https://doi.org/10.3390/foods10112639

Chang, H. C. (2018). Healthy and safe Korean traditional fermented foods: Kimchi and chongkukjang. *Journal of Ethnic Foods, 5*(3), 161–166. https://doi.org/10.1016/j.jef.2018.08.003

Chaudhary, A., Sharma, D. K., & Arora, A. (2018). Prospects of Indian traditional fermented foods agricultural sciences. *International Food Research Journal, 88*(10), 1496–1501.

Comitini, F., Agarbati, A., Canonico, L., & Ciani, M. (2021). Yeast interaction and molecular mechanism in wine fermentation: A comprehensive review. *International Journal of Molecular Sciences, 22*(14), 7754. https://doi.org/10.3390/ijms22147754

Dahiya, D., & Nigam, P. S. (2022). Probiotics, prebiotics, synbiotics, and fermented foods as potential biotics in nutrition improving health via microbiome-gut-brain axis. *Fermentation, 8*(7), 303. https://doi.org/10.3390/fermentation8070303

Dimidi, E., Cox, S. R., Rossi, M., & Whelan, K. (2019). Fermented foods: Definitions and characteristics, impact on the gut microbiota and effects on gastrointestinal health and diseases. *Nutrients, 11*, 1806. https://doi.org/10.3390/nu11081806

Gour, S., Mathur, N., Singh, A., & Bhatnagar, P. (2015). Environment Molecular Microbiology Lab., Department of Zoology, University of Rajasthan, Jaipur, India 2Dean, Life sciences, IIS University, Jaipur, India.

Gupta, V., & Nagar, R. (2014). Minerals and antinutrients profile of rabadi after different traditional preparation methods. *Journal of Food Science and Technology, 51*(8), 1617–1621. https://doi.org/10.1007/s13197-012-0667-2

Hui, W., Hou, Q., Cao, C., Xu, H., Zhen, Y., Kwok, L. Y., Sun, T., Zhang, H., & Zhang, W. (2017). Identification of microbial profile of Koji using single molecule, real-time sequencing technology. *Journal of Food Science, 82*(5), 1193–1199. https://doi.org/10.1111/1750-3841.13699

Jang, C. H., Oh, J., Lim, J. S., Kim, H. J., & Kim, J. S. (2021). Fermented soy products: Beneficial potential in neurodegenerative diseases. *Food, 10*(3), 636. https://doi.org/10.3390/foods10030636

Joshi, V., Senapati, A., & Raj, D. (2019). 4 traditional fermented foods composition and nutritive value. *Fermentation, 7*(4), 289.

Jung, S. J., Chae, S. W., & Shin, D. H. (2022). Fermented foods of Korea and their functionalities. *Fermentation, 8*, 645. https://doi.org/10.3390/fermentation8110645

Kallscheuer, N. (2018). Engineered microorganisms for the production of food additives approved by the European Union—A systematic analysis. *Frontiers in Microbiology, 9*, 1746. https://doi.org/10.3389/fmicb.2018.01746

Kameshwari, V., Selvaraj, S., & Sundaramoorthy, S. (2020). Department of Microbiology, St. Peters Institute of Higher Education and Research, Avadi, Chennai-600054, India.

Kok, C. R., & Hutkins, R. (2018). Yogurt and other fermented foods as sources of health-promoting bacteria. *Nutrition Reviews, 76*(1), 4–15. https://doi.org/10.1093/nutrit/nuy056

Maicas, S. (2020). The role of yeasts in fermentation processes. *Microorganisms, 8*, 1142. https://doi.org/10.3390/microorganisms8081142

Mannaa, M., Han, G., Seo, Y. S., & Park. (2021). Evolution of food fermentation processes and the use of multi-omics in deciphering the roles of the microbiota. *Food, 10*, 2861. https://doi.org/10.3390/foods10112861

Materia, V. C., Linnemann, A. R., & Smid, E. J. (2021). Contribution of traditional fermented foods to food systems transformation: Value addition and inclusive entrepreneurship. *Food Security, 13*, 1163–1177. https://doi.org/10.1007/s12571-021-01185-5

Mathur, H., Beresford, T. P., & Cotter, P. D. (2020). Health benefits of lactic acid bacteria (LAB) fermentates. *Nutrients, 12*(6), 1679. https://doi.org/10.3390/nu12061679

Nagaoka, S. (2019). Yogurt production. *Methods in Molecular Biology, 1887*, 45–54. https://doi.org/10.1007/978-1-4939-8907-2_5

Peluzio, M. D., Dias, M. M. E., Martinez, J. A., & Milagro, F. I. (2021). Kefir and intestinal microbiota modulation: Implications in human health. *Frontiers in Nutrition, 8*, 638740. https://doi.org/10.3389/fnut.2021.638740

Quigley, L., O'Sullivan, O., Stanton, C., Beresford, T. P., Ross, R. P., Fitzgerald, G. F., & Cotter, P. D. (2013). The complex microbiota of raw milk. *FEMS Microbiology Reviews, 37*(5), 664–698. https://doi.org/10.1111/1574-6976.12030

Ray, C. R., Sheikha, A. E., & Kumar, S. (2014). Oriental fermented functional (probiotics) foods. *Journal of Advanced Pharmaceutical Technology & Research, 12*(4), 335–339. https://doi.org/10.13140/2.1.2898.4006

Reyes-Gavilán, C., María Fernández, M., Hudson, J. A., & Korpela, R. (2015). Role of microorganisms present in dairy fermented products in health and diseases. *BioMed Research International, 2015*, 204173. https://doi.org/10.1155/2015/204173

Romulo, A., & Surya, R. (2021). Tempe: A traditional fermented food of Indonesia and its health benefits. *International Journal of Gastronomy and Food Science, 26*, 100413. https://doi.org/10.1016/j.ijgfs.2021.100413

Rosa, D., Dias, M., Grzeskowiak, L., Reis, S., Conceicao, L., & Peluzio, M. (2017). Milk kefir: Nutritional, microbiological and health benefits. *Nutrition Research Reviews, 30*(1), 82–96. https://doi.org/10.1017/S0954422416000275

Sarma, U., & Gupta, S. (2022). An overview on ethnic fermented food and beverages of India: Interplay of microbes, immunity and nutrition. *Nutrition and Health, 28*(3), 331–339. https://doi.org/10.1177/02601060221085138

Sharifi, M., Moridnia, A., Mortazavi, D., Salehi, M., Bagheri, M., & Sheikhi, A. (2017). Kefir: A powerful probiotics with anti-cancer properties. *Medical Oncology, 34*(11), 183. https://doi.org/10.1007/s12032-017-1044-9

Sharma, R., Diwan, B., & Singh, B. P. (2022). Probiotic fermentation of polyphenols: Potential sources of novel functional foods. *Food Production, Processing and Nutrition, 4*, 21. https://doi.org/10.1186/s43014-022-00101-4

Sharma, R., Garg, P., Kumar, P., Bhatia, S. K., & Kulshrestha, S. (2020). Microbial fermentation and its role in quality improvement of fermented foods. *Fermentation, 6*, 106. https://doi.org/10.3390/fermentation6040106

Shi, C., & Maktabdar, M. (2022). Lactic Acid Bacteria as biopreservation against spoilage molds in dairy products - A review. *Frontiers in Microbiology, 12*, 819684. https://doi.org/10.3389/fmicb.2021.819684

Shurtleff, W., & Aoyagi, A. (2018). *The book of miso* (2nd ed.). Soyinfo Centre.

Singh, R., Kumar, M., & Mittal, A. (2016). Microbial enzymes: Industrial progress in 21st century. *3 Biotech, 6*, 174. https://doi.org/10.1007/s13205-016-0485-8

Singh, R. P., Shadan, A., & Ma, Y. (2022). Biotechnological applications of probiotics: A multifarious weapon to disease and metabolic abnormality. *Probiotics and Antimicrobial Proteins, 14*, 1184–1210. https://doi.org/10.1007/s12602-022-09992-8

Singh, V. P. (2018). Recent approaches in food bio-preservation - A review. *Open Veterinary Journal, 8*(1), 104–111. https://doi.org/10.4314/ovj.v8i1.16

Soemarie, Y. B., Milanda, T., & Barliana, M. I. (2021). Fermented foods as probiotics: A review. *Journal of Advanced Pharmaceutical Technology & Research, 12*(4), 335–339. https://doi.org/10.4103/japtr.japtr_116_21

Tamang, J. P. (2015). Naturally fermented ethnic soybean foods of India. *Journal of Ethnic Foods, 2*(1), 8–17. https://doi.org/10.1016/j.jef.2015.02.003

Tamang, J. P. (2021). 'Ethno-microbiology' of ethnic Indian fermented foods and alcoholic beverages. *Journal of Applied Microbiology, 133*(1), 15382. https://doi.org/10.1111/jam.15382

Tamang, J. P., & Lama, S. (2022). Probiotic properties of yeasts in traditional fermented foods and beverages. *Journal of Applied Microbiology, 132*(5), 3533–3542. https://doi.org/10.1111/jam.15467

Thapa, N. (2016). Ethnic fermented and preserved fish products of India and Nepal. *Journal of Ethnic Foods, 3*(1), 69–77. https://doi.org/10.1016/j.jef.2016.02.003

Virdis, C., Sumby, K., Bartowsky, E., & Jiranek, V. (2021). Lactic acid bacteria in wine: Technological advances and evaluation of their functional role. *Frontiers in Microbiology, 11*, 612118. https://doi.org/10.3389/fmicb.2020.612118

Yildirum, H. K. (2021). Insight into role of yeast in alcoholic beverages. In *Microbial biotechnology in food and health*. Elsevier.

Young, T. W. (2022). *Beer*. Retrieved from https://www.britannia.com/topic/beer

Zheng, X., Shi, X., & Wang, B. (2021). A review on the general cheese processing technology, flavor biochemical pathways and the influence of yeasts in cheese. *Food Microbiology, 12*, 703284. https://doi.org/10.3389/fmicb.2021.703284

# Chapter 5
# Benefaction of Probiotics for Human Health

**Maryam Fatima, Sultan Mehmood Ghani, Samia Khanum, Abdel Rahman Mohammad Said Al-Tawaha, Arun Karnwal, Abhijit Dey, Abdel Razzaq M. Al Tawaha, and Iftikhar Ali**

**Abstract** The human microbiota is made up of a diverse and vibrant set of bacteria that are only found in humans. Over 100–1000 microbes in the human gut essentially regulate the host's internal environment and, thus, significantly impact host health. This excellent symbiotic interaction has drawn much attention from researchers. Live microbes, called probiotics, are helpful bacteria that provide dietary and clinical benefits due to their various beneficial characteristics. Probiotic strains have a substantial impact on improving human health. Bacteria, yeast, and the bacteria and yeast species *Lactobacillus* and *Bifidobacterium* are among the most often utilized probiotic strains. Microbiology and biotechnology researchers are increasingly interested in probiotics because of the potential benefits they may provide to human health in the face of infection. Numerous high-quality clinical investigations have shown that using probiotics is beneficial. Inflammatory bowel disease, aller-

M. Fatima · S. M. Ghani
National Institute of Food Science and Technology, University of Agriculture Faisalabad, Faisalabad, Pakistan

S. Khanum
Department of Botany, University of the Punjab, Lahore, Pakistan

A. R. Mohammad Said Al-Tawaha (✉)
Department of Biological Sciences, Al Hussein Bin Talal University, Ma'an, Jordan
e-mail: Abdel-al-tawaha@ahu.edu.jo

A. Karnwal
Department of Microbiology, School of Bioengineering and BioSciences, Lovely Professional University, Phagwara, India

A. Dey
Department of Life Sciences, Presidency University, Kolkata, India

A. R. M. Al Tawaha
Faculty of Agriculture, Department of Crop Science, Universiti Putra Malaysia, Serdang, Selangor, Malaysia

I. Ali
School of Life Science, Chinese University of Hong Kong, Hong Kong, Hong Kong

Centre for Plant Science and Biodiversity, University of Swat, Charbagh, Pakistan


A. Karnwal, A. R. Mohammad Said Al-Tawaha (eds.), *Food Microbial Sustainability*, https://doi.org/10.1007/978-981-99-4784-3_5

gies, urogenital infections, *Helicobacter pylori* infections, and gastrointestinal problems are only some of the conditions that probiotics have shown promise in treating. This chapter intends to highlight the benefaction of probiotics on human health, the medical field, and improved living.

**Keywords** Lactobacillus · Bifidobacterium · Health safety · Probiotic · LAB · Pharmaceutical

## 5.1 Introduction

A term from the contemporary age, probiotics means "for life" and describes bacterial associations that are good for human and animal health. Probiotics are "live, non-pathogenic microorganisms" that positively affect the host's health when supplied adequately. There are also nutritional and medical advantages to eating them. Maintaining a proper ratio of helpful bacteria in the gut flora requires a diet rich in foods containing probiotics microflora. The type and amount of probiotic bacteria, such as Lactobacillus and Bifidobacterium strains, present in GIT and dietary supplements vary widely. A crucial role of the gut microbiota is maintaining human health (Dallal et al., 2015). Probiotic microbial treatments can improve the body's resident microbes' structure, composition, and function.

Additionally, it reduces pathogen invasion and colonization additional to these. It has been shown that the availability of vitamins in the human host can be increased, that probiotic bacteria can create vitamins, and that this improves GIT transit overall. The food, feed, dairy, and fermentation sectors increasingly turn to probiotics as a natural alternative to conventional medical therapies to better serve the healthcare system's needs (Anandharaj & Sivasankari, 2014). Foods that contain probiotics may serve a specific purpose in the digestive system of humans, particularly in decreasing the likelihood of and treating established human diseases. The most common method is to consume probiotic organisms through dietary items. By activating certain genes in localized host cells, probiotics also activate, alter, and control the host's immunological response. As a component of the gut–brain axis, they even control the release of gastrointestinal hormones and control brain function through bilateral neural transmission. Probiotics substantially impact the development of inflammatory bowel disease by stimulating intestinal angiogenesis through vascular endothelial growth factor receptor signalling, which controls both acute and chronic inflammation in the intestinal mucosa (Chen et al., 2013).

Products containing probiotics may serve a specific purpose in the human digestive system, especially in lowering the risk of and treating established human diseases. The most common method is to consume probiotic organisms through dietary items. Probiotics may help lessen objective and subjective symptoms of particular diseases, lowering the chance of developing those conditions. Probiotic microorganisms are often sold as dried or deep-frozen culture concentrations that may be added to a food matrix. Probiotics' role in improving the immune system overall is well researched. The natural flora in the GIT is crucial in maintaining health (Pacini & Ruggiero, 2017). These include enteric toxin reduction, metabolic processes, defensive functions that rely on microorganisms, and structure and histological function. Due to their versatility and long shelf life, probiotics are frequently used in meals, sauces, and beverage fermentation. Mechanisms may be delivered in various ways, each optimized for a certain age range. It may be taken orally as a supplement or as part of a probiotic meal. Probiotic cells can survive within the harsh conditions of the gastrointestinal tract (GIT) and exert their health benefits because they are always carried in food.

However, it must remain stable throughout the gastrointestinal system to obtain a total plate count of at least $10^6$ CFU/g. Although it is often believed that dairy products are the primary sources of probiotics, many other food items outside dairy also contain probiotics (Nagpal et al., 2012). A vast range of bacteria and other microorganisms, including yeast, fall under the umbrella term "probiotics," each performing a somewhat different function. Thus, not all probiotics will necessarily share a given medical benefit just because one strain has a such advantage. It is necessary to evaluate the effects of each strain individually (Kerry et al., 2018). The goals of this chapter are to encourage further study into probiotics, review the most current discoveries on their effect on human health, and increase overall awareness of them.

## 5.2 Prebiotics, Synbiotics, Probiotics, and Postbiotics

Probiotics, prebiotics, and synbiotics have been given many different definitions, but the one that best describes them is a microbiome or a collection of microorganisms living in the gastrointestinal tract that internally provide nutrition to the host. They are frequently ingested as formulations containing live, active microbial cultures that include bacterium isolated from the environment, such as *lactobacilli*, *lactococci*, or *bifidobacteria* (Bongaerts & Severijnen, 2016).

### *5.2.1 Postbiotics*

In the absence of living organisms, bacterial metabolites may nevertheless affect signal transduction pathways and barrier functions. These probiotic microorganisms produce postbiotics, which are generally understood to be non-viable bacterial metabolites or metabolic by-products with biologic functions in the host (Patel & Denning, 2013). However, it has been discovered that some heat-killed probiotics can also retain significant bacterial formations that may have biological activities in host cells. Bacteriocins, organic acids, ethanol, diacetyl, acetaldehydes, and hydrogen peroxide are examples of postbiotics, which are generally defined as remnants of bacterial metabolism. Such metabolic products can be employed as a substitute for antibiotics since they have a wide-ranging inhibitory effect towards harmful bacteria. Postbiotics are non-viable bacterial metabolites from probiotics that are not harmful or pathogenic and immune to degradation by enzymes present in mammals. In some cases, postbiotics can also promote angiogenesis in epithelial cells and barrier function against microorganisms like Saccharomyces boulardii (Giorgetti et al., 2015).

### 5.2.2 *Prebiotics*

Prebiotics are the specific substances that alter the gastrointestinal microflora but are difficult for humans to digest. They were created due to further research into probiotics. They have a specific function in promoting the development or activities of beneficial species of bacteria in the gut. Prebiotics can easily provide these needs because they are easily fermented by gut microbiota, including such bifidobacteria, which is a primary energy source for epithelial cells in the colon. Some different intestinal microbes ferment these non-digestible oligosaccharides in addition to bifidobacteria (Rastall & Gibson, 2015). Prebiotics may be derived naturally from foods like grains, fruits, and vegetables daily. Prebiotics have numerous beneficial properties in addition to their role as an energy source, including a decrease in the frequency and length of diarrhoea, relief from inflammatory and other gastrointestinal illnesses' symptoms, and protection against colon cancer. Prebiotics may also reduce certain cardiovascular disease risk factors, increase mineral bioavailability and absorption, and promote fullness and weight reduction (Egan & Van Sinderen, 2018).

### 5.2.3 *Synbiotics*

Synbiotics are the combination of prebiotic substances and probiotics that aid in boosting the well-being and transplantation of live microbial nutritional supplementation in the intestinal tract, resulting from advancements in microbial research. When the probiotic and prebiotic cooperate in the living system, the synergistic effects are promoted more effectively. According to growing scientific research, the symbiotic link among prebiotics and probiotics has a major positive impact on health. Due to the rising public knowledge of the advantages for gut health, illness prevention, and treatment, there has been a constant rise in the financial importance of functional foods containing synbiotics. Synbiotics' probiotic and prebiotic composition offers a range of possible health advantages, including as enhanced immunity, decreased inflammation, and better digestive function. Prebiotics can help feed the beneficial bacteria in the gut, while probiotics in synbiotics can aid digestion and enhance nutrient absorption (Tufarelli & Laudadio, 2016). Synbiotics can assist in increasing the growth of beneficial bacteria while decreasing the number of bad bacteria in the stomach, which can reduce inflammation and enhance general health. It is crucial to remember that the exact health advantages of synbiotics might change based on the probiotic and prebiotic strains used by the user's lifestyle and general health. A healthcare professional should be consulted before beginning a synbiotic supplement since certain people may have particular medical issues that may be affected by the usage of probiotics (Pandey et al., 2015).

## 5.3 History of Probiotics

Elie Metchnikoff, the recipient of the Nobel Prize, developed the word "probiotics," which refers to lactic acid bacteria, a group of highly advantageous microbes. "The non-digested substances and faeces inside the colon are the major reasons for the generation of toxic substances that decrease the average human life", this was the basic idea offered by Metchnikoff in his masterwork "The Prolongation of Life", which clarified and identified the association of regular intake of lactic acid fermented milk by Bulgarian and Caucasian people and their general well-being and average life span was improved (Zommiti et al., 2020). Scientific research in the probiotics and their potential benefits has increased significantly throughout recent years. Probiotics have demonstrated strong potential in clinical practice, despite their positive impacts on human health as beneficial microorganisms. There are solid indications that probiotic bacteria help prevent or treat a wide range of diseases and infections, especially when they are directly related to gastrointestinal issues in both children and adults. Many scientific researchers have studied the critical part of probiotic bacteria and their characteristics to treat various diseases in recent decades, suggesting a strong correlation with probiotics and immune responses (Kang & Im, 2015).

## 5.4 Sources of Probiotic Strains and Microbial Species as Probiotics

The primary supplies may come from human origin like breast milk, the human intestinal tract, or both. Animal and other dietary substances like raw milk or foods that have undergone fermentation are the major sources of probiotics. Isolation of these strains from human microbiota is more susceptible due to their higher adhesion to the normal gastrointestinal epithelium than other strains. However, a number of probiotic food supplements also include unique microbes that have never been used safely in either animals or humans. The bacterial strains utilized in probiotic foods play a major role in decreasing the cholesterol level, improving metabolic functions, decreasing the onset of cancer, reducing yeast and vaginal infection, preventing the symptoms of PMS by stimulating the immune system (Sanders, 2008).

The only strains of lactic acid bacteria significant in nutrition are those belonging to the genus Lactococcus and Bifidobacterium. These species have the most significant features in a practical setting (Naeem et al., 2012). Gram-positive, catalase-negative bacteria known as lactic acid bacteria can produce lactic acid as the primary by-product of the fermentation of carbohydrates. Because they utilize a diverse mechanism for its metabolism, the genus Bifidobacterium is consequently placed among them more conventionally than phylogenetically. *From a commercial standpoint, Streptococcus thermophilus and Lactococcus lactis are significant lactic acid bacteria,* two additional species that play a significant part in the food business, especially dairy products (Puniya et al., 2016).

## 5.5 Mode of Action of Probiotics

Investigations on probiotic microorganisms have advanced steadily over the past 20 years, generally in terms of selecting probiotic cultures. Their characteristics, their possible applications, and their effects on improving human health are studied nowadays. Probiotics play a key role in forming the microflora in the gastrointestinal tract (GIT), which is essential for maintaining the homeostasis between pathogens and beneficial bacteria. There are now three main probiotic activity mechanisms that have been identified. The first one is currently managed by the local anaerobic flora, which limits intestinal infections. By preventing attachment of pathogens, probiotics can directly affect neighbouring bacteria. This distinct defensive pathway is utilized for maintaining inner health condition. By colonizing harmful bacteria and then engaging in antagonism against gastrointestinal pathogens, lactobacilli and bifidobacteria have indeed been found to suppress a wide variety of infections. This concept is frequently essential to prevent and treat infections and restore the microbial balance in the gut (Oelschlaeger, 2010).

The second mechanism aids in the production of anti-microbial compounds, such as bacteriocins, toxins, organic acids, short-chain fatty acids, and a decrease in gut pH. In the GIT ecosystem, these compounds prevent the growth of other hazardous bacteria, including food-borne pathogenic and spoilage microorganisms. By generating a hostile environment, this activity causes the pathogen to die and may also cause toxins to become inactive. Probiotic effects are produced by microbial products, which are used to pinpoint a potential probiotic function and apply it effectively to prevent a particular disease by killing the target cells. The third method involves T-cell activation, cytokine generation, and overall immune regulation, which results in triggering phagocytosis and IgA secretion, altering T-cell response, promoting $Th_1$ response, and minimizing $Th_2$ response. The mechanism is crucial to treating and preventing infectious diseases (Kechagia et al., 2013). Probiotics' ability can play an important role in the modulation of the immune system. Probiotics improve immunity in several ways, including altering the balance of pro- and anti-inflammatory cytokines, activating local macrophages, affecting local and systemic

IgA production, and regulating the body's reaction to food antigens (Ghadimi et al., 2008; Kabeerdoss et al., 2011).

The probiotics use a three-step action mechanism.

- It boosts and controls immunological response.
- It maintains colonization resistance, normalizes intestinal microbiota, and manages irritable-bowel syndrome and other inflammatory bowel illnesses.
- The last path supplies nutrients to the colon epithelium. Still, it also has metabolic effects, including lactose breakdown, bile salt de-conjugation and release, and a reduction in mutagenic and toxic reactions in the gut (Jain & Sharma, 2012).

Additionally, extensive molecular, bioengineered, and genetic research allowed for the direct engagement of four pathways to reveal the fundamental idea of the advantageous impact of good bacteria known as "probiotics". (1) Microbial conflicts caused by the action of anti-microbial substances, (2) pathogenic bacteria competing with each other for epithelial adherence and nutrition, (3) host immune system regulation, and (4) reduction in the generation of bacterial toxins (Zommiti et al., 2020).

## 5.6 Desirable Probiotic Properties

A prospective probiotic strain is anticipated to possess a number of desirable traits in order to utilize its beneficial effects. For example, the capacity to tolerate acid and bile, which also appears to be extremely important for orally administered, adhesion to mucosal and epithelial surfaces, and bile salts hydrolase activities are all important. However, there are still questions about the relevance and inconsistencies of clinical investigations, and a lack of operational technique uniformity that must be taken into account, thus, their usefulness is still up for dispute (Mercenier et al., 2008).

The colony-forming unit per gram of the finished product is a crucial criterion, and probiotic dosage levels should be based on those that have been shown effective in human research. It is widely accepted that probiotics must have lowest threshold of $10^6$ CFU/mL and a maximum of $10^8$–$10^9$ CFU/ml probiotic strains must be ingested daily for the probiotic impact transmitted to the consumers, even though knowledge regarding minimum efficacious concentration levels is still absent. Additionally, the strains have to be capable of flourishing in manufacturing and commercial settings and should maintain viability throughout routine storage. As it enhances processes like adherence, decreases gut permeability, and modulates the immune system, viability is a need for probiotic functioning and poses a commercial hurdle. However, several research have shown that not all probiotic effects require viability since all clinical advantages or pathways are not directly tied to it and probiotics DNA may have important health impacts (Mercenier et al., 2008). Therefore, for certain probiotic strains, optimum development during the first manufacturing processes could be acceptable, and they are not required to maintain excellent viability throughout their storage (Kosin & Rakshit, 2006).

## 5.7 Probiotics and the Available Food Products

The variety of dietary items that include probiotic strains is substantial and continues to expand. Dairy-based goods such as fermented milks, cheese, ice cream, buttermilk, milk powder, and yoghurt make up the majority of the market's offerings. Yoghurt sales account for the greatest percentage of total sales. Soy-based products, protein bars, cereals, and a range of juices are examples of non-dairy food applications that are suitable for delivering probiotics to consumers. In addition to safety, the compatibility of the product with the microbes and the maintenance of its suitability for food processing, packaging, and storage conditions are factors that must be taken into consideration while determining the efficacy of the inclusion of the probiotic strain in these substances (Ewe et al., 2010). The pH level plays a crucial role in the success of incorporating probiotics into a product, and it can impact the survival and growth of the probiotics. This is why cheese is seen as a better delivery system for probiotics compared to yoghurt, as it provides more favourable conditions for the probiotics to thrive. New technological advancements offer solutions to the stability and viability challenges faced in incorporating probiotics. This allows for the creation of new options that satisfy the increased consumer demands for probiotic-rich substances. Developing microencapsulation technology has enabled the protection of microorganisms from adverse environmental conditions. This technology has enabled beverage producers to deliver probiotics to consumers through a straw delivery device that contains a dried form of the probiotic. Market-available probiotic fragments are also being utilized to provide benefits during processing. Furthermore, bifidobacteria's production of lantibiotics, which are antibacterial substances, is being explored for potential application in the food industry (O'Sullivan, 2012; Sheehan et al., 2007).

## 5.8 Health Benefits of Probiotics

Lactic acid bacteria (LAB), primarily comprising various species of Lactobacillus and Enterococcus, have a long history of use in dairy-based fermented foods. The ingestion of these bacteria has been a regular part of the human diet for centuries. It is important to note that not all strains of LAB have the same health benefits. The probiotic effects of each strain are unique and cannot be generalized. Studies have shown that certain LAB strains can positively impact various health conditions. However, it is crucial to understand that the health benefits of probiotics are highly dependent on the specific strain in question. The beneficial effects of probiotics are closely linked to several disease conditions. These microorganisms have significant importance and potential in mitigating various microbial infections (Nagpal et al., 2012). It has been shown through scientific studies that certain strains of probiotics can modulate the gut microbiome and improve immune function, leading to a reduction in the incidence and severity of several diseases, including gastrointestinal disorders, allergies, and infections. Probiotics have a strong and growing scientific basis, with numerous studies demonstrating their effectiveness in promoting human health and preventing various disease conditions. The importance of these microorganisms cannot be overstated, and their continued research and development holds great promise for the future of human health.

### 5.8.1 The Anti-Pathogenic Effect of Probiotics

One of the main advantages of probiotics is their ability to limit pathogenic activity since, unlike traditional antibiotics, they do not modify or disrupt the diverse gut microbiota community. Probiotics or a probiotic mixture's anti-pathogenic efficacy has been the subject of extensive investigation. Probiotics suppress infectious pathogens by producing short-chain fatty acids, including acetic, propionic, butyric, and lactic acids. Short-chain fatty acids aid in maintaining the colonic lumen's proper pH, which is essential for the production of a variety of bacterial enzymes as well as the digestion of foreign substances and toxins removal in the gut (Ammor et al., 2006). Many probiotics create a variety of anti-microbial substances, including bacteriocins, ethanol, organic acids, acetaldehydes, $H_2O_2$, and peptides. Proteins and bacteriocins are primarily responsible for raising the target cells' membrane permeability which causes the membrane potential to depolarize and eventually lead to cell death (Kareem et al., 2014). The generation of $H_2O_2$ by these groups of bacteria leads to the oxidizing of sulfhydryl groups, inactivation of numerous enzymes, and peroxidation of membrane lipids which will increase membrane permeability of the harmful microbes that will ultimately lead to cell death. Several substances might work by releasing lactic acid and acetic acid to reduce pH. Probiotics generate cationic anti-microbial peptides called defences in various cells, including Paneth cells in the intestinal epithelium (Figueroa-González et al., 2011). Besides generating bioactive compounds directly influencing harmful microbes, probiotic bacteria also play an important role in stimulating host defensive mechanisms. Probiotics can also exhibit anti-pathogenic action by fighting pathogens for nutritional growth, pathogen adsorption, and receptor sites.

### 5.8.2 Probiotics and Bacterial Vaginitis

According to the Centers for Disease Control and Prevention, more than a billion women worldwide suffer from nonsexually transmitted urogenital infections such as bacterial vaginosis, urinogenital infections, and various additional illnesses caused by different yeast species (Waigankar & Patel, 2011). *Gardnerella vaginalis*, *Ureaplasma urealyticum*, and *Mycoplasma hominis* are the organisms frequently linked to BV. Globally, sexually transmitted infections (STDs) constitute a major source of morbidity. Gonorrhoea and Chlamydia are the two bacterial STDs that have been reported the most frequently in several industrialized nations (Chan et al., 2016; Hanson et al., 2016). When lactobacilli concentrations in the vaginal environment alter or disappear, urogenital infections can result. The main microbial variables that control the existence, expansion, colonization, and persistence of non-endogenous microbes in the vagina include Lactobacillus spp. The protection offered by Lactobacillus spp. against uropathogens declines as their number does (Ya et al., 2010). Despite having advanced medications to cure medical ailments, the main problem of the present decade is that pathogenic bacteria simultaneously develop resistance to the available medicines. Therefore, at this time, our attention must be on creating living supplement, such as harmless microorganisms that may work to prevent or destroy pathogenic microbes rather than producing new medications.

Additionally, it has been hypothesized that lactobacilli create bio-films that coat the urogenital cells. Positive outcomes from clinical trials provided strong support to the use of lactobacilli in bacterial vaginosis. For the avoidance of recurring bacterial vaginosis, probiotic capsules containing *Lactobacillus rhamnosus*, *Lactobacillus crispatus*, *Lactobacillus gasseri*, *Lactobacillus vaginalis*, *Lactobacillus acidophilus*, *Lactobacillus reuteri*, and *Streptococcus thermophilus* are helpful (Shamshu et al., 2017; Siroli et al., 2017).

Lactobacilli play a crucial role in maintaining urogenital health through several key mechanisms:

1. Boosting the immune response.
2. Competing with other microbes for essential resources and space on the cells in the vaginal epithelium, urinary tract, and vagina.
3. Lowering the pH of the vagina by producing organic acid, generally lactic acid.
4. Secreting anti-microbial compounds and inhibitors, such as bacteriocins and hydrogen peroxide, prevents harmful bacteria growth.

#### 5.8.2.1 Anti-diabetic Activities of Probiotics

425 million people globally, including 78 million in the Southeast Asian area, have diabetes, based on the International Diabetic Federation (IDF) of Southeast Asia. In addition, if nothing is done, this number is predicted to increase to 629 million by 2045 (Atlas, 2015). Despite the fact that there is no permanent treatment for diabetes, this illness is managed using a variety of drugs. However, pharmacological and bimolecular scientists have made advances in comprehending the significance of synbiotics in treating the illness. Two distinct bacterial phyla that predominate in the gut microbiome are the gram negative bacteria and the gram-positive firmicutes (Iqbal et al., 2014). Recent studies have shown that obesity is linked to an increase

in bacteroidetes over time along with a decrease in firmicutes. More particular, type-2 diabetic individuals have considerably less firmicute species leads to increase the bacteroidetes/firmicutes ratio, that increasing the average with plasma glucose levels. Autoimmune diseases such as type 1 diabetes have been linked to a comparable developmental pattern (Kobyliak et al., 2016; Le Barz et al., 2015). Changes in the microbiome also promote the intrusion of opportunistic diseases, which can simultaneously reduce sulfates and decrease the development of bacteria that produce butyrate while being resistant to oxidative stress. Type-2 diabetes is controlled by modifying gut hormones like glucagon-like peptide-1 and gastric inhibitory polypeptide using probiotic and prebiotic therapies. In this situation, hormones are thought to be involved in glucose homeostasis, which neutralizes the problem brought on by insulin resistance or the inability of β-cell to make insulin (Hartstra et al., 2015). Currently, research is concentrated on developing novel prebiotics, such as arabin-oxylan oligosaccharide, which potentially prevents associated anabolic and catabolic illnesses. Both carbohydrates have also been important in lowering the adiposity (Grover et al., 2012; Kerry et al., 2018).

### *5.8.3 Allergies and Probiotics*

There is a high economic and social cost to society at large due to the rising incidence of allergy illnesses brought on by immunological disorders. To monitor and prevent these illnesses, it is critical to understand the underlying molecular mechanisms that contribute to the pathophysiology of allergic disorders along with novel therapeutic approaches. Allergies are immune system responses to particles (that should be nontoxic), which cause these reactions. A hypersensitive response triggered by immunological processes is an allergy. Food antigens cause food allergies, which also induce gastrointestinal inflammation. Probiotics cure allergies by restoring health to your digestive system that reduces inflammation, restores immune functioning, and fortifies the lining of your stomach (Kim et al., 2013). Probiotics are beneficial in lessening the effects of food allergies because they improve gut defence in two ways (non-immunologic and immunologic). First, the gut bacteria are returned to normal, and membrane permeability is reduced. The second process includes enhancing the host's immune defence system by increasing IgA activity. As a result, food allergies are decreased and food antigens degrade more quickly. Probiotics alter how antigens are built, decreasing their immunogenicity, intestine permeability, and the production of anti-inflammatory cytokines which are prominent in people with a variety of allergy illnesses. The positive impact of probiotics in managing and preventing allergic illnesses has enhanced our understanding of their causes and prevention in recent years (Goyal et al., 2012; Licciardi et al., 2013). Food allergy symptoms are reduced by *Lactobacillus* GG and *L. rhamnosus* GG, which also substantially lower the chance of acquiring allergic illness (Abatenh et al., 2018).

### 5.8.4 Probiotics and Cancer

Around the world, there have been almost 14 million new cases of the terrible disease and 8.2 million additional deaths from cancer up to 2012. The continents of Asia, Africa, and America present more than 70% of all cancer-related issues worldwide. Genomic, proteomic, and molecular pathology-based research on cancer has increased public awareness of the disease and our understanding of it during the past 10 years. Numerous novel medications with intriguing luminous features have been developed concurrently employing nanotechnology and biotechnology (nanocapsules), but tolerance to their load and side effects remains a significant barrier (Gayathri & Rashmi, 2016). Natural sources with anti-carcinogenic properties, such as probiotics, have recently received much attention. These have generated great interest from nutritionists, researchers, and industry professionals who want to collaborate to fight the condition and create a medicine that works well and has few to no adverse effects. Through the synthesis of SCFAs, in vitro investigations have shown that the probiotic strains *Lactobacillus fermentum* are particularly effective in inhibiting colorectal cancer cells and encouraging normal epithelial colon cell development (ferulic acid). The capability was also contrasted with two additional probiotics, *L. acidophilus and L. rhamnosus*, which had both been previously identified as having tumorigenic activity (So et al., 2017; Vafaeie, 2006).

Animal studies demonstrated the protective benefits of LAB against rodent colon cancer. Human studies also indicate that some LAB types may be anti-carcinogenic due to their capacity to reduce the activity of an enzyme called glucuronidase (which can cause the production of chemicals that might cause cancer in the digestive system). According to demographic research, dairy consumers have a lower incidence of colon cancer than other groups. To establish this impact, however, there is still much work to be done (Iqbal et al., 2014). Probiotics may be crucial in fighting cancer, but thus far, only in vitro studies have been done. Therefore, in vivo models must demonstrate the anti-cancer potential of probiotics before moving on to animal and human studies.

### 5.8.5 Probiotics and Hypertension

Increased blood cholesterol has become more common in adults, kids, and adolescents. Obesity, hypercholesterolemia, and lipid abnormalities are the main causes of hypertension. One of first document to demonstarte the hypocholesterolemic effects of milk fermented by lactobacillus. Recent studies have revealed that, in addition to Lactobacilli's hypocholesterolemic effects, Bifidobacteria may also significantly lower blood cholesterol levels when cholesterol levels are increased. Since most cholesterol is produced and absorbed in the gut, intestinal microflora has been found to affect blood cholesterol levels. In addition, studies have demonstrated probiotics to help reduce blood pressure by lowering cholesterol levels and raising LDL cholesterol's resistance to oxidation (Guo et al., 2011). Probiotics increase total and low-density lipoprotein cholesterol, which lowers hypertension. High blood or serum cholesterol levels reduce glucose and insulin resistance, regulate the renin–angiotensin system, and lower cholesterol. Therefore, probiotic supplements may reduce hypertension in individuals blood pressure (Patel et al., 2010).

### 5.8.6 *Probiotics and their Anti-inflamatory Activities*

Inflammatory bowel disease, a group of chronic inflammatory illnesses of the gastrointestinal tract, includes Crohn's disease and ulcerative colitis. Any area of the gastrointestinal tract, including the mucosa, submucosa, and serosa, can be affected by Crohn's disease, and the inflammation may even extend across the whole GIT. In contrast, ulcerative colitis often affects the big bowel, more especially the colon's mucosa and submucosa. According to research, an imbalance in the gut microbiota is a key pathophysiological factor in controlling inflammatory bowel disease. It is also known that probiotic, prebiotic, and synbiotic supplements may impact the disease. The formation of SCFAs, notably acetate, butyrate, and propionate, is linked to inflammatory bowel disease (Cammarota et al., 2015).

Furthermore, it has been established that these SCFAs are essential for preserving colonic homeostasis. They also have anti-inflammatory properties and enhance intestinal propulsive performance. Thus, it is fair to assume that increasing the synthesis of SCFAs by supplementing with non-digestible carbohydrates and fibre (prebiotic) alone or in conjunction with probiotics might be beneficial therapeutic treatments (Currò et al., 2017). The field's progress is focused primarily on creating genetically modified probiotic bacterial strains that can produce and release immunomodulators or lipoteichoic acid, a key component of Gram-positive bacteria's cell walls, which can affect the host immune system and restore the population of barrier protection mutualistic species of bacteria. The most common probiotics utilized in foods include *Lactobacillus*, *Bifidobacterium*, *Enterobacter*, and *E. coli* (Domingo, 2017).

## 5.9 Conclusion

The majority of our fundamental therapeutic and nutritional supplementing needs might be satisfied by probiotics, which have significant functional properties. These microorganisms have responded well to therapeutic therapy for various illnesses and conditions, including food intolerances, IBS, and diarrhoea brought on by rotavirus. Additionally, the role of probiotics in preventing and treating disorders linked to harmful bacteria, such as cancer, diabetes, and obesity, is an intriguing and quickly developing study field. Dietary supplements often include dairy products. However, probiotics can also be added to fermented food products that are not dairy-based, offering a different and better source for assessing novel probiotic strains. Additionally, recent clinical and dietary analyses have effectively revealed some astonishing abilities of some probiotic strains. Particularly, control of energy in a variety of metabolic processes, tolerance to acid and bile, capacity to cling to gut epithelial cells, and ability to fight off pathogens, along with some other properties, such as their ability to increase safety, suitability as food, and usefulness as probiotics for human health. To pave a new way for investigating and exploiting probiotics to enhance human health, the current focus is assessing novel strains of probiotic microbes and their application in biomedical/clinical research.

## References

Abatenh, E., Gizaw, B., Tsegay, Z., Tefera, G., & Aynalem, E. (2018). Health benefits of probiotics. *Journal of Bacteriology and Infectious Diseases, 2*(1), 8–27.

Ammor, S., Tauveron, G., Dufour, E., & Chevallier, I. (2006). Antibacterial activity of lactic acid bacteria against spoilage and pathogenic bacteria isolated from the same meat small-scale facility: 1—Screening and characterization of the antibacterial compounds. *Food Control, 17*(6), 454–461.

Anandharaj, M., & Sivasankari, B. (2014). Isolation of potential probiotic Lactobacillus oris HMI68 from mother's milk with cholesterol-reducing property. *Journal of Bioscience and Bioengineering, 118*(2), 153–159.

Atlas, D. (2015). *International diabetes federation. IDF Diabetes Atlas* (7th ed., p. 33). International Diabetes Federation.

Bongaerts, G. P., & Severijnen, R. S. (2016). A reassessment of the PROPATRIA study and its implications for probiotic therapy. *Nature Biotechnology, 34*(1), 55–63.

Cammarota, G., Ianiro, G., Cianci, R., Bibbò, S., Gasbarrini, A., & Currò, D. (2015). The involvement of gut microbiota in inflammatory bowel disease pathogenesis: Potential for therapy. *Pharmacology & Therapeutics, 149*, 191–212. https://doi.org/10.1016/j.pharmthera.2014.12.006

Chan, P. A., Robinette, A., Montgomery, M., Almonte, A., Cu-Uvin, S., Lonks, J. R., & Hardy, E. J. (2016). Extragenital infections caused by Chlamydia trachomatis and Neisseria gonorrhoeae: A review of the literature. *Infectious Diseases in Obstetrics and Gynecology, 2016*, 5758387.

Chen, X., Yang, G., Song, J.-H., Xu, H., Li, D., Goldsmith, J., & Kelly, C. P. (2013). Probiotic yeast inhibits VEGFR signaling and angiogenesis in intestinal inflammation. *PLoS One, 8*(5), e64227.

Currò, D., Ianiro, G., Pecere, S., Bibbò, S., & Cammarota, G. (2017). Probiotics, fibre and herbal medicinal products for functional and inflammatory bowel disorders. *British Journal of Pharmacology, 174*(11), 1426–1449.

Dallal, M. M. S., Mojarrad, M., Baghbani, F., Raoofian, R., Mardaneh, J., & Salehipour, Z. (2015). Effects of probiotic Lactobacillus acidophilus and Lactobacillus casei on colorectal tumor cells activity (CaCo-2). *Archives of Iranian Medicine, 18*(3), 167–172.

Domingo, J. J. S. (2017). Review of the role of probiotics in gastrointestinal diseases in adults. *Gastroenterología y Hepatología, 40*(6), 417–429.

Egan, M., & Van Sinderen, D. (2018). Carbohydrate metabolism. In *Bifidobacteria the bifidobacteria and related organisms* (pp. 145–164). Elsevier.

Ewe, J.-A., Wan-Abdullah, W.-N., & Liong, M.-T. (2010). Viability and growth characteristics of Lactobacillus in soymilk supplemented with B-vitamins. *International Journal of Food Sciences and Nutrition, 61*(1), 87–107.

Figueroa-González, I., Cruz-Guerrero, A., & Quijano, G. (2011). The benefits of probiotics on human health. *Journal of Microbial & Biochemical Technology, 1*, 1948–5948.

Gayathri, D., & Rashmi, B. (2016). Anti-cancer properties of probiotics: a natural strategy for cancer prevention. *EC Nutrition, 5*(4), 1191–1202.

Ghadimi, D., Fölster-Holst, R., De Vrese, M., Winkler, P., Heller, K. J., & Schrezenmeir, J. (2008). Effects of probiotic bacteria and their genomic DNA on TH1/TH2-cytokine production by peripheral blood mononuclear cells (PBMCs) of healthy and allergic subjects. *Immunobiology, 213*(8), 677–692.

Giorgetti, G., Brandimarte, G., Fabiocchi, F., Ricci, S., Flamini, P., Sandri, G., & Lecca, P. G. (2015). Interactions between innate immunity, microbiota, and probiotics. *Journal of Immunology Research, 2015*, 501361.

Goyal, M., Bansal, M., Yadav, S., & Malhotra, K. (2012). Probiotics: The friendly microbes. *Journal of Clinical Practice, 23*(3), 126–130.

Grover, S., Rashmi, H. M., Srivastava, A. K., & Batish, V. K. (2012). Probiotics for human health–new innovations and emerging trends. *Gut Pathogens, 4*(1), 1–14.

Guo, Z., Liu, X., Zhang, Q., Shen, Z., Tian, F., Zhang, H., & Chen, W. (2011). Influence of consumption of probiotics on the plasma lipid profile: A meta-analysis of randomized controlled trials. *Nutrition, Metabolism, and Cardiovascular Diseases, 21*(11), 844–850.

Hanson, L., VandeVusse, L., Jerme, M., Abad, C. L., & Safdar, N. (2016). Probiotics for treatment and prevention of urogenital infections in women: A systematic review. *Journal of Midwifery & Women's Health, 61*(3), 339–355.

Hartstra, A. V., Bouter, K. E., Bäckhed, F., & Nieuwdorp, M. (2015). Insights into the role of the microbiome in obesity and type 2 diabetes. *Diabetes Care, 38*(1), 159–165.

Iqbal, M. Z., Qadir, M. I., Hussain, T., Janbaz, K. H., Khan, Y. H., & Ahmad, B. (2014). Probiotics and their beneficial effects against various diseases. *Pakistan Journal of Pharmaceutical Sciences, 27*(2), 405.

Jain, P., & Sharma, P. (2012). Probiotics and their efficacy in improving oral health: A review. *Journal of Applied Pharmaceutical Science, 2*(11), 151–163.

Kabeerdoss, J., Devi, R. S., Mary, R. R., Prabhavathi, D., Vidya, R., Mechenro, J., & Ramakrishna, B. S. (2011). Effect of yoghurt containing Bifidobacterium lactis Bb12® on faecal excretion of secretory immunoglobulin A and human beta-defensin 2 in healthy adult volunteers. *Nutrition Journal, 10*(1), 1–4.

Kang, H.-J., & Im, S.-H. (2015). Probiotics as an immune modulator. *Journal of Nutritional Science and Vitaminology, 61*(Supplement), S103–S105.

Kareem, K. Y., Hooi Ling, F., Teck Chwen, L., May Foong, O., & Anjas Asmara, S. (2014). Inhibitory activity of postbiotic produced by strains of Lactobacillus plantarum using reconstituted media supplemented with inulin. *Gut Pathogens, 6*(1), 1–7.

Kechagia, M., Basoulis, D., Konstantopoulou, S., Dimitriadi, D., Gyftopoulou, K., Skarmoutsou, N., & Fakiri, E. M. (2013). Health benefits of probiotics: A review. *International Scholarly Research Notices, 2013*, 481651.

Kerry, R. G., Patra, J. K., Gouda, S., Park, Y., Shin, H.-S., & Das, G. (2018). Benefaction of probiotics for human health: A review. *Journal of Food and Drug Analysis, 26*(3), 927–939.

Kim, H.-J., Kim, H. Y., Lee, S.-Y., Seo, J.-H., Lee, E., & Hong, S.-J. (2013). Clinical efficacy and mechanism of probiotics in allergic diseases. *Korean Journal of Pediatrics, 56*(9), 369.

Kobyliak, N., Conte, C., Cammarota, G., Haley, A. P., Styriak, I., Gaspar, L., & Kruzliak, P. (2016). Probiotics in prevention and treatment of obesity: A critical view. *Nutrition and Metabolism, 13*(1), 1–13.

Kosin, B., & Rakshit, S. K. (2006). Microbial and processing criteria for production of probiotics: A review. *Food Technology and Biotechnology, 44*(3), 371–379.

Le Barz, M., Anhê, F. F., Varin, T. V., Desjardins, Y., Levy, E., Roy, D., & Marette, A. (2015). Probiotics as complementary treatment for metabolic disorders. *Diabetes & Metabolism Journal, 39*(4), 291–303.

Licciardi, P. V., Ismail, I. H., Balloch, A., Mui, M., Hoe, E., Lamb, K., & Tang, M. L. (2013). Maternal supplementation with LGG reduces vaccine-specific immune responses in infants at high-risk of developing allergic disease. *Frontiers in Immunology, 4*, 381.

Mercenier, A., Lenoir-Wrjnkoop, I., & Sanders, M. E. (2008). Physiological and functional properties of probiotics. *Journal of Milk Science and Biotechnology, 26*(1), 53–57.

Naeem, M., Ilyas, M., Haider, S., Baig, S., & Saleem, M. (2012). Isolation characterization and identification of lactic acid bacteria from fruit juices and their efficacy against antibiotics. *Pakistan Journal of Botany, 44*(323), 8.

Nagpal, R., Kumar, A., Kumar, M., Behare, P. V., Jain, S., & Yadav, H. (2012). Probiotics, their health benefits and applications for developing healthier foods: A review. *FEMS Microbiology Letters, 334*(1), 1–15.

O'Sullivan, D. J. (2012). Exploring the potential to utilize lantibiotic-producing bifidobacteria to create dairy ingredients with increased broadspectrum anti-microbial functionalities yields encouraging results. *Food Technology, 66*(6), 45–50.

Oelschlaeger, T. A. (2010). Mechanisms of probiotic actions–A review. *International Journal of Medical Microbiology, 300*(1), 57–62.

Pacini, S., & Ruggiero, M. (2017). Description of a novel probiotic concept: Implications for the modulation of the immune system. *American Journal of Immunology, 13*(2), 107–113.

Pandey, K. R., Naik, S. R., & Vakil, B. V. (2015). Probiotics, prebiotics and synbiotics-A review. *Journal of Food Science and Technology, 52*, 7577–7587.

Patel, A. K., Singhania, R. R., Pandey, A., & Chincholkar, S. B. (2010). Probiotic bile salt hydrolase: Current developments and perspectives. *Applied Biochemistry and Biotechnology, 162*, 166–180.

Patel, R. M., & Denning, P. W. (2013). Therapeutic use of prebiotics, probiotics, and postbiotics to prevent necrotizing enterocolitis: What is the current evidence? *Clinics in Perinatology, 40*(1), 11–25.

Puniya, M., Ravinder Kumar, M., Panwar, H., Kumar, N., & Ramneek, A. K. P. (2016). Screening of lactic acid bacteria of different origin for their probiotic potential. *Journal of Food Processing & Technology, 7*(1), 545.

Rastall, R. A., & Gibson, G. R. (2015). Recent developments in prebiotics to selectively impact beneficial microbes and promote intestinal health. *Current Opinion in Biotechnology, 32*, 42–46.

Sanders, M. E. (2008). Probiotics: Definition, sources, selection, and uses. *Clinical Infectious Diseases, 46*(2), 58–61.

Shamshu, R., Vaman, J., & Nirmala, C. (2017). Role of probiotics in lower reproductive tract infection in women of age group 18 to 45 years. *International Journal of Reproduction, Contraception, Obstetrics and Gynecology, 6*(2), 671.

Sheehan, V. M., Ross, P., & Fitzgerald, G. F. (2007). Assessing the acid tolerance and the technological robustness of probiotic cultures for fortification in fruit juices. *Innovative Food Science & Emerging Technologies, 8*(2), 279–284.

Siroli, L., Patrignani, F., Serrazanetti, D. I., Parolin, C., Ñahui Palomino, R. A., Vitali, B., & Lanciotti, R. (2017). Determination of antibacterial and technological properties of vaginal lactobacilli for their potential application in dairy products. *Frontiers in Microbiology, 8*, 166.

So, S. S., Wan, M. L., & El-Nezami, H. (2017). Probiotics-mediated suppression of cancer. *Current Opinion in Oncology, 29*(1), 62–72.

Tufarelli, V., & Laudadio, V. (2016). An overview on the functional food concept: Prospectives and applied researches in probiotics, prebiotics and synbiotics. *Journal of Experimental Biology and Agricultural Sciences, 4*(3), 273–278.

Vafaeie, F. (2006). Critical review on probiotics and its effect on cancer. *Journal of Applied Microbiology, 100*(6), 1171–1185.

Waigankar, S. S., & Patel, V. (2011). Role of probiotics in urogenital healthcare. *Journal of Mid-life Health, 2*(1), 5–10.

Ya, W., Reifer, C., & Miller, L. E. (2010). Efficacy of vaginal probiotic capsules for recurrent bacterial vaginosis: A double-blind, randomized, placebo-controlled study. *American Journal of Obstetrics and Gynecology, 203*(2), e121–e126.

Zommiti, M., Feuilloley, M. G., & Connil, N. (2020). Update of probiotics in human world: A nonstop source of benefactions till the end of time. *Microorganisms, 8*(12), 1907.

# Chapter 6
# Algal Protein: Future of Sustainable Food

**Komal, Rohan Samir Kumar Sachan, Rahul Kumar Kashodhan, Inderpal Devgon, Nisha, Khushboo, Abdel Rahman Mohammad Said Al-Tawaha, and Arun Karnwal**

**Abstract** Due to factors including population growth, rising incomes, increased urbanization, and aging populations, the need for protein components is continuously expanding on a global scale. Traditionally, foods produced from animals (such as dairy, eggs, and meat) provide most of the dietary protein needs of people. However, the identification of sustainable alternative protein sources is thus required to replace animal proteins. Based on exceptional and widespread ecological adaption, microalgae act as of the edible proteins. Microalgae have a high photosynthetic efficiency and can flourish in disadvantaged environments using non-potable effluent. The purity of the amino acids (AAs) and the technological functional characteristics of microalgae proteins are comparable to those of reference proteins. Recalcitrant cell walls, however, make it difficult to digest and use the microalgae proteins effectively. Furthermore, the palatability and low sensory scores limits the microalgal biomass in the food and feed industries. Meat analogues, emulsifiers, and bioactive peptides are a few novels uses for microalgae proteins. The development of low-cost growing techniques, wet downstream processing based on biomass, with bio-refinery strategies would increase the sustainability ability as human and animal feed applications.

**Keywords** Microalgae · Essential amino acids · Single-cell protein · Eukaryotes · Photosynthesis · Proteins supplements · Animal feed · Dietary proteins

---

Komal · R. K. Kashodhan
Department of Biotechnology, School of Bioengineering and Biosciences, Lovely Professional University, Phagwara, Punjab, India

R. S. K. Sachan · I. Devgon · Nisha · Khushboo
Department of Microbiology, School of Bioengineering and Biosciences, Lovely Professional University, Phagwara, Punjab, India

A. R. Mohammad Said Al-Tawaha
Department of Biological Sciences, Al Hussein Bin Talal University, Ma'an, Jordan

A. Karnwal (✉)
Department of Microbiology, Lovely Professional University, Phagwara, Punjab, India



A. Karnwal, A. R. Mohammad Said Al-Tawaha (eds.), *Food Microbial Sustainability*, https://doi.org/10.1007/978-981-99-4784-3_6

## 6.1 Introduction

Proteins, as we know are regarded as building blocks of life and provide energy, structurally these are macromolecules that helps to build body tissue and muscles. The daily intake of protein is around 0.8 g/kg of body weight. The other beneficial effects of proteins are increase intestinal digestion, provide omega-3s with essential fatty acids, and protection against oxidants by generating antioxidants (Shen et al., 2021).

With increasing population and lifestyle, daily requirement of protein has compromised our bodily functions. To cope up with the rapid changing lifestyle, proteins supplements are available that completes the daily requirements and contributes to better health. The exploration for protein as supplements has been carried out since the early 50s (Kårlund et al., 2019). Many approved protein supplements such as whey-based, soy-based, and plant-based are available in the market. However, these supplements need to be used as per advisory as people may have allergic reactions. Apart from the protein components, many other micronutrients might be added to the supplements from other sources that can be a potential allergen (Hertzler et al., 2020). People are forced to have an extra precautions or thoughts while choosing protein supplements for diet preferences that should not interfere with health. However, historically, algae have been used as protein supplements over more than thousands of years. Since then, algal-based protein has gained prominence and healthier choice among the population.

Algae are a diverse group of unicellular and multicellular photosynthetic eukaryotes that are mostly aquatic and autotrophs (Jan Stevenson, 2022). These have chlorophyll *a* as primary photosynthetic pigment that helps to generate food internally (Kuczynska et al., 2015). Algae has proven its usage in various anthropogenic sectors (industrial and traditional) because of broader range types. For instance, East Asia cultivate seaweeds or often regard as seaweed farming for food. Apart from dietary functions, algae are also used in bioremediation processes for controlling pollution. A noteworthy approach of algae has been utilized in generation of biofuel (Vanthoor-Koopmans et al., 2013) because of light entrapping mechanism by chlorophyll. Algae provide varied human applications in terms of remediation of the environment to nutrients for humans and animals (Fig. 6.1). However, algal protein as sustainable food, the chapter focuses on diversity of algae and its protein production. The chapter also focuses on aspects of sustainable application as functional and nutritional protein products.

## 6.2 Algal Diversity and Classification

Despite the algae being a polyphyletic category, the different algal lineages seem to be monophyletic groups. Compared to the three major algal lineages, four small algal lineages contain fewer species and less morphological variation. which

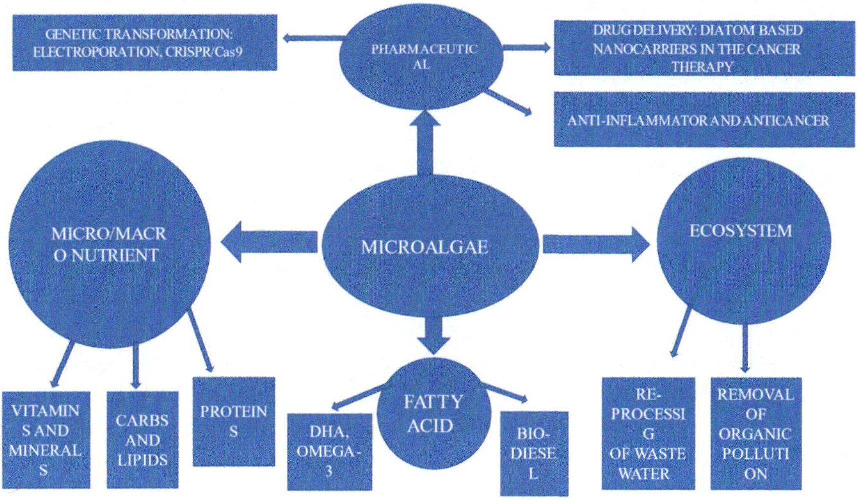


**Fig. 6.1** Beneficial effects of algae to humans

contain many species and significant morphological diversity. The three main lineages of algae are chromophyte, green, and red algae. The cryptophytes, dinoflagellates, glaucophytes, and euglenophytes are the minor lineages.

### 6.2.1 *Red Algae*

These photosynthetic species have chloroplasts with a girdling thylakoid but no external endoplasmic reticulum covering along with chlorophyll A, D, and phycobilipigments. Floridian starch, deposited in the cytoplasm rather than the chloroplast, is the main storing substance (Dorrell & Smith, 2011). The primary chain is of glucan carbohydrate linked by 1,4-1inked and 1,6-1inked as cross chains. There are no flagella at all. Apical cells start the growth process, and the thallus is basically filamentous. Most of the species of red algae have intricated reproductive mechanisms including various types of spores and multifarious life backgrounds that include clearly distinct plants. The cause of this is linked to the fact that fertilization "is fundamentally inefficient in the absence of gamete motility, a restriction that has been partially compensated for by improved spore formation associated with biphasic and triphasic life cycles." Red algae or red seaweed are most abundant species as these were formerly macrophytes at some time in their evolutionary past (Balasubramaniam et al., 2022). From the tropics to the poles, red seaweeds are found along the world's coastlines, albeit the number of species and individuals is lowest there. Although many species are thought to be connected forms, completely unattached species can survive in comfortable surroundings in significant numbers. Freshwater, especially clear streams, can also contain red algae. These freshwater algae may be constrained in correspond to the ecological requirements to cultivate algae by spore dispersal. Sheath and Handbook (1988) show that there are numerous species in old, glacier-free regions, in glaciated regions. An explicit association engaging the species enumeration and the number of years following glacial retreat has well documented (Sheath & Handbook, 1988). Seaweeds are widespread in streams that flow through rain forests in Australia and North America (Andersen, unpublished observations), although many rain forest locations lack even basic surveys. Excellent summary of the ecological study on red algae may be found in two recent reviews. Many red algae are photosynthetic, and many kinds of invertebrate and vertebrate species feed on the thalli. Grazing animals frequently have limited access to algal species, making the survival of the alga essential to the life of the animal. The red alga thallus is consumed in enormous volumes by several grazers, unlike many plants that build up biomass. Iridaea was nibbled by snails in one investigation at the same rate as new tissue was being made. In such circumstances, the dried biomass does not account for an accurate representation of the yearly production. The alga provide significantly higher energy to the ecosystem than inflicted by instantaneous biomass measurements (Chan et al., 2022; David et al., 2022).

### *6.2.2 Green Algae*

The term "green algae" refers to a broad group of photosynthesis-dependent organisms, including the embryophytes (land plants). Although they are mostly found in freshwater ecosystems, several varieties can also be found in marine, brackish, and terrestrial habitats. The group contains macroscopic thallus with well-differentiated tissues, multicellular filamentous forms, colonial and unicellular planktons, and unicellular planktons. Chlorophylls a and b are found in photosynthetic forms, and some species also possess chlorophyllides that resemble chlorophyll c as well as the light-harvesting carotenoid fucoxanthin. Their chloroplasts lack an exterior endoplasmic reticulum coat and a girdle lamella. Starch including amylopectin and amylose is the main product for storage, and it is kept in grains inside the chloroplast. Flagellate vegetative cells, zoospores, and gametes have moving structures called flagella; there is an absence of lack tripartite tubular hairs on these structures. There are also three primary types of basal apparatuses, each of which is unique for a sizable taxonomic group, typically a class. The green algae are the lineage of eukaryotic algae that can be found in the widest range of settings. These algal bodies are very common in fresh water lakes and rivers, can be seen as open ocean phytoplankters and coastal seaweeds, and can grow on a variety of hard substrates including topsoil, vegetation, wildlife, rocks, and other hard substrates. Even in the lower atmosphere, they thrive (Douglas et al., 2003; Torres & De-la-Torre, 2022; Wang et al., 2014).

### 6.2.3 Chromophyte Algae

These classes of algae have chlorophyll A with different variables of chlorophyll C. Carotenoids that capture light through photosynthetic activity are many and diverse. Their exterior endoplasmic reticulum coat and girdle lamella are both common features of their chloroplasts. Among the various classes, the main storage product varies, but typically it has three glucan chain 1,3 linked with three other glucan chains by 1,6 branches. It is constantly kept outside of the chloroplast. Flagella are heterodynamic, paired structures, with the longer one frequently having two rows of tripartite tubular hairs. They are present on flagellate vegetative cells, gametes, and zoospores. Each class has a unique basal apparatus, and during phagotrophic feeding, flagella may be active in addition to swimming. The chromophyte lineage has the most taxonomic classes of algae. Brown algae are the largest individuals, and some kelps can grow to 60 m in just a few weeks. In terms of the number of species and individuals, diatoms are the most numerous. Chrysophytes, synurophytes, xanthophytes, silicoflagellates, and other classifications are among the surviving groups. Although they are widely dispersed as a main group, several of the classes may only appear seldom (Duckett et al., 1993).

### 6.2.4 Dinoflagellates

Dinoflagellates are tiny, unicellular protists or creatures that belong to the class Dinoflagellate. They can be found in most aquatic habitats, including equatorial, arctic, and open ocean conditions as well as lakes. Dinoflagellates can generate "red tides" when they bloom because of the extremely high cell density in the surface water, which causes a shift in color. A few types of dinoflagellates can produce neurotoxins that can be bioconcentrated by filtering organisms, most notably shellfish, and become toxic and harmful to the health of animals that consume them.

Dinoflagellates have chlorophyll A and other varieties of chlorophyll C in their photosynthetic forms. There are many light-harvesting carotenoids that are actively photosynthetic, particularly peridinin. The girdle lamella is absent with extra membrane surrounding the plastids. Dinophyceae starch is the primary storage substance which composed of glucan monomers linked by 1,4 and 1,6 linkages. Most of the storage bodies in algae resides in chloroplast as these are synthesized in them, however, in Dinophyceae storage bodies reside outside the chloroplast. Major algal species are biflagellate, one is having a dangled posterior flagellum and the other has encircled girdle flagellum. The basal apparatus differ within the group. In phagocytosis, the flagella and related structures are active (Hoppenrath, 2016).

### 6.2.5 *Euglenophytes*

It is likely that Antony van Leeuwenhoek, a pioneer in the study of protists, first described Euglenophyta, a division of highly differentiated algal flagellates, in the seventeenth century. However, it was not until Ehrenberg's 1838 description of Euglena, for which the division is named and whose features characterize the attributes of Euglenophyta, that a comprehensive account of the euglenoids. Usually having two flagella for propulsion, an undifferentiated cell wall, and chloroplasts, these unicellular flagellates are cylindrical, ovoid to fusiform, microscopic plants of eukaryotic organization (possessing a genuine nucleus and other membrane-bound organelles); colorless variations are known.

Although they have photosynthesis-related compounds like chlorophyll A and B, they do not have photosynthesis-related light-harvesting carotenoids. The chloroplasts are enveloped by an extra single membrane without girdle lamella. Paramylon, a filamentous structured glucan chain present as grains in the cytoplasm, is the main storage bodies in this class of algae. Vegetative cells of biflagellate are most frequently found, but one quadriflagellate has just been identified. Phagotrophic is frequent, in which basal apparatus perform role in feeding. The basal apparatus has a variety of morphologies (Wang et al., 2021).

### 6.2.6 *Cryptomonads*

Small (between 5 and 50 mm) biflagellate protists known as cryptomonads can be found in a variety of freshwater, brackish, and marine settings. They stand out for having a characteristic cellular asymmetry and ejectosomes, which are extrusive organelles. Numerous cryptomonads use photosynthetic processes, and the pigments and colors of their plastids vary widely. The presence of a "nucleomorph," a remnant nucleus of secondary endosymbiotic origin, in plastid-bearing cryptomonads is notable. These contain chlorophyll A, chlorophyll C2, and other types of phycobiliproteins. Despite the absence of a girdle lamella, the chloroplast has an exterior endoplasmic reticulum coat. The starch that is produced from 1,4 glucans and is the main storage component which is stored linking chloroplastic ER to envelope of the chloroplast. Periplasts, which are highly complex, cover cells. The life cycle is also complicated, as every species has biflagellate monads to some extent with different patterns of bipartite hairs (Hoef-Emden & Archibald, 2016).

### 6.2.7 *Glaucophytes*

Only four recognized genera—glaucocystis, cyanophora, gloeochaete, and cyanoptyche—and 15 species make up the Glaucophyta, which is by far the Archaeplastida phylum with the least number of species. However, more unique lineages than previously thought exist in glaucophytes, according to recent molecular and morphological investigations, proving that the group is not as species impoverished as previously believed. The remaining peptidoglycan wall in their envelope is one of the numerous ancestral plastids features that glaucophytes, freshwater phototrophs of moderate to low abundance, still possess. These traits come from the cyanobacterial donor of this organelle. When its shared ancestry with other Archaeplastida was discovered, the name "cyanelles" for these plastids was changed to "muroplasts." Chlorophyll *a* and phycobilin pigments are two examples of photosynthetic pigments. Although they lack real chloroplasts, they do have "cyanelles," which are endosymbionts made of blue-green algae; the thylakoids are found singly. The storage substance resembles starch and could be floridian starch. Little further about the organization has been released (Jackson et al., 2015).

## 6.3 Composition and Production of Algal Protein

The overly increasing population around the globe and limited land and water resources along with awareness in healthy lifestyle has derived the demand for possible substitutes of protein sources. Algae, an aquatic organism is rich in number of bio compounds that are beneficial for human health, majorly proteins and amino acids, lipids, fatty acids, carbohydrates, polysaccharides, and pigments such as carotenoid, chlorophyll, and phycocyanin along with polyphenols, vitamins, and minerals (Demarco et al., 2022).

Microalgae is now approved as capable source for edible protein due to its excellent adaptation ecologically. It can grow in marginal areas by utilizing non-potable wastewater and shows great photosynthetic efficacy. Microalgae has shown similarities with reference protein in terms of quality of amino acid and techno-functional properties. Formerly, to produce single-cell protein (SCP), microalgae species (*Arthospira* and *Chorella*) have been used having limited application in pharmaceutical industries (Kumar et al., 2022). On the other hand, red seaweed, also known as macroalgae, with 6500 species are assumed to be the oldest eukaryotic algae. Its functions are analogous to plant proteins (pea proteins and soy protein) because of versatility in structure that offers bunch of applications. According to studies from 221 macroalgae species, 145 species have been used for food applications purpose and 101 species for hydrocolloids production (Rawiwan et al., 2022).

Both macroalgae and microalgae have shown potential in the production of biofuels as renewable sources of energy to reduce the carbon emission as algae absorbs carbon dioxide to produce algae biomass through photosynthesis. Algal biomass has higher photosynthetic efficiency and energy content therefore it has become evident feedstock to produce biofuel (Chen et al., 2022).

Availability of essential amino acids and their digestibility decides the quality of proteins. Animal sources of protein which the human body cannot biosynthesize consist of essential amino acids including lysine, methionine, phenylalanine, threonine, valine, tryptophan, histidine, leucine, and isoleucine. These proteins are considered complete proteins whereas on the other hand plant sources of protein lack some of these essential amino acids and are therefore incomplete proteins. Due to the presence of polysaccharides, plant-based proteins are difficult to digest (Bleakley & Hayes, 2017). According to studies 15 artic red algae species are useful in medicine and food industries, are rich sources of minerals and various metabolites. Many filamentous ceramialean algae including *Ceramium virgatum, Polysiphonia stricta, Savoiea arctica*, have 20–32% protein content of dry weight. It has a high quantity of macronutrients, pigments, and ascorbic acid. *C. virgatum* and *P. palmata* consist of nonproteogenic β-alanine amino acid (Yanshin et al., 2021).

The four genera of microalgae *Chlorella*, *Nannochloropsis*, and *Phaeodactylum* consist of crude protein as main content after that ether extract and crude ash (Table 6.1). Therefore, commercial products of microalgae are great alternative for feed and food products (Wild et al., 2018).

### 6.3.1 Production Process

The algae being highly efficient in photosynthesis and growth rate showed higher productivities for carbohydrates, lipids, and protein. Additionally, microalgae farming can be done in areas with unsuitable soil, climate, and water where conventional crops cannot grow. Hence microalgae do not compete with the food crops directly and beneficial in purifying water for agriculture. The steps involved in microalgae protein production includes biomass recovery, disruption of microalgae cells, and separation of proteins (Amorim et al., 2021).

**Table 6.1** Given below give the information about lipid and protein content of some of the species of both macroalgae and microalgae on dry matter basis

| Algal biomass | Lipid content (%) | Protein content (%) | References |
|---|---|---|---|
| *Spyridia filamentosa* | 1.10 | 2.20 | Polat and Ozogul (2008) |
| *Polysiphonia* | 4.3 | 28.2 | Kulikova et al. (2022) |
| *Spirogyra* sp. | 11–21 | 6–20 | Milano et al. (2016) |
| *Porphyridium cruentum* (red algae) | 9–14 | 28–40 | Milano et al. (2016) |
| *Nannochloropsis oceanica* (Eustigmatophyte) | 19.10 | 24.80 | Cheng et al. (2014) |
| *Dunaliella salina* (Chlorophyta) | 11.47 | 8.57 | Pirwitz et al. (2016) |

For harvesting microalgae coagulation/flocculation techniques are typically used. These techniques enable the formation of large aggregates in microalgae that increases the size, and the process of separation becomes much easier. Typically, microbial cells' surface repels other cells as a result they remain as colloidal suspension in water. To avoid this condition chemical flocculation, in which polyelectrolytes such as polyacrylamide polymers and metallic salts neutralizes the negative charge present on the surface of the cells, and auto flocculation in which cells are induced with the addition of calcium or magnesium at pH 9, techniques are used (Vandamme et al., 2013).

As most of the important components are inside the thick resistant cell walls of microalgae therefore high-end technologies including enzymatic, mechanical (for e.g., high pressure homogenization, bead milling and ultrasonication), and chemical treatments are used for cell disruption (Li et al., 2022). In chemical treatment cells are treated with chemicals such as surfactants, acids, and bases, that change the permeability of the cell wall by degrading the chemical linkages on the cell walls. As the permeability of the cells increases up to a certain level it leads to the disruption of the cell wall. Alkaline treatment is the most widely used amongst all the chemicals (Phong et al., 2018).

In the case of mechanical methods, emulsification process is well suited for high pressure homogenizers (HPHs) in which cell disruption is maximized by various valve-seat arrangements. The cell suspension radially passes through a valve, colloids with an impact ring, it then departs the valves and it either discharges or goes to second valve. Then the shear forces of accelerated jet fluid acts upon valve surface and hydrodynamic cavitation from pressure drop induced shear causes the cell breakage. Whereas bead milling includes the mechanism of mechanical compaction and shear stress and cavitation and free radical formation in the case of ultrasonication (Günerken et al., 2015).

Once the cells are disrupted, they undergo the process of membrane filtration. It includes the process of liquid separation into two different fractions by selectively passing of the compounds through semi-permeable membrane depending upon the molecular weight. Ultrafiltration, nanofiltration, reverse osmosis, and microfiltration are the most used membrane technologies. Algal proteins with molecular weight more than 200kDa can be separated with the help of microfiltration. Through the process of ultrafiltration, macromolecules, and proteins of molecular weight 1–200 kDa is used and nanofiltration helps in the minimization of osmotic pressure. RO (reverse osmosis) is applied to concentrate the fluid volume (Bleakley & Hayes, 2017).

Apart from being the protein source, bioactive peptides are formulated using algal proteins with other proteinaceous compounds which positively modulate human health as anti-diabetic, anti-microbial properties antioxidant, treatment of atherosclerosis, hypersensitivity, inflammation, and unwanted tissue proliferation (Geada et al., 2021).

## 6.4 Extraction and Processing Procedures of Algal Protein

### 6.4.1 Ultrasound Assisted Extraction

Ultrasonication is frequently used in a variety of processes, including dispersion, catalysis, extraction, nanoparticle synthesis, and the creation of graphene. Through cavitation, which is the emergence, expansion, and dissolution of bubbles in a solution, ultrasonication can increase the yield of chemicals. Researchers are interested in ultrasonication because of its green technology that has applications in the food and related industries. Ultrasonication works by compressing and releasing air by sound waves with a frequency greater than 20 kHz. There are several ways for how ultrasound works, including matrix breakdown and detexturization in plant cell walls, as well as fragmentation, erosion, the sonocapilary effect, sonoporation, and local shear stress. Cell wall disruption is a result of these mechanisms taken collectively. However, depending on the type of biomass and process variables, a given mechanism's proportionate level of contribution may change (Braspaiboon et al., 2022; Carreira-Casais et al., 2021; Liu et al., 2022; Sankaran et al., 2018; Vernès et al., 2019). According to a study by Tomšik et al. (2016), 1.60 g of gallic acid equivalent/100 g of dry weight of total phenolic content may be extracted from wild garlic using 79.8 min of irradiation time (RSM as a mathematical tool) (Tomšik et al., 2016). In addition, it has been claimed that lemon balm and peppermint extracts can produce substantial levels of total phenol content (891.76 mg $l^{-1}$) and antioxidant activity (2.17 mmol TE $l^{-1}$) with only a 25-min irradiation period. Consequently, ultrasonication was used in the current investigation due to the method's enormous potential as a means of cell disintegration. By using ultrasonication, larger extraction yields can be obtained with less effort and frequently at lower temperatures, making it more suited for chemicals that are thermolabile (Chia et al., 2020; Mittal et al., 2017). The mechanical mixing and disruption caused by ultrasonication may result in a greater mass transfer between the biomass or sample and the solvent. Contrarily, the cavitation bubbles burst causes shock waves that cause the cell to rupture, improving the release of intracellular components into the solution (Chia et al., 2020; Mittal et al., 2017).

### 6.4.2 Pulsed Electric Field

PEF technology is a non-thermal, minimally intrusive, and environmentally benign method that has several uses in the food, biotechnology, medical, and environmental industries to treat biological tissues and biomaterials. PEF technology is a non-thermal, minimally intrusive, and environmentally benign method that has several uses in the food, biotechnology, medical, and environmental industries to treat biological tissues and biomaterials. Because the PEF can accelerate the mass transfer of intracellular components, its inclusion in the extraction process has a favorable impact on the extraction. Furthermore, because it makes specific intracellular components more accessible, this non-thermal method lowers the temperature and solvent concentration utilized in the extraction process. Thus, it is possible to use fewer natural resources and solvents. As a result, this cutting-edge technology illustrates the fundamentals of green chemistry because it has beneficial effects on both society and the environment. The PEF technology encourages minimal changes to the goods' sensory and nutritional qualities while also boosting the yield of the extraction process. This fact is directly related to the low temperatures—between 30 and 40 °C—that were employed during the procedure. These support minimizing thermosensitive chemical toxicity in the food matrix (Carullo et al., 2020, 2022; Prabhu et al., 2019; t'Lam et al., 2017).

## 6.5 Functional and Nutritional Properties of Algal Proteins

### *6.5.1 Functional Properties*

Polysaccharides, proteins, lipids, pigments, fibers, and polyphenols are just a few of the substances that make up micro- and macroalgae. Algal proteins are a source of bioactive peptides, also known as cryptides, which, when released from their parent proteins where they are inactive, can have immediate physiological effects. In addition to having the potential to serve as an alternative protein source, bioactive peptides (BAPs) and other proteinaceous compounds with biological value and favorable effects on health, such as antioxidant, anti-proliferative, anti-inflammatory, anti-hypertensive, anti-diabetic, anti-atherosclerotic, anti-coagulant, and antimicrobial properties, can be produced from algae proteins.

Hypertension is a substantial risk factor for cardiovascular disease (CVD), which is currently one of the top causes of death worldwide. Renin-angiotensin system plays a major role in blood pressure regulation. Blood pressure rises because of the very potent vasoconstrictor angiotensin II being created when the angiotensin converting enzyme (ACE) binds and cleaves angiotensin I. ACE is a target that is quite amenable to medication, and some commonly given anti-hypertensive drugs, such captopril, are ACE inhibitors. An increase in blood pressure is prevented by these inhibitors by blocking the ACE-mediated conversion of angiotensin I into angiotensin II. From protein extracts of the microalgae, many peptides having ACE inhibitory activities have been discovered (Bleakley & Hayes, 2017; de Amorim et al., 2022; Thiviya et al., 2022; Wells et al., 2017).

Free radicals are chemical entities that have unpaired electrons and are extremely reactive and short-lived. Reactive oxygen species (ROS) can develop through exposure to external elements like ionizing radiation and ultraviolet light in addition to the typical metabolic mechanisms that produce them. Proteins, lipids, carbohydrates, and nucleic acids can all be oxidatively modified by free radicals. The risk of several human diseases has been linked to an increase in the presence of these modified versions. Most organisms have antioxidant systems that work to lower or completely remove the amounts of typical ROS. Numerous algal peptides have been found to work as antioxidants by directly chelating metals, scavenging ROS, and preventing lipid peroxidation cascades (Chen et al., 2019; Rezayian et al., 2019; Zuluaga et al., 2017).

Peptides can pass through cell membranes because of their small size and chemical makeup, which prevents hazardous levels from building up as they do with proteins and antibodies. These substances have demonstrated great specificity and affinities with little interactions with other medicinal therapies. The use of peptides is restricted by a few limitations, such as the fact that they are less effective than conventional cancer medication therapies. Numerous issues with peptide therapies are mostly brought about by the medications' lack of specificity, which prevents them from being able to distinguish between cancerous and healthy cells. Furthermore, the efficacy of the peptides may be hampered if the bonds between the chemotherapeutic drugs and the transport molecules are broken.

### *6.5.2 Nutritional Value*

Vegetable proteins are frequently regarded as an incomplete protein supply since they lack some EAAs, however, animal-based proteins are rich sources of EAAs. Seasonal variations in protein content and AA balance have been observed in marine macroalgae. Therefore, the best time to harvest is when the protein content is higher. Brown seaweeds may have less protein than other types, but it seems that they are abundant in Thr, Val, Leu, and Lys. While Trp is sometimes impossible to quantify due to protein breakdown during the employed extraction procedures, Methionine, another essential amino acid is absent in both macroalgae and microalgae. Although macroalgae tend to have a more complete EAA profile than microalgae, the potential of microalgae as a protein provider is larger (in terms of protein content). However, overall, both algae appeared to be a source that might help healthy persons meet the WHO's EAA requirements.

It is well known that algae have an intriguing nutritional label. Vitamins, minerals, and dietary fiber abound in macroalgae. In contrast, microalgae are touted as "healthy foods" since they are high in protein and bioactive substances. However, there is still debate regarding the precise protein composition of algae. The total organic nitrogen concentration is analyzed using the Kjeldahl method which is the most popular technique for measuring protein content. The value of organic nitrogen-protein is then calculated by multiplying this amount by a conversion factor. The best conversion factors to use for each type of macroalgae are still up for debate, though, because macroalgae contain considerable amounts of non-protein nitrogen.

Brown macroalgae often have low protein contents (3–15% (w/w) of dry biomass), as compared to red macroalgae is up to 47% and for green macroalgae is about 9–26% For instance, *Porphyra* sp. possesses protein levels like those in soybean meal; according to some sources, the protein concentration can reach up to 44%. The conversion factor of nitrogen-to-protein in microalgae is still a hot topic since using the usual value (6.25) can lead to findings that are either under- or over-estimated. Through literature survey, 6.25 is the conversion factor value to calculate nitrogen-to-protein in microalgae research, resulting in larger and incorrect percentages, even though an average ratio of 4.78 is frequently used and even advised (Geada et al., 2021; Kusmayadi et al., 2021; Yang et al., 2023).

## 6.6 Algal Protein Application as Sustainable Food

Algae offers variety of application as shown in Fig. 6.2, however, their role as food supplement has been explored more. Proteins are component of humans that promotes and enhances growth. Structurally, proteins are composed of building blocks called amino acids linked by peptide bonds. Because different species of microalgae have a high protein content, microalgae have been used as a source of protein since the 1950s. Algae have been used by humans as a food source and supplement for thousands of years. Algae not only acted as a natural carbon source that reduced global warming, but they also helped relieve the growing strain on the freshwater and arable land resources needed for land-based food production. Because of their diverse biochemical compositions, microalgae can act as one of future alternatives

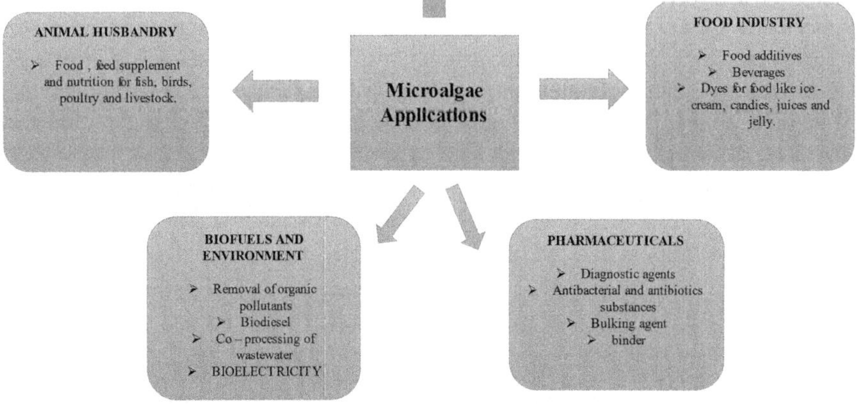


**Fig. 6.2** General application of algae

for sustainable food source in terms of both functionality and nutritional values. Algae as food sources for human consumption is due to higher essential amino acids (EAAs) values and other beneficial elements. Also, more than 72,500 different species of algae can be found in both freshwater and saltwater. Macroalgae, which are the largest ones, account for 20% of all species. Microalgae make up the remaining 80%. In lab flasks with a liquid culture, these single-celled microalgae start their trip to becoming a food source. The nutritional value of each microalga varies. Important amino acids, essential fatty acids, such as omega-3, omega-6, and omega-7, and vitamins A, D, and E are some of the nutrients they include. Chlorella and spirulina are two of the most well-liked for human ingestion (Caporgno & Mathys, 2018; Ibañez & Cifuentes, 2013; Kumar et al., 2022).

A future scope can be research on beneficial algae to be use in human diet and animal feed due to the wide spectrum of nutrients that different microalgae strains supply. A typical alga possesses fatty acids, proteins and amino acids, pigments, vitamins, and carbohydrates as its chemical and structural elements along which can be processed for their health advantages. Microalgal in diets for people and animals is a biggest market now-a-days and its current market are increasing exponentially. In some impoverished nations, the inclusion of nutrient-rich microalgae in staple foods like bread aids in enhancing food security and addressing malnutrition problems. Microalgal biomass, like spirulina, added to food and soft drinks offers consumers a healthier option.

With global regulation enforcing industrial process to be more sustainable and greener process, circular economy and zero-waste concepts have been formularized. For instance, biomass and metabolite productivity improvement with minimal industrial wastewater treatment, agricultural wastes byproducts need to be utilized as nutritional needs for algal growth. Algal proteins are known to be susceptible to overproduction when grown on nitrogen- or phosphorus-rich substrates, making the underutilized streams potentially useful for boosting the protein content of algae. People are now being urged to consume "superfoods," which are foods that are incredibly nutrient-dense, improve health, and help avoid several chronic ailments. The World Health Organization (WHO) has approved spirulina as one of the superfoods. Also, spirulina accounts for both natural colorants and cosmetics (Kusmayadi et al., 2021). This has inspired businesses and academics to investigate microalgae's potential as functional meals. Spirulina produces 12,000 tons annually, Dunaliella salina 3000 tons annually, and Aphanizomenon flos-aquae 1500 tons annually, all of which are used to make products that are derived from algae. Credence Research estimates that the market for products manufactured from algae will be valued US$44.6 billion by 2023, with a cumulative annual growth rate (CAGR) of over 5.2%.

## 6.7 Conclusion and Challenges Ahead

Algal protein can be a sustainable protein source for the next generation. In doing so, the need for new techniques to produce and commercialize is still a hindrance. As we know, algal growth is favored by $CO_2$ rich environment, an approach for novel PBR that will optimize the algal growth and retain $CO_2$ in the chamber. However, $CO_2$ in the medium if present for extended time causes unintended culture acidification that endangering the entire growth process. In addition to the financial and environmental benefits, processing options may permit the simultaneous or sequential refinement of multiple streams. These can direct other added value products with improved functionality and purity. Microalgal products now have numerous intriguing advantages in terms of health benefits, practical qualities, nutritional value, etc. thanks to the development of microalgal biotechnology. The algal biomass can also be used in food ingredients that can satisfy customer demands for environment-friendly food. As both microalgae and macroalgae (seaweed) have proteins, soluble fibers, polysaccharides, lipids, polyunsaturated fatty acids, pigments, vitamins, and minerals are present in abundance to be used as ingredients. However, one of the major obstacles to integrating algae in food systems is connected to their sensory palatability, since their biomass may contain several odor-active volatile chemical compounds that may not be acceptable for some algal-food products. Their food industry is slowly being penetrated by microalgal products such as non-animal proteins, high-value chemicals, and microalgal dry biomass. However, there are still a lot of issues that need to be resolved before microalgae-based food may go in the direction of the future food industry. Problems including high production costs, poor product acceptance, and unreliable safety are now the hurdles preventing widespread use. The advancement of food made from microalgae requires the collaboration of experts from a variety of sectors.

## References

Amorim, M. L., Soares, J., Coimbra, J. S. D. R., Leite, M. D. O., Albino, L. F. T., & Martins, M. A. (2021). Microalgae proteins: Production, separation, isolation, quantification, and application in food and feed. *Critical Reviews in Food Science and Nutrition, 61*(12), 1976–2002.

Balasubramaniam, A., Arumugham, I., Nathan, P. S., Santhosh Kumar, M. P., Murugesan, K., Dharmaraj, S., Thangavelu, L., Yadalam, P. K., Ramadoss, R., & Ashokkumar, V. (2022). Emerging technologies and potential applications of algae in dentistry – A critical review. *Journal of Biotechnology, 360*, 1–10. https://doi.org/10.1016/J.JBIOTEC.2022.09.021

Bleakley, S., & Hayes, M. (2017). Algal proteins: Extraction, application, and challenges concerning production. *Food, 6*(5), 1–34. https://doi.org/10.3390/FOODS6050033

Braspaiboon, S., Osiriphun, S., Surawang, S., Jirarattanarangsri, W., Kanha, N., & Laokuldilok, T. (2022). Ultrasound-assisted alkaline extraction of proteins in several algae and their nutritional characteristics. *International Journal of Food Science & Technology, 57*(9), 6143–6154. https://doi.org/10.1111/IJFS.15975

Caporgno, M. P., & Mathys, A. (2018). Trends in microalgae incorporation into innovative food products with potential health benefits. *Frontiers in Nutrition, 5*, 58. https://doi.org/10.3389/FNUT.2018.00058/BIBTEX

Carreira-Casais, A., Otero, P., Garcia-Perez, P., Garcia-Oliveira, P., Pereira, A. G., Carpena, M., Soria-Lopez, A., Simal-Gandara, J., & Prieto, M. A. (2021). Benefits and drawbacks of ultrasound-assisted extraction for the recovery of bioactive compounds from marine algae. *International Journal of Environmental Research and Public Health, 18*(17), 9153. https://doi.org/10.3390/IJERPH18179153

Carullo, D., Abera, B. D., Scognamiglio, M., Donsì, F., Ferrari, G., & Pataro, G. (2022). Application of pulsed electric fields and high-pressure homogenization in biorefinery cascade of C. vulgaris microalgae. *Food, 11*(3), 471. https://doi.org/10.3390/FOODS11030471

Carullo, D., Pataro, G., Donsì, F., & Ferrari, G. (2020). Pulsed electric fields-assisted extraction of valuable compounds from arthrospira platensis: Effect of pulse polarity and mild heating. *Frontiers in Bioengineering and Biotechnology, 8*, 551272. https://doi.org/10.3389/FBIOE.2020.551272

Chan, S. S., Low, S. S., Chew, K. W., Ling, T. C., Rinklebe, J., Juan, J. C., Ng, E. P., & Show, P. L. (2022). Prospects and environmental sustainability of phyconanotechnology: A review on algae-mediated metal nanoparticles synthesis and mechanism. *Environmental Research, 212*, 113140. https://doi.org/10.1016/J.ENVRES.2022.113140

Chen, H., Xia, A., Zhu, X., Huang, Y., Zhu, X., & Liao, Q. (2022). Hydrothermal hydrolysis of algal biomass for biofuels production: A review. *Bioresource Technology, 344*, 126213.

Chen, Y., Chen, J., Chang, C., Chen, J., Cao, F., Zhao, J., Zheng, Y., & Zhu, J. (2019). Physicochemical and functional properties of proteins extracted from three microalgal species. *Food Hydrocolloids, 96*, 510–517. https://doi.org/10.1016/J.FOODHYD.2019.05.025

Cheng, J., Huang, R., Yu, T., Li, T., Zhou, J., & Cen, K. (2014). Biodiesel production from lipids in wet microalgae with microwave irradiation and bio-crude production from algal residue through hydrothermal liquefaction. *Bioresource Technology, 151*, 415–418.

Chia, S. R., Chew, K. W., Leong, H. Y., Manickam, S., Show, P. L., & Nguyen, T. H. P. (2020). Sonoprocessing-assisted solvent extraction for the recovery of pigment-protein complex from Spirulina platensis. *Chemical Engineering Journal, 398*, 125613. https://doi.org/10.1016/J.CEJ.2020.125613

David, S. R., Baharulnizam, N. B., & Rajabalaya, R. (2022). A review on biological assays of red algae marine compounds: An insight into skin whitening activities. *Journal of Herbal Medicine, 35*, 100585. https://doi.org/10.1016/J.HERMED.2022.100585

de Amorim, A. P., da Silva, G. H., Brandão, R. M. P., Porto, A. L. F., & Bezerra, R. P. (2022). Algae as a source of peptides inhibitors of the angiotensin-converting enzyme: a systematic review. *Anais da Academia Brasileira de Ciências, 94*(2), 20201636. https://doi.org/10.1590/0001-3765202220201636

Demarco, M., de Moraes, J. O., Matos, Â. P., Derner, R. B., de Farias Neves, F., & Tribuzi, G. (2022). Digestibility, bioaccessibility and bioactivity of compounds from algae. *Trends in Food Science & Technology, 121*, 114–128.

Dorrell, R. G., & Smith, A. G. (2011). Do red and green make brown? Perspectives on plastid acquisitions within chromalveolates. *Eukaryotic Cell, 10*(7), 856. https://doi.org/10.1128/EC.00326-10

Douglas, S. E., Raven, J. A., & Larkum, A. W. D. (2003). *The algae and their general characteristics* (pp. 1–10). Springer. https://doi.org/10.1007/978-94-007-1038-2_1

Duckett, J. G., Dransfield, J., Schmid, R., Cutler, D. F., Ferguson, I. K., Dalby, D. H., Vaughan, J. G., van Emden, H. F., Jury, S. L., Rutherford, R. W., Sands, M. J. S., & Field, D. V. (1993). The chromophyte algae: Problems and Perspectives, the structural biology of palms, morphology of flowers and inflorescences, anatomy of European woods, Nephelieae Pollen (Sapindaceae): Form, Function and Evolution, Saltmarsh Ecology, Atlas of Microscopy of Medical Plants, Culinary Herbs and Spices, Plant Nematodes and their Control, Biological Invasions in Europe and the Mediterranean, John Lightfoot: his work and travels, with a biographical introduction and a catalogue of the Lightfoot Herbari. *Botanical Journal of the Linnean Society, 111*(3), 399–410. https://doi.org/10.1111/J.1095-8339.1993.TB01912.X

Geada, P., Moreira, C., Silva, M., Nunes, R., Madureira, L., Rocha, C. M. R., Pereira, R. N., Vicente, A. A., & Teixeira, J. A. (2021). Algal proteins: Production strategies and nutritional and functional properties. *Bioresource Technology, 332*, 125125. https://doi.org/10.1016/J.BIORTECH.2021.125125

Günerken, E., D'Hondt, E., Eppink, M. H. M., Garcia-Gonzalez, L., Elst, K., & Wijffels, R. H. (2015). Cell disruption for microalgae biorefineries. *Biotechnology Advances, 33*(2), 243–260.

Hertzler, S. R., Lieblein-Boff, J. C., Weiler, M., & Allgeier, C. (2020). Plant proteins: Assessing their nutritional quality and effects on health and physical function. *Nutrients, 12*(12), 1–27. https://doi.org/10.3390/NU12123704

Hoef-Emden, K., & Archibald, J. M. (2016). Cryptophyta (cryptomonads). In *Handbook of the Protists* (pp. 1–41). Springer. https://doi.org/10.1007/978-3-319-32669-6_35-1

Hoppenrath, M. (2016). Dinoflagellate taxonomy — A review and proposal of a revised classification. *Marine Biodiversity, 47*(2), 381–403. https://doi.org/10.1007/S12526-016-0471-8

Ibañez, E., & Cifuentes, A. (2013). Benefits of using algae as natural sources of functional ingredients. *Journal of the Science of Food and Agriculture, 93*(4), 703–709. https://doi.org/10.1002/JSFA.6023

Jackson, C., Clayden, S., & Reyes-Prieto, A. (2015). The glaucophyta: The blue-green plants in a nutshell. *Acta Societatis Botanicorum Poloniae, 84*(2), 149–165.

Jan Stevenson, R. (2022). Algae and primary production of streams and rivers ecosystems. *Encyclopedia of Inland Waters, 2022*, 312–322. https://doi.org/10.1016/B978-0-12-819166-8.00134-1

Kårlund, A., Gómez-Gallego, C., Turpeinen, A. M., Palo-Oja, O. M., El-Nezami, H., & Kolehmainen, M. (2019). Protein supplements and their relation with nutrition, microbiota composition and health: Is more protein always better for sportspeople? *Nutrients, 11*(4), 829. https://doi.org/10.3390/NU11040829

Kuczynska, P., Jemiola-Rzeminska, M., & Strzalka, K. (2015). Photosynthetic pigments in diatoms. *Marine Drugs, 13*(9), 5847. https://doi.org/10.3390/MD13095847

Kulikova, Y., Sukhikh, S., Kalashnikova, O., Chupakhin, E., Ivanova, S., Chubarenko, B., & Babich, O. (2022). Assessment of the resource potential of Baltic sea macroalgae. *Applied Sciences, 12*(7), 3599.

Kumar, R., Hegde, A. S., Sharma, K., Parmar, P., & Srivatsan, V. (2022). Microalgae as a sustainable source of edible proteins and bioactive peptides – Current trends and future prospects. *Food Research International, 157*, 111338. https://doi.org/10.1016/J.FOODRES.2022.111338

Kusmayadi, A., Leong, Y. K., Yen, H. W., Huang, C. Y., & Chang, J. S. (2021). Microalgae as sustainable food and feed sources for animals and humans – Biotechnological and environmental aspects. *Chemosphere, 271*, 129800. https://doi.org/10.1016/J.CHEMOSPHERE.2021.129800

Li, Y., Kiani, H., Tiwari, B. K., & Halim, R. (2022). Unit operations applied to cell disruption of microalgae. In *3rd generation biofuels* (pp. 225–248). Woodhead Publishing.

Liu, Y., Liu, X., Cui, Y., & Yuan, W. (2022). Ultrasound for microalgal cell disruption and product extraction: A review. *Ultrasonics Sonochemistry, 87*, 106054. https://doi.org/10.1016/J.ULTSONCH.2022.106054

Milano, J., Ong, H. C., Masjuki, H. H., Chong, W. T., Lam, M. K., Loh, P. K., & Vellayan, V. (2016). Microalgae biofuels as an alternative to fossil fuel for power generation. *Renewable and Sustainable Energy Reviews, 58*, 180–197.

Mittal, R., Tavanandi, H. A., Mantri, V. A., & Raghavarao, K. S. M. S. (2017). Ultrasound assisted methods for enhanced extraction of phycobiliproteins from marine macro-algae, Gelidium pusillum (Rhodophyta). *Ultrasonics Sonochemistry, 38*, 92–103. https://doi.org/10.1016/J.ULTSONCH.2017.02.030

Phong, W. N., Show, P. L., Le, C. F., Tao, Y., Chang, J. S., & Ling, T. C. (2018). Improving cell disruption efficiency to facilitate protein release from microalgae using chemical and mechanical integrated method. *Biochemical Engineering Journal, 135*, 83–90.

Pirwitz, K., Rihko-Struckmann, L., & Sundmacher, K. (2016). Valorization of the aqueous phase obtained from hydrothermally treated Dunaliella salina remnant biomass. *Bioresource Technology, 219*, 64–71.

Polat, S., & Ozogul, Y. (2008). Biochemical composition of some red and brown macro algae from the Northeastern Mediterranean Sea. *International Journal of Food Sciences and Nutrition, 59*(7-8), 566–572.

Prabhu, M. S., Levkov, K., Livney, Y. D., Israel, A., & Golberg, A. (2019). High-voltage pulsed electric field preprocessing enhances extraction of starch, proteins, and ash from marine macroalgae Ulva ohnoi. *ACS Sustainable Chemistry & Engineering, 7*(20), 17453–17463. https://doi.org/10.1021/ACSSUSCHEMENG.9B04669/SUPPL_FILE/SC9B04669_SI_001.PDF

Rawiwan, P., Peng, Y., Paramayuda, I. G. P. B., & Quek, S. Y. (2022). Red seaweed: A promising alternative protein source for global food sustainability. *Trends in Food Science and Technology, 123*, 37–56.

Rezayian, M., Niknam, V., & Ebrahimzadeh, H. (2019). Oxidative damage and antioxidative system in algae. *Toxicology Reports, 6*, 1309. https://doi.org/10.1016/J.TOXREP.2019.10.001

Sankaran, R., Manickam, S., Yap, Y. J., Ling, T. C., Chang, J. S., & Show, P. L. (2018). Extraction of proteins from microalgae using integrated method of sugaring-out assisted liquid biphasic flotation (LBF) and ultrasound. *Ultrasonics Sonochemistry, 48*, 231–239. https://doi.org/10.1016/J.ULTSONCH.2018.06.002

Sheath, R. G., & Handbook, J. A. (1988). Mechanical adaptations to flow in freshwater red algae1. *Journal of Phycology, 24*(1), 107–111. https://doi.org/10.1111/J.1529-8817.1988.TB04462.X

Shen, Y., Levin, A., Kamada, A., Toprakcioglu, Z., Rodriguez-Garcia, M., Xu, Y., & Knowles, T. P. J. (2021). From protein building blocks to functional materials. *ACS Nano, 15*(4), 5819–5837. https://doi.org/10.1021/ACSNANO.0C08510/ASSET/IMAGES/LARGE/NN0C08510_0008.JPEG

t'Lam, G. P., Postma, P. R., Fernandes, D. A., Timmermans, R. A. H., Vermuë, M. H., Barbosa, M. J., Eppink, M. H. M., Wijffels, R. H., & Olivieri, G. (2017). Pulsed electric field for protein release of the microalgae Chlorella vulgaris and Neochloris oleoabundans. *Algal Research, 24*, 181–187. https://doi.org/10.1016/J.ALGAL.2017.03.024

Thiviya, P., Gamage, A., Gama-Arachchige, N. S., Merah, O., & Madhujith, T. (2022). Seaweeds as a source of functional proteins. *Phycology, 2*(2), 216–243. https://doi.org/10.3390/PHYCOLOGY2020012

Tomšik, A., Pavlić, B., Vladić, J., Ramić, M., Brindza, J., & Vidović, S. (2016). Optimization of ultrasound-assisted extraction of bioactive compounds from wild garlic (Allium ursinum L.). *Ultrasonics Sonochemistry, 29*, 502–511. https://doi.org/10.1016/J.ULTSONCH.2015.11.005

Torres, F. G., & De-la-Torre, G. E. (2022). Green algae as a sustainable source for energy generation and storage technologies. *Sustainable Energy Technologies and Assessments, 53*, 102658. https://doi.org/10.1016/J.SETA.2022.102658

Vandamme, D., Foubert, I., & Muylaert, K. (2013). Flocculation as a low-cost method for harvesting microalgae for bulk biomass production. *Trends in Biotechnology, 31*(4), 233–239.

Vanthoor-Koopmans, M., Wijffels, R. H., Barbosa, M. J., & Eppink, M. H. M. (2013). Biorefinery of microalgae for food and fuel. *Bioresource Technology, 135*, 142–149. https://doi.org/10.1016/J.BIORTECH.2012.10.135

Vernès, L., Abert-Vian, M., el Maâtaoui, M., Tao, Y., Bornard, I., & Chemat, F. (2019). Application of ultrasound for green extraction of proteins from spirulina. Mechanism, optimization, modeling, and industrial prospects. *Ultrasonics Sonochemistry, 54*, 48–60. https://doi.org/10.1016/J.ULTSONCH.2019.02.016

Wang, L., Wang, X., Wu, H., & Liu, R. (2014). Overview on biological activities and molecular characteristics of sulfated polysaccharides from marine green algae in recent years. *Marine Drugs, 12*(9), 4984–5020. https://doi.org/10.3390/MD12094984

Wang, Y., Feng, J., Lv, J., Liu, Q., Nan, F., Liu, X., & Xie, S. (2021). Phylogenetic and morphological evolution of green euglenophytes based on 18S rRNA. *Journal of Eukaryotic Microbiology, 68*(1), e12824. https://doi.org/10.1111/JEU.12824

Wells, M. L., Potin, P., Craigie, J. S., Raven, J. A., Merchant, S. S., Helliwell, K. E., Smith, A. G., Camire, M. E., & Brawley, S. H. (2017). Algae as nutritional and functional food sources: revisiting our understanding. *Journal of Applied Phycology, 29*(2), 949. https://doi.org/10.1007/S10811-016-0974-5

Wild, K. J., Steingaß, H., & Rodehutscord, M. (2018). Variability in nutrient composition and in vitro crude protein digestibility of 16 microalgae products. *Journal of Animal Physiology and Animal Nutrition, 102*(5), 1306–1319.

Yang, Y., Hassan, S. H. A., Awasthi, M. K., Gajendran, B., Sharma, M., Ji, M. K., & Salama, E. S. (2023). The recent progress on the bioactive compounds from algal biomass for human health applications. *Food Bioscience, 51*, 102267. https://doi.org/10.1016/J.FBIO.2022.102267

Yanshin, N., Kushnareva, A., Lemesheva, V., Birkemeyer, C., & Tarakhovskaya, E. (2021). Chemical composition and potential practical application of 15 red algal species from the White Sea Coast (the Arctic Ocean). *Molecules, 26*(9), 2489.

Zuluaga, M., Gueguen, V., Pavon-Djavid, G., & Letourneur, D. (2017). Carotenoids from microalgae to block oxidative stress. *BioImpacts: BI, 7*(1), 1. https://doi.org/10.15171/BI.2017.01

# Chapter 7
# Microbial Biofactories: A Promising Approach Towards Sustainable Omega-3 Fatty Acid Production

**Megha Rawat, Shreya Mundepi, Manu Pant, Gaurav Pant, and Gaurav Kumar**

**Abstract** Omega-3 fatty acids or polyunsaturated fatty acids known as are indispensable to human health and are more commonly referred to as "healthy fats." These include improving neuronal functioning and promoting cardiac health by decreasing triglycerides, as well as controlling blood pressure, reducing hypertension, improving mental health, and checking microbial proliferation in the body. Only meals rich in n-3 FA can provide us with these essential FA as the human body is unable to generate them. Plants are a rich source of ALA, while seafood is known for containing significant levels of DHA and EPA. These food sources have their own drawbacks, such as the potential for heavy metal contamination with mercury, disagreeable taste and odor, and difficulties in being accepted by vegans and vegetarians. Plants are not thought to be a trustworthy supply of these fatty acids due to their slow growth, inconsistent synthesis, and complicated enzyme system. Consequently, research has been done on microbial sources as prospective sources of the crucial omega-3 fatty acids. Commercially produced n-3FAs from cultivated microorganisms under controlled conditions have shown improved FA synthesis and quick development. Moreover, modifications can be done in microorganisms by genetic engineering much more easily than higher organisms. In this context, the significance of microalgae has also been investigated and confirmed. This paper talks about the significance of n-3 FA and the possibility for bacteria to produce them in greater and more widespread amounts.

M. Rawat · S. Mundepi · M. Pant (✉)
Department of Biotechnology, Graphic Era Deemed to be University, Dehradun, Uttarakhand, India
e-mail: manupant@geu.ac.in

G. Pant
Department of Microbiology, Graphic Era Deemed to be University, Dehradun, India

G. Kumar (✉)
Department of Microbiology, School of Bioengineering and Biosciences, Lovely Professional University, Punjab, India



A. Karnwal, A. R. Mohammad Said Al-Tawaha (eds.), *Food Microbial Sustainability*, https://doi.org/10.1007/978-981-99-4784-3_7

**Keywords** Omega-3 fatty acids · EPA · DHA · Microalgae · Fermentation · Biosynthesis · n-3FA Production

## 7.1 Introduction

Fatty acids (FAs) are aliphatic carbon chains that end with a carboxylate group. n-3 FA are essential components of cell membranes, particularly those of neurons (Brunner, 2006). The docosahexaenoic acid (22:6) and the eicosapentaenoic acid (EPA) are the two essential acids in this category that are required (eicosapentaenoic acid 20:5). Alpha-linolenic acid, found in plants, may be transformed into DHA and then EPA, the latter of which is necessary in higher proportions for the appropriate functioning of the eye and brain. These fatty acids are regarded as a crucial nutrient that must be obtained from outside sources because organisms are unable to produce n-3 FAs on their own (Kralovec et al., 2011). DHA and EPA have been shown to help reduce inflammatory, neurological, and cardiovascular disorders (Sijtsma & De Swaaf, 2004). The endocrine and immunological systems, blood vessels, lungs, and heart are all able to reap the benefits of fatty acids' capacity to generate energy for the body (Rubio-Rodriguez et al., 2010; Gong et al., 2014). The primary source of DHA and EPA for the diet of humans at the moment is marine fatty fish. These sources have a number of drawbacks, including the fact that vegans do not approve of them, a disagreeable odor, and the possibility of pollution from water sources, such as mercury. As a result, researchers have been looking for alternate sources of DHA and EPA from microorganisms. It has been reported that algae oils contain 35–75% DHA (wt/wt). All microorganisms produce and use lipids as an energy source, as well as fatty acids and sterols for cell membrane biogenesis and maintenance.

Plants need rich soil, longer growing seasons, and low enzymatic activity to create DHA and EPA, if they are genetically altered; fungi, on the other hand, need an organic carbon supply and typically longer growing seasons (Barclay et al., 1994). Microalgae are considered as major source of DHA and EPA. They may naturally develop quickly and have a high capacity for synthesizing FAs under a vast range of autotrophic, mixotrophic, and heterotrophic growth circumstances (Li et al., 2009). This makes them an excellent choice for the yield of n-3FAs.

Microbial oil has several advantages, such as the presence of squalene or phytosterols, which are favorable for the human well-being (Gong et al., 2014). Furthermore, natural antioxidants in these microbial oils prevent their oxidative damage. Additionally, these microbes are readily cultivable, available all year round, and seasonally independent (Deeba et al., 2016). Because of these qualities, the generation of omega-3 PUFA by microorganisms is a desirable and cost-effective option for the industrial sector. The future pursuit of industrial application will benefit greatly from the metabolic engineering of oleaginous bacteria for improved n-3 FA production. In microbial species with increased n-3FA synthesis, Yuan and Alper (2019) have looked into the use of metabolic engineering. The focus of the present review is on the function of bacteria as biofactories in the synthesis of n-3FAs and its related components.

## 7.2 Applications of Omega-3 Fatty Acids

Omega-3 fatty acids have a wide variety of therapeutic effects, including anti-inflammatory, anti-arrhythmic, anti-atherosclerotic, anti-thrombotic, anti-fibrotic, and endothelium relaxing qualities. Because of these benefits, n-3 FAs are highly valued. They are believed to be essential for mental health and have a number of therapeutic benefits, particularly when used in conjunction with antidepressant medications due to their ability to treat depression and blood levels. This is especially true when considering the fact that antidepressants themselves have the ability to treat depression. Regular consumption can help prevent the risk of myocardial infarction, cardiac arrhythmias, hypertension, thrombosis, which is important for maintaining a healthy cardiovascular system. This is because n-3 FA's increase lipoprotein ratio while lowering total cholesterol to high-density lipoprotein ratio (Horrocks & Yeo, 1999).

For the fetal brain to develop properly, pregnant women must consume enough EPA and DHA (Damude & Kinney, 2008). ARA and DHA are also necessary for a baby's healthy growth and functional development (Dyerberg et al., 1995).

n-3FAs have been widely studied to treat several disorders. Some of them are discussed as follows:

*Cardiovascular diseases*: These include myocarditis, an inflammation of the heart muscle that leads to atherosclerosis (plaque buildup of calcium or cholesterol in the arteries). Because of this, there is an increased risk of blood clots forming, which can lead to cardiovascular problems such as heart attacks and strokes. FAs have the ability to modify inflammatory cascades because of their interactions with inflammatory signaling pathways, their regulation of inflammatory genes via peroxisome proliferator-activated receptor, and their capacity to suppress inflammatory cytokine production (Nagy & Tiuca, 2017). Thus, consuming food that contains FA greatly lowers the likelihood of such inflammation. Additionally, n-3 FA supplements have been shown to reduce smokers' dependence on nicotine, which is a key contributor to the rise in heart diseases. FA supplements provide a lot of advantages, including ease of use, low cost, and a lower chance of negative effects (Domenichiello et al., 2014).

*Neurodegenerative disorders*: It is well established that FAs improve membrane flexibility, ion permeability, and membrane shape. This facilitates the quick uptake of glucose into the brain, which in turn enhances neuronal function. The benefits of n-3 FA in treating neurodegenerative diseases are widely recognized. High concentrations of unsaturated fatty acids in synaptic nerve terminals contribute to the modification of vascular and immunological processes, which has a direct influence on central nervous system health. Memory loss, disorganized thinking, speech difficulties, mental incapacity, and other significant neurological disease including Alzheimer's and dementia may all be increased by a deficiency in n-3 FA. Consuming foods high in dietary DHA and EPA reduces the risk of acquiring neuropsychiatric disorders, as well as the occurrence of dementia and neurological diseases. There is evidence that boosting one's diet with both Vit B12 and n-3 FA can help minimize the incidence of neurodevelopmental problems (Rathod et al., 2016).

*Hypertension*: The word "hypertension" is frequently used to describe elevated blood pressure in the arteries caused by endothelial dysfunction, systemic vascular resistance, and enhanced insulin resistance. Therefore, taking the proper steps to control and avoid it is crucial. It is well established that ALA consumption from a variety of sources can significantly lower blood pressure (Das, 2010).

*Antimicrobial activity*: Studies indicate that EPA and DHA possess antibacterial, antiviral, and antifungal activities. They have been shown to act against *Plasmodium* sp., yeast growth, and hepatitis C virus. Additionally, n-3 FAs can stop infectious diseases from creating cytotoxins, which helps keep antibiotics working effectively against germs. Some of the antimicrobial processes that are triggered by n-3 fatty acids include internal protein leakage, preventing or delaying cell-to-cell communication, disturbing membrane hydrophobicity, impeding ATP generation, disrupting with the electron transport chain, altering cell permeability, and ending FA synthesis by impeding the FsbI enzymes activity (Hamilton et al., 2015).

*Biomarkers*: FAs have been discovered to be a valuable biomarker for a variety of illnesses, allowing researchers to understand the molecular process, assess risk, and make the correct diagnosis. FAs are preferred as biomarkers because of their high precision, medium sensitivity, and low cost; this characteristic also allows for simple biomarker screening (Nagy & Tiuca, 2017).

*Probiotics*: High quantities of n-3 FA are maintained inside the intestinal tract by co-microencapsulating probiotic bacteria with omega-3 rich tuna oil, which has been demonstrated in tests to promote probiotic bacterial viability and adhesion to the intestines. Omega-3 FA-supplemented probiotics have been shown to have favorable effects on inflammation, adipose tissue function, and hepatic lipid metabolism, all of which aid to prevent the onset of hepatic sclerosis (Hibbelna & Gowb, 2015).

## 7.3 Microbial Sources of Omega-3 PUFA

Various diatoms (*Phaeodactylum tricornutum*), microalgae (*Chlorella minutissima, Crypthecodinium cohnii, Pavlova lutheri, Aurantiochytrium*), and fungi (*Mortierella alpina, Candida glabrata, Rhodosporidium toruloides*) are reported to generate high quantities of n-3 FAs like ARA, DHA, and EPA (Amjad Khan et al., 2017;

Abedi & Sahari, 2014; Shimiziu et al., 1988; Mironov et al., 2018; Hamilton et al., 2016; Guiheneuf et al., 2009). Genetic engineering was employed to boost omega-3 FA production in microorganisms. *Yarrowia lipolytica*, a yeast, has been genetically altered to generate more EPA, DGLA, and ARA (Xue et al. 2013). Fungi *Rhodosporidium toruloides* was also genetically modified, with three desaturase (FAD3) and 12 desaturase (FAD2) genes introduced, resulting in a 49% increase in ALA synthesis (Liu et al., 2018). Table 7.1 shows PUFA content in different microorganisms:

## 7.4 Docosahexaenoic Acid (DHA) Production by Microorganisms

DHA is made up of 22-carbon atoms and has 6-cis-double bonds. It is a primary component of the CNS, skin, heart, and retina. DHA is an essential dietary component due to its wide range physiological effects and health benefits which includes reducing blood pressure, arthritis, hypertension, and thrombosis. DHA is mostly found in marine fatty fish. DHA is mainly consumed from fish food, making microbes a preferred source (Sijtsma & De Swaaf, 2004).

**Table 7.1** PUFA content in various microorganisms

| Microorganisms | PUFA composition | Reference(s) |
|---|---|---|
| *Aurantiochytrium* sp. ATCC PRA-276 | Docosahexaenoic acid (61.3–70.5%) | Furlan et al. (2017) |
| *Candida glabrata* UCP1566 | ALA (90.5%) | Lima et al. (2015) |
| *Chlorella minutissima* UTEX 2341 | EPA (31.8%) | Yongmanitchai and Ward (1991) |
| *Chlorella minutissima* | EPA (45%) | Seto et al. (1984) |
| *Crypthecodinium cohnii* | DHA (50%) | Mendes et al. (2007) |
| *Mortierella alpina* | ARA (50%) | Higashiyama et al. (2002) |
| *Mortierella*. alpine CFR-GV15 | ARA (56.82%) + EPA (3.46%) + DHA (4.30%) | Vadivelan and Venkateswaran (2014) |
| *Phaeodactylum tricornutum* | EPA + DHA (36.5% + 23.6%) | Hamilton et al. (2016) |
| *Phaeodactylum tricornutum* UTEX 640 | EPA (30.5%) | Yongmanitchai and Ward (1991) |
| *Thraustochytrium* sp. ONC-T18 | DHA (31.5%) | Burja et al. (2006) |

Source: Adarme-Vega et al. (2012), Deeba et al. (2016)

### 7.4.1 Potential Microorganisms for DHA Production

*Aurantiochytrium* spp., *Crypthecodinium cohnii, Thraustochytrium* spp., *Ulkenia* spp. are considered important producers of DHA (Mendes et al., 2007). The species *T. aurem, T. roseum,* and *T. striatum* of the genus *Thraustochytrium* accumulate a large amount of DHA, with levels ranging from 42 to 52% (Singh & Ward, 1996). Many algal classes including Dinophyceae, Cryptophyceae, and Haptophyceae are considered as good source of DHA (Singh & Ward, 1996). *Schizochytrium* spp. produces DHA at the highest levels. *Schizochytrium spp.* and *Crypthecodinium cohnii* are employed in the commercial synthesis of DHA by Martek Corporation (USA) (Sijtsma & De Swaaf, 2004). Production of DHA from bacteria is not considered good as it requires specialized growing conditions such as 2 °C and extremely high pressure (Ratledge et al., 1997) (Table 7.2).

### 7.4.2 Biosynthesis of DHA

Under conditions of carbon and nutritional limitation, a constant availability of acetyl-CoA and NADPH is necessary for FA synthesis (Ochsenreither et al., 2016). The isocitrate dehydrogenase enzyme of the TCA cycle is inhibited by an increase in ATP inside the mitochondria, which results in an enlarged pool of citrate that is transported to the cytoplasm (Patel et al., 2016). ATP: citrate lyase (ACL) enzyme transforms citrate into oxaloacetate and acetyl Co-A. Acetyl-CoA carboxylase then catalyzes the conversion of acetyl Co-A into malonyl Co-A (ACC). Additionally, fatty acid synthase (FAS) uses malonyl Co-A to produce palmitic acid (16:0) or stearic acid (18:0). The saturated fatty acid is then changed into other PUFAs.

The creation of n-3 FA can be caused by either an PKS pathway (polyketide synthase pathway) or aerobic desaturase system can be involved in n3 FA production. Both of these systems are known as polyketide synthase pathways (Fig. 7.1).

**Table 7.2** Percentage of DHA in various microorganisms

| DHA content | Microorganisms | Reference(s) |
|---|---|---|
| 61.3–70.5% | *Aurantiochytrium* sp. ATCC PRA-276 | Furlan et al. (2017) |
| 31.5% | *Thraustochytrium* sp. ONC-T18 | Burja et al. (2006) |
| 50% | *Crypthecodinium cohnii* | Mendes et al. (2007) |
| 46% | *Ulkenia* sp. | Kiy (2005) |
| DHA + EPA (23.6% + 36.5%) | *Phaeodactylum tricornutum* | Hamilton et al. (2016) |
| EPA + DHA (62.9%) | *Thraustochytrium aureum* | Ward and Singh (2005) |
| EPA + DHA (26.7%) | *Nannochloropsis* sp. | Hu and Gao (2003) |
| EPA + DHA (41.5%) | *Pavlova lutheri* | Guiheneuf et al. (2009) |

Source: Adarme-Vega et al. (2012), Deeba et al. (2016)

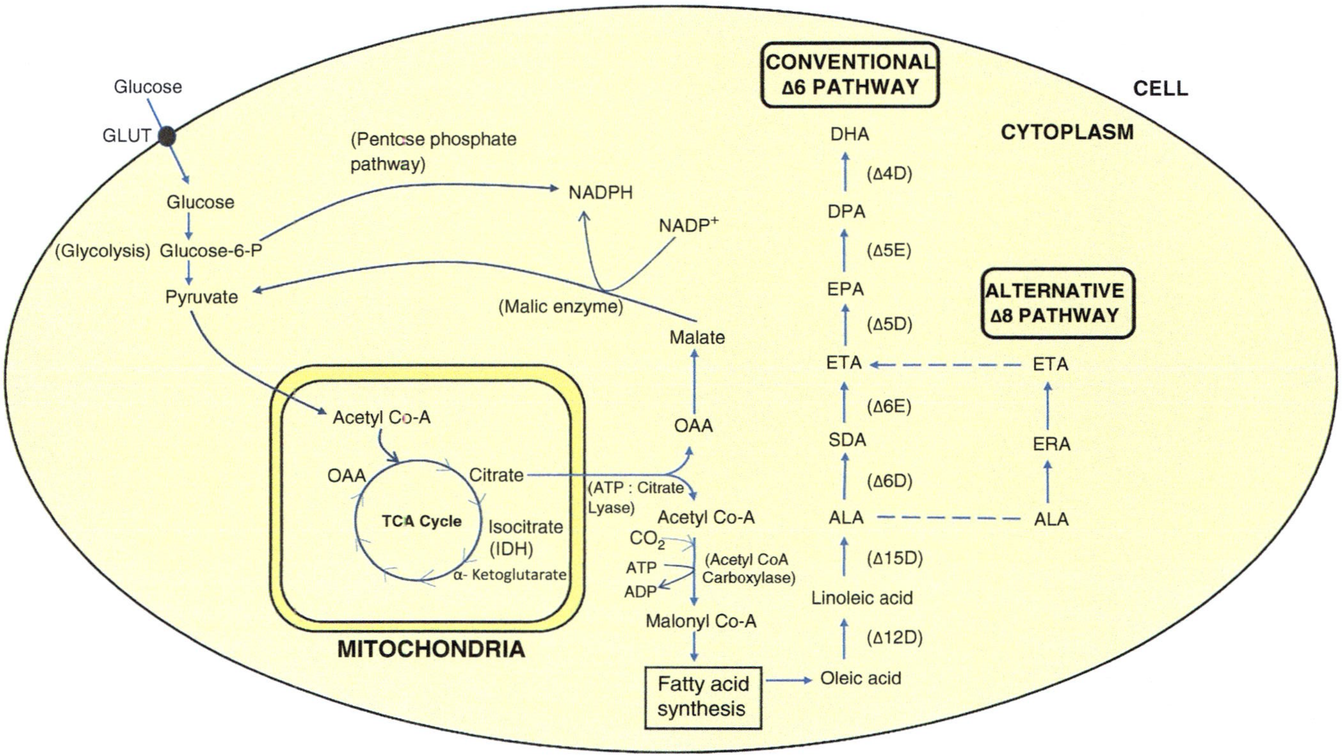


**Fig. 7.1** Overview of biosynthesis of ω-3PUFA in microbial cell. *ALA* α-linolenic acid, *SDA* stearidonic acid, *ERA* eicosatrienoic acid, *ETA* eicosatetraenoic acid, *EPA* eicosapentaenoic acid, *DPA* docosapentaenoic acid, *DHA* docosahexaenoic acid

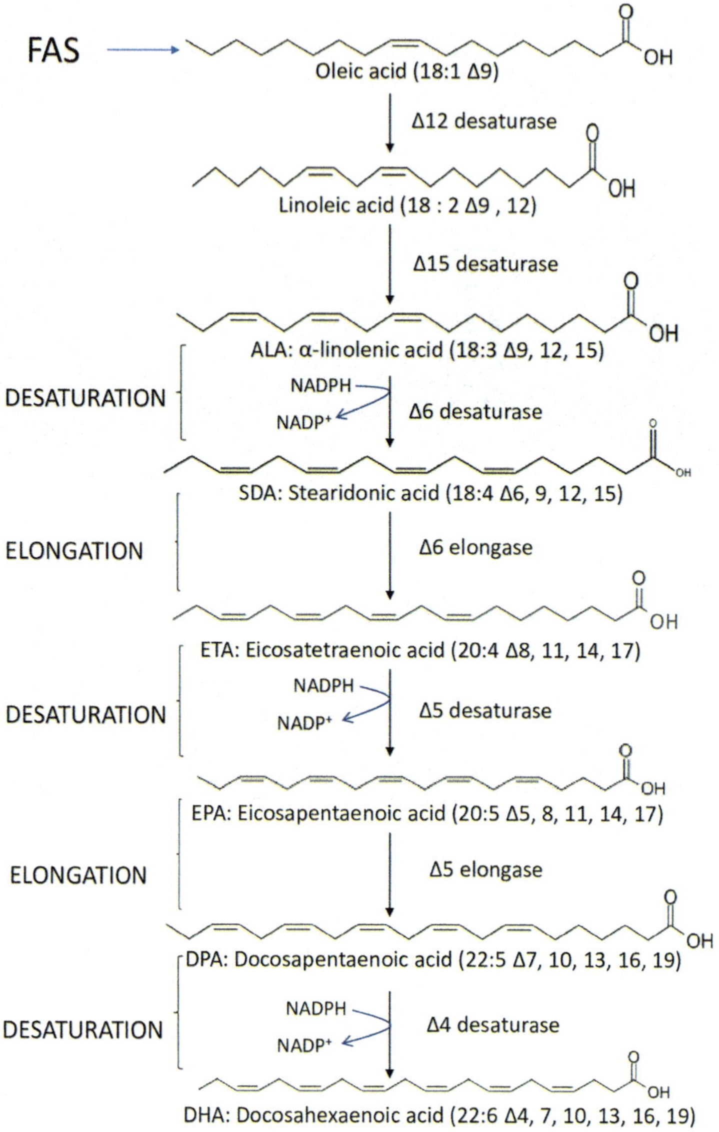


**Fig. 7.2** Pathway for the production of docosahexaenoic acid (DHA)

#### 7.4.2.1 Aerobic Pathway

The enzymes for desaturation and elongation are involved for synthesis of DHA via this particular pathway. While the elongase enzyme is responsible for fatty acid chain extension, Desaturase enzyme forms a double bond between two pre-existing double bonds in a fatty acid substrate. Figure 7.2 provides a general description of the process that leads to the creation of DHA.

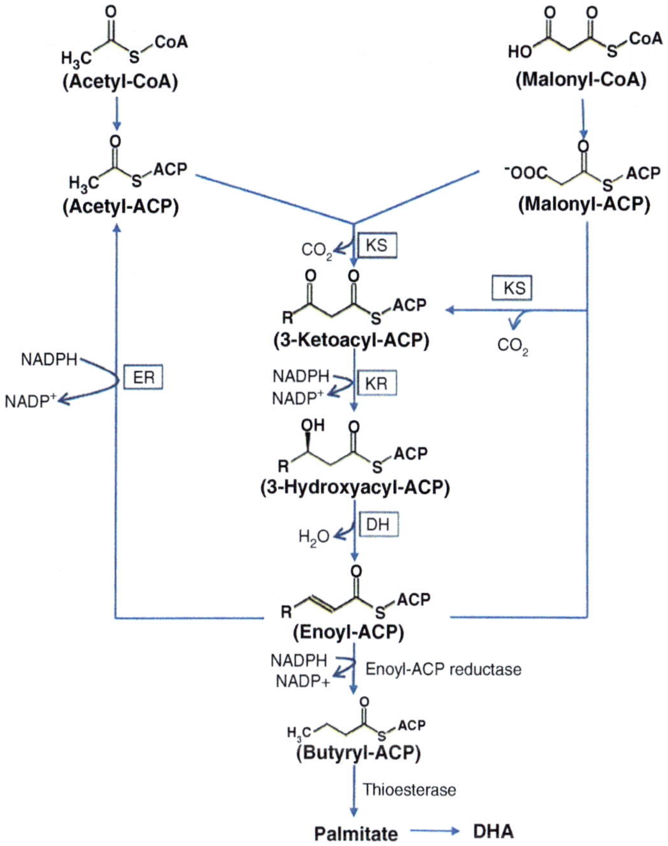


**Fig. 7.3** PKS pathway for the biosynthesis of DHA. *KS* keto synthase, *KR* keto reductase, *DH* dehydratase, *ER* enoyl reductase

### 7.4.2.2 Anaerobic Pathway (PKS)

Some fungal species like *C. cohnii*, *Schizochytrium* spp., and *Ulkenia* spp. synthesize DHA by PKS pathway (Ratledge, 2004) which was initially discovered in marine bacteria, *Shewanella pneumatophore* strain SCRC-2378 (Yazawa, 1996). In this pathway dehydratase-isomerase enzyme is used for the double bond insertion (Gong et al., 2014).

Microbes use either or both of these pathways (*Schizochytrium* and *Thraustochytrium* species) for long chain PUFAs biosynthesis (Qiu et al., 2001). The condensation, reduction, and dehydration steps in the PKS pathway involved in the synthesis of DHA is described in Fig. 7.3.

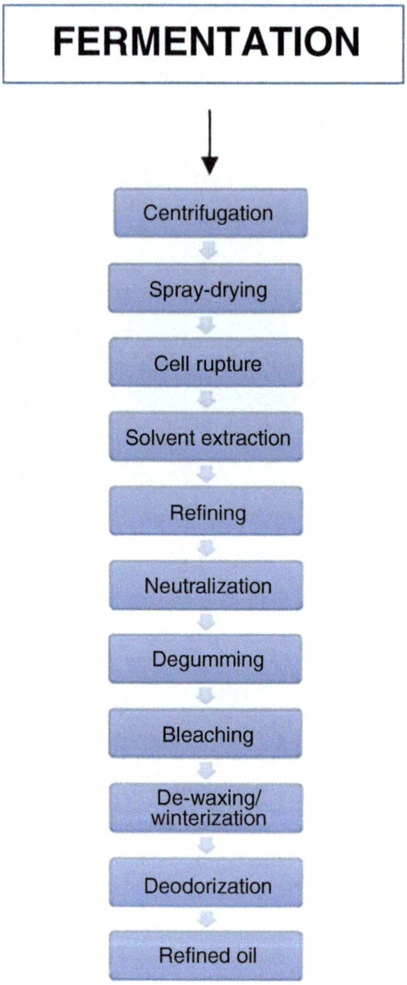


**Fig. 7.4** Flowchart representing Downstream Processing of DHA

## *7.4.3 Downstream Processing*

The methods used for DHA extraction and purification are referred to as downstream processing (Fig. 7.4). This process starts by the centrifugation for separating biomass. The harvested material is dried, frozen, and stored until extraction under nitrogen. It is very important to minimize exposure of DHA to oxygen (after the cell wall is ruptured), as DHA is exposed to potential oxidation (Wynn et al., 2005). To rupture the cell for the extraction of intracellular DHA, the biomass is sent through a homogenizer for disruption. For extraction, the biomass is mixed with organic solvent like hexane, ethanol, and chloroform (solvent extraction). Super critical fluid extraction process can also be used which produces highly purified extracts. The oil containing DHA is processed by removing hexane solvent through evaporation. The resulting crude oil is then kept at freezing temperatures, usually under nitrogen to

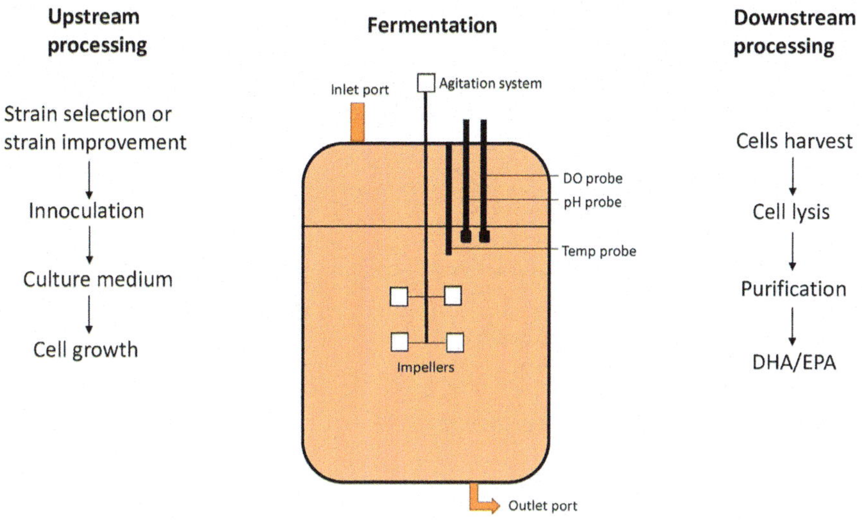


**Fig. 7.5** Laboratory production of DHA/EPA

prevent exposure to oxygen. Several steps such as refining, neutralization, degumming, bleaching, winterization, and deodorization are done for processing crude oil. Chemical refining is done to remove free fatty acids. Degumming involves addition of water to remove phosphatides, sterols, etc. Oil is then washed with water for removing soap. For the removal of color pigments, trace metals, and oxidation products bleaching is done. De-waxed bleached oil is then delivered to a deodorizer. In the processed oil, antioxidants such as tocopherols are added which is mixed with high-oleic sunflower oil to get DHA (40% w/w) (Fig. 7.5).

Mendes et al. (2007) reported the three-step process for DHA extraction. Firstly, ethanol-KOH is used for the direct solvent extraction of PUFA which is then transmethylated. For separating saturated FA from unsaturated ones on the basis of melting point, methyl esters are subjected to winterization technique, i.e., cooling of oil for the crystallization of saturated FAs and separation of unsaturated FAs. Urea is used to purify PUFA because of its capacity to solidify complexes with saturated free FAs.

The use of microorganisms to create DHA, a key n-3 FA, is a developing area of research that has the potential to be both environmentally friendly and economically viable.

## 7.5 Microbial Production of EPA (Eicosapentaenoic Acid)

Eicosapentaenoic acid (EPA) is another important n-3 FA having 20 carbons with five-cis-double bonds. It has anti-inflammatory and potential antineoplastic and chemo-preventive properties. Eicosanoids generated from EPA are believed to be beneficial in treating heart disease and other inflammatory chronic illnesses.

**Table 7.3** Percentage of EPA in various microorganisms

| EPA content (%) | Microorganisms | Reference(s) |
|---|---|---|
| 31.8 | *Chlorella minutissima* UTEX 2341 | Yongmanitchai and Ward (1991) |
| 30.5 | *Phaeodactylum tricornutum* UTEX 640 | Yongmanitchai and Ward (1991) |
| 45 | *Chlorella minutissima* | Seto et al. (1984) |
| 56.6 | *Yarrowia lipolytica* Y4184 | Xue et al. (2013) |
| 28 | *Nannochloropsis salina* | Van Wagenen et al. (2012) |
| 40 | *Shewanella putrefaciens* | Yazawa (1996) |
| 24 | *Alteromonas putrefaciens* | Yazawa et al. (1988) |
| 20 | *Mortierella* | Jareonkitmongkol et al. (1993) |

Source: Adarme-Vega et al., (2012), Deeba et al., (2016)

### *7.5.1 Potential Microorganisms*

The primary sources of EPA generation are fungi and algae (Table 7.3). Though some marine bacteria species are known to produce EPA but relatively in smaller proportions such as *Alteromonas*, *Shewanella*, *Flexibacter*, and *vibrio*. As per a study, *Shewanella putrefaciens* produced a significant quantity of EPA, near to 40% of total FAs (Bajpai & Bajpai, 1993). E. coli and other bacterial strains have been modified using genetic engineering techniques, and several laboratory tests have been conducted. The research showed that certain species of fungi, such as *Mortierella alpina* and *Mortierella hygrophila* produces maximum yield of EPA, i.e., 29 and 41 mg EPA per gram of cells, respectively. Another fungal class known as *Pythium* was found to yield 25–34 mg EPA per g cells (Ratledge et al., 1997).

Several kinds of marine algae are identified for producing high amounts of EPA (Bajpai & Bajpai, 1993). However, the cultivation of macroscopic algae is not so easy therefore they are not preferred for commercial production. Researchers discovered that phototrophic algae like Nannochloropsis oculata and Chlorella minutissima produce 45% of the total FAs' EPA (Ratledge et al., 1997).

EPA (55%) rich oils may now be produced commercially using a strain of oleaginous yeast called Yarrowia lipolytica that DuPont created using a PUFA engineering process (Xue et al., 2013). For enhanced EPA levels, Δ5-desaturase gene has been introduced in *Thraustochytrium* sp. (Kobayashi et al., 2011). Therefore, modified strains with efficient transformation mechanisms have been designed to produce high EPA levels.

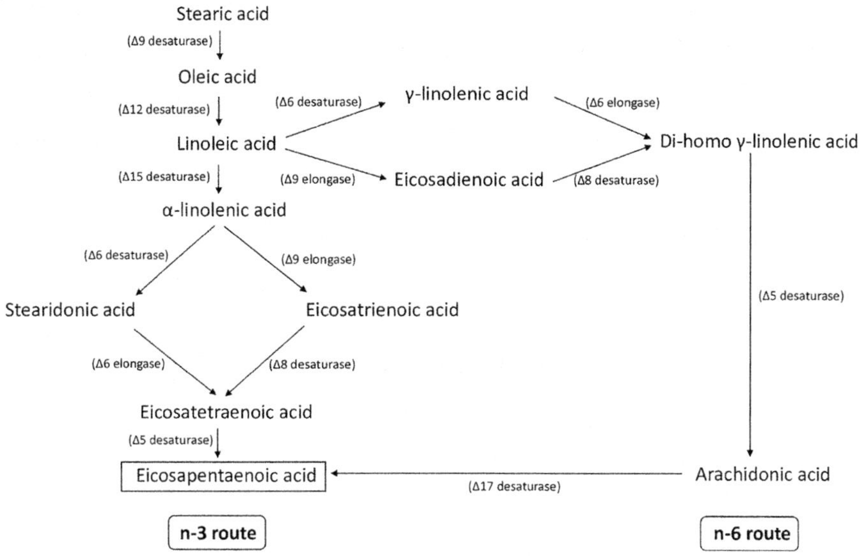


**Fig. 7.6** Desaturation and elongation pathway for EPA production

## 7.5.2 *Biosynthesis of EPA*

### 7.5.2.1 Aerobic Pathway

The series of reactions including desaturation and elongation leads to the formation of EPA like other PUFAs. The saturated Stearic acid (18:0) is first formed by condensation reactions which is then converted into Oleic acid (18:1) which is finally desaturated to form ALA (18:3). LA and ALA now act as parent chain and EPA is synthesized either from LA (n-6 route) or ALA (n-3 route). The pathway is represented in the Fig. 7.6.

### 7.5.2.2 PKS Pathway

The initial steps of this pathway are similar to those of FAS and involve the condensation of malonyl-CoA and acyl-CoA. This step is then followed by several iterations of reactions involving reduction, dehydration, reduction, and condensation, which ultimately result in fatty acyl chain extension through the addition of two carbon units at each cycle (Huang et al., 2004).

#### 7.5.2.3 Cultivation Systems

The cultivation systems for EPA production depends on the nature of microorganisms. Phototrophic cultivation systems are required for growing phototrophic algae while heterotrophic algae which depend on the externally supplied carbon source uses heterotrophic cultivation system (culturing microorganisms in the absence of light) for their growth and EPA production.

#### 7.5.2.4 Downstream Processing

The downstream processing of EPA extraction from harvested broth involves the separation of water from the biomass via centrifugation. The biomass is processed using a combination of hexane and methanol, and is then reacted with acetyl chloride to create esters. Then it is centrifuged and chromatographic methods such as silver silica gel chromatography is used for purifying EPA from crude extract, thus allowing the recovery and recycling of the solvent hexane and acetone. The presence of other polyunsaturated acids reduces the effectiveness of the process.

## 7.6 Microalgae as Omega-3 Producer

Through the process of photosynthesis, the great majority of aquatic environments are home to microalgae, which transform the light energy and $CO_2$ into various forms of biomass. When resources are limited but light is abundant, photosynthetic byproducts such as lipids and carbohydrates accumulate (Wynn et al., 2005). Microalgae are the main source of DHA and EPA, and these substances are progressively gathered at various trophic levels. Numerous fish, mollusc, and zooplankton species can have their development and diets impacted by changes in microalgae concentration (Brown, 2002).

As a consequence of this, the responsibility for preserving the structural soundness of aquatic food webs all around the world falls on microalgae (Sajjadi et al., 2018). According to the stage of development an organism is in, the ratio of polar to non-polar lipids in the cell membrane fluctuates. Polar lipids, also known as structural lipids, can be exemplified by glycolipids and phospholipids, respectively. Non-polar lipids include things such as triglycerides, diglycerides, monoglycerides, and free fatty acids, among other things. There are several strains of Phaeodactylum, Nannochloropsis, Thraustochytrium, and Schizochytrium that may produce substantial amounts of DHA and EPA (between 30% and 40% total fatty acid content). Controlling a number of variables, such as pH, temperature, and growing conditions, enables for large-scale synthesis of EPA and DHA (Adarme-Vega et al., 2012).

Due to their health-promoting qualities, microalgae are presently in great demand in the pharmaceutical and nutraceutical businesses. Infant formulae, a $10 billion market annually, are enriched with microalgal-derived PUFA like ARA and DHA.

For aquaculture, selection of the best microalgae species is important. A suitable species must have high rates of development and lipid contents (Patil et al., 2007). It must also be able to be mass-produced. Furthermore, the microalgae/species employed in aquaculture must be easily digested and of suitable size for ingestion (e.g., 1–15 m for filter feeders; 10–100 m for grazers). The majority of microalgae use $CO_2$ for photoautotrophic growth, however, some can also grow heterotrophically using solely organic molecules for carbon and energy (Fernandez et al., 2021). When compared to photoautotrophs, they have higher cell biomass and lipid productivities since they don't need light and grow at their fastest rate possible (Hossain & Mahlia, 2019).

## 7.7 Cultivation of Microorganisms and Influence of Various Conditions for Improved Omega-3 FA Synthesis

The n-3FA synthesis from microorganisms is optimized by a number of parameters. It is also vital to keep the circumstances stable since adverse conditions might lead to the accumulation of neutral lipids. The kind of bioreactors to employ for production depends on the kind and strain of microorganisms (Hossain & Mahlia, 2019). Some of the crucial factors are discussed in the following.

### *7.7.1 Temperature*

Temperature requirements are different for different organisms under study. For the growth and production of DHA, product accumulation increased with decrease in temperature. Most bacteria are thought to create additional PUFAs as a response to the low temperature circumstances. This change helps them to maintain membrane fluidity, as they are an integral part of cell membrane. At low temperatures, oxygen solubility increases. Desaturase enzyme production is accelerated, resulting in more effective catalysis (Singh & Ward, 1996). However, maintaining low temperatures requires a powerful cooling system, which is both expensive to construct and challenging to operate. As a result, in order to combat this, microorganisms have been cultivated at higher temperatures for a certain amount of time in order to achieve an optimal cell concentration before switching to lower temperatures. These temperature change experiments led to higher DHA content in a variety of microorganism groups. Compared to cultures kept at 25 °C for 72 h, C. cohnii cultivated at 25 °C for 48 h and then moved to 12 °C for 24 h showed a 20% increase in DHA content (Jiang & Chen, 2000a, b). Similar results were observed with the temperature shift trials among *Mortierella* spp., *Entomophthora exitalis,* and *Isochrysis galbana* which improved the DHA accumulation (Singh & Ward, 1996). However, this temperature shift strategy does not offer significant insights for scaling up the process.

C. cohnii has also been found to grow at temperatures ranging from 15 to 30 °C (Mendes et al., 2009).

Fungi and algae need low temperatures in order to produce EPA. The temperature of 25 °C was determined to be the most conducive for the development of several types of fungi and algae. After conducting experiments with temperature shifts, researchers discovered that many fungal species, with the exception of M. alpina, and algal species, including Porphyridium cruentum, Nannochloropis spp., and Pythium irregulare, produced more EPA when they were incubated at temperatures as low as 12 °C. The elongation and denaturation processes may include thermolabile enzymes, as evidenced by the fact that *Chlorella minutissima* produced more EPA at low temperatures (Seto et al., 1984). It has been proposed that lower temperatures enhance the solubility of oxygen in the medium, which facilitates microorganisms' ability to perform out elongation and desaturation activities, eventually increasing the synthesis of omega-3 fatty acids (Ratledge et al., 1997).

Temperatures that are either lower than or higher than the optimum temperature for microalgae may limit or hinder their development, which ultimately affects how much biomass is generated overall. Reports indicate that the majority of microalgae reach their full potential between 20 and 300 °C, depending on the environment in which they are situated (Brindhadevi et al., 2021). As opposed to that, thermophilic algae are able to survive at temperatures as high as 800 °C (Khoo et al., 2020).

## 7.7.2 *Nutrients*

It is crucial to supply all nutrients at the right pH and temperature (Khoo et al., 2020). Microalgae require certain nutritional ratio in order to flourish. Autotrophs, for example, require sodium bicarbonate and $CO_2$ as their primary carbon source. For heterotrophs, supplies of nitrogen, glycerol, and glucose are necessary, but in case mixotrophs, mixed carbon sources are employed. Different nitrogen sources have been investigated for C. cohnii development, including yeast extract, glutamic acid, waste molasses, maize steep liquor, ammonium chloride, and potassium nitrate (Mendes et al., 2009). Organic nitrogen sources yeast extract and glutamic acid are mainly used for cultivation because of the inability of the *C. cohnii* to grow in inorganic salts (De Swaaf et al., 2003; Pleissner & Eriksen, 2012). Due to its abundance in micro- and macro-nutrients and ability to lower production costs, yeast extract has been widely employed by Martek Corporation as a nitrogen source in their commercial production of DHA (Kyle et al., 1995).

### 7.7.2.1 Nitrogen

It is the main nutrient that supports the growth of biomass (Khoo et al., 2020). Nitrogen in various forms is required by microalgal species at concentrations ranging from 1% to 10%. Inadequate nitrogen availability, like other variables, may impede growth and total output. The two nitrogen sources that are most frequently used are nitrate and urea. According to reports, *Chlorella vulgaris* grew the fastest when exposed to potassium nitrate, producing 0.57 g/L/day of biomass and 47.1 g/L/day of lipids (Khoo et al., 2020). Kinds of algae such as *Botryococcus* sp., *Dunaliella bardawal,* and *D. salina* yielded high EPA under low nitrogen concentration on the contrary *Scenedesmus* and *Chlorella* (fresh water algae) produced PUFAs at higher nitrogen concentrations (Bajpai & Bajpai, 1993). *Phaedactylum tricornutum* and *N. laevis* showed optimum growth when nitrate and urea is present while the presence of ammonium decreased the growth and EPA yield. For the synthesis of EPA, yeast extract and peptone are suitable (Shimiziu et al., 1988). Despite using meat extract, tryptone, and amino acids, the biomass output was high but the EPA concentration was low (Bajpai & Bajpai, 1993; Shimiziu et al., 1988).

Micro-nutrients are as important as macronutrients for the growth of microorganisms and EPA production. In *Phaedactylum tricornutum*, the EPA production is increased by 65% in the presence of vitamin B12 (Ratledge et al., 1997). With silicate-limited cultures, diatoms such as N. laevis produced high levels of EPA (Wen & Chen, 2010). The reduction in EPA production in Euglena is brought on by the absence of certain metal ions, including manganese (Constantopoulos, 1970).

#### 7.7.2.2 Carbon

The most common form of carbon source utilized by *Cohnii* throughout the process of DHA synthesis is glucose. Many investigations have been conducted to discover the ideal levels of glucose essential for the development of *C. cohnii*. De Swaaf et al., (1999) discovered that when the glucose levels exceeded 25 g/L, the growth rate halted. According to some reports, however, the glucose content for industrial-scale manufacturing is approximately 50 g/L (Kyle et al., 1995). Other carbon sources have also been evaluated to determine their suitability for the growth of C. cohnii. It was revealed that glucose provided the best-quality DHA and is hence best suited for commercial DHA production (Mendes et al., 2009). When compared to an autotrophic culture, Chlorella cells generated in an autotrophic culture accumulated 28% less lipids and 45% less carbs than those produced in a heterotrophic culture using 10 g/L of glucose as the carbon source. A recent study found that all microalgae, with the exception of Nannochloropsis oculata, are capable of growing heterotrophically using 5 g of glucose per liter of water. A recent study found that glucose was the most efficient carbon source for the synthesis of FA from Schizochytrium limacinum (Shene et al., 2010).

Synthesis of EPA can be accomplished using a variety of oils, including linseed oil and corn oil, among others (Ratledge et al., 1997). It was discovered that linseed oil, which has 58% ALA (a precursor for the production of EPA), is the most effective carbon source for the fungus class Mortierella (Ratledge et al., 1997). M. elongata NRRL 5513 produced 29.5% mg/g of biomass when it was cultivated for 8 days at 15 °C in a media with 2% weight-per-volume (w/v) linseed oil (Ratledge et al., 1997).

The requirement of vitamin is important if the microorganism is auxotrophic like *C. cohnii*. Though vitamins are not required if yeast extract is added. It has been reported that a vitamin mixture including thiamine, biotin, and cyanocobalamine was applied to large-scale culture of Schizochytrium sp., and that extremely high volumetric productivity was produced (Ren et al., 2010). It was also shown that thiamine-HCl and cyanocobalamine boosted biomass production in Thraustochytrium species (Singh & Ward, 1996). But to keep the production cost low, it is advised to use yeast extract.

### 7.7.3 pH

pH, like other variables, plays an important function in development. From pH 6 to 8.76, most microalgal species grew rapidly (Andrade et al., 2021). *Thraustochytrid* strains may be cultivated at pH levels ranging from 4 to 8. With adaption of pH from 4 to 10, *Chlorella vulgaris* has shown its maximum growth at pH 9–10. Fungal infection can be stopped under situations of acidity at pH 3–4 (Khoo et al., 2020). The studies confirm that pH is critical for cultivating microalgal cells.

When *C. cohnii* was cultivated, different reports about the optimum pH came. The optimal pH was reported as 6.636 by Tuttle and Loeblich (1975) and for the purpose of producing DHA on a wide scale, Jiang and Chen (2000a, b) found that C. cohnii had the maximum specific growth rate at pH 7.2, pH range between 7.0 and 7.8 was reported by Kyle et al., (1995). Ammonium hydroxide is employed in Schizochytrium sp. for the industrial manufacture of DHA and serves as both a nitrogen supply and a pH stabilizer (Bailey et al., 2003). The usage of ammonium hydroxide results in a reduction in the pH since the cells take up nitrogen in the form of ammonia, which leaves more hydrogen ions in the medium.

### 7.7.4 Salinity

Salinity is an important factor for marine species because the majority of DHA-producing microorganisms are marine, it is necessary to use a saline environment similar to natural seawater for cultivating them. Several different salinities (0–5% NaCl) were investigated on the phycomycetes genus, the ideal concentration of NaCl was discovered to be 2.5% (Singh & Ward, 1996). Many species cannot thrive in either extremely high (5% NaCl) or extremely low (0–1% NaCl) salt conditions. For C. cohnii, it was discovered that higher salinities result in the production of more oleic acid, whereas lower salinities result in higher DHA content (Singh & Ward, 1996). The ideal salinity of *C. cohnii* ATCC 30772 was found to be 17.8 g/L which was half the salinity of sea water (De Swaaf et al., 1999).

### 7.7.5 Activators

Advanced fermentation techniques have been used in microalgal strains to improve the synthesis of lipids and DHA. Earlier research has also primarily focused on greater DHA production because there have been several clinical trials incorporating a variety of nutrient-based therapies that have led to improved cognitive health (Charles et al., 2019). The use of alcohols in the fermentation medium has been reported to increase the amount of Astaxanthin and lipids (Zhang et al., 2017; Du et al., 2019).

In a prior investigation, a mixture of chemical modulators consisting of ethanolamine, napthoxyacetic acid, and butylated hydroxyanisole induced *C. cohnii* cells to collect lipids that were anywhere from 8 to 20% greater than normal (Wang et al., 2018). In prior testing, the low-cost carbon source glycerol resulted in increased lipid content (Saenge et al., 2011; Sengmee et al., 2017). Geng et al., (2019) found that gibberellin is an excellent tool for boosting lipid and DHA synthesis in Aurantiochytrium spp.

### 7.7.6 Mixing/Agitation

In fermentation processes, agitation plays an important mixing and shearing role. Impellers are used in fermenters which helps in the continuous mixing of nutrients, oxygen, and heat in the fermentation broth. Agitation depends on the viscosity of the medium. Very high agitation speed can lead to heterogenous mixing and shear forces can damage fragile microorganisms while very low agitation speed can increase the viscosity of fermentation broth causing reduction in mass transfer efficiency. The length between the impellers is two times greater in a low-viscosity medium than it is in a high-viscosity one (Larsson et al., 2011).

Homogeneous mixing of nutrients, temperature, and pH in a culture is another important aspect of microalgae cultivation. This prevents sedimentation and clumping and also provides light where it is needed (Khoo et al., 2020).

### 7.7.7 Aeration Rate

Aeration determines the oxygenation of the fermentation process. Different DHA-producing microorganisms require different air saturation levels. During fermentation, the oxygen supply (one of the key-limiting factor of microbial growth) is monitored via DO (dissolved oxygen) probe. *Schizochytrium* sp. And *C. cohnii* have been produced commercially using a variety of air saturation levels. The commercial synthesis of DHA by C. cohnii is planned to use DO levels between 10% and 50% of the air saturation level (Kyle et al., 1995). *Schizochytrium sp.*, on the other hand, ferments with relatively low oxygen saturation values in the air because it uses the oxygen-free PKS route to create DHA (Bailey et al., 2003). The oxygen partial pressure is also impacted by the bioreactor's height (Larsson et al., 2011).

## 7.8 Conclusion

Utilizing microorganisms as a source of n-3 FA provides a solution that seems to be more environmentally friendly, dependable, and of superior quality. Single cell oils (SCOs), commonly referred to as microbial lipids, are a possible substitute for traditional sources of edible oils. These oils are derived from microbes. They provide additional benefits, such as ease of production and maintenance, and antioxidant properties of microbial oils, while overcoming the drawbacks of seafoods, such as chemical contamination and rapid declination of stocks. Several microorganisms have been used to produce n-3FAs on a commercial scale. Microalgae have been identified as one of these with the greatest potential for producing EPA and DHA. Microalgae are known to accumulate photosynthetic bioproducts including lipids and carbohydrates when the resources are scarce. These oils have been utilized in newborn formula, and they are also seeing growing application in dietary supplements as well as goods enriched with nutrients. Researchers have looked at the possibility of using metabolic engineering as a stand-in for the external stress that is often used. This entails optimizing nutrients, temperature, pH, agitation, and other factors for various strain types that are employed to produce FA. Additionally, genetic alterations are being made to improve wild strains with increased PUFA production for commercial production. The production of specific PUFAs can be done by manipulating fatty acid biosynthetic pathways or using novel pathways. Overall, the microbial synthesis of omega-3 FA is a promising and developing area that may offer a reliable and affordable source of these important n-3 FA. There is scope to explore other microbes and develop improved strains and standardize their cultivation methods for efficient n-3FA production.

## References

Abedi, E., & Sahari, M. A. (2014). Long-chain polyunsaturated fatty acid sources and evaluation of their nutritional and functional properties. *Food Science & Nutrition, 28*(36), 49833–49851.

Adarme-Vega, T. C., Lim, D. K., Timmins, M., Vernen, F., Li, Y., & Schenk, P. M. (2012). Microalgal biofactories: A promising approach towards sustainable omega-3 fatty acid production. *Microbial Cell Factories, 11*(1), 1.

Amjad Khan, W., Chun-Mei, H., Khan, N., Iqbal, A., Lyu, S. W., & Shah, F. (2017). Bioengineered plants can be a useful source of omega-3 fatty acids. *BioMed Research International, 2017*, 7348919.

Andrade, D. S., Amaral, H. F., Gavilanes, F. Z., Morioka, L.R., Nassar, J. M., de Melo, J. M., Silva, H.R., Telles, T.S. (2021) Microalgae: Cultivation, biotechnological, environmental, and agricultural applications. In Advances in the Domain of Environmental Biotechnology. Maddela, N. R., Garcia Cruzatty, L.C., Charaborty, S.Spinger. pp 635-701. https://doi.org/10.1007/978-981-15-8999-7_23

Bailey, R. B., DiMasi, D., Hansen, J. M., Mirrasoul, P. J., & Ruecker, C. M. (2003). *Enhanced production of lipids containing polyenoic fatty acid by very high density cultures of eukaryotic microbes in fermentors*. Patent NumberUS Patent: 6607900.

Bajpai, P., & Bajpai, P. K. (1993). Eicosapentaenoic acid (EPA) production from microorganisms: A review. *Journal of Biotechnology, 30*(2), 161–183.

Barclay, W. R., Meager, K. M., & Abril, J. R. (1994). Heterotrophic production of long chain omega-3 fatty acids utilizing algae and algae-like microorganisms. *Journal of Applied Phycology, 6*(2), 123–129.

Brindhadevi, K., Mathimani, T., Rene, E. R., Shanmugam, S., Chi, N. T., & Pugazhendhi, A. (2021). Impact of cultivation conditions on the biomass and lipid in microalgae with an emphasis on biodiesel. *Fuel, 284*, 119058.

Brown, M. R. (2002). Nutritional value and use of microalgae in aquaculture. In: Cruz-Suárez, L. E., Ricque-Marie, D., Tapia-Salazar, M., Gaxiola-Cortés, M. G., Simoes, N. (Eds.). *Avances en Nutrición Acuícola VI. Memorias del VI Simposium Internacional de Nutrición Acuícola*. 3 al 6 de Septiembre del 2002. Cancún, Quintana Roo, México.

Brunner, E. (2006). Oily fish and omega 3 fat supplements. *BMJ, 332*(7544), 739–740. https://doi.org/10.1136/bmj.38798.680185.47. Epub 2006 Mar 24. PMID: 16565094; PMCID: PMC1420720.

Burja, A. M., Radianingtyas, H., Windust, A., & Barrow, C. J. (2006). Isolation and characterization of polyunsaturated fatty acid producing Thraustochytrium species: Screening of strains and optimization of omega-3 production. *Applied Microbiology and Biotechnology, 72*(6), 1161–1169.

Charles, C. N., Msagati, T., Swai, H., & Chacha, M. (2019). Microalgae: An alternative natural source of bioavailable omega-3 DHA for promotion of mental health in East Africa. *Scientific African, 6*, e00187.

Constantopoulos, G. (1970). Lipid metabolism of manganese-deficient algae: Effect of manganese deficiency on the greening and the lipid composition of *Euglena gracilis* Z. *Plant Physiology, 45*(1), 76–80.

Damude, H. G., & Kinney, A. J. (2008). Enhancing plant seed oils for human nutrition. *Plant Physiology, 147*(3), 962–968.

Das, U. N. (2010). Essential fatty acids and their metabolites in the context of hypertension. *Hypertension Research, 33*, 782–785.

Deeba, F., Pruthi, V., & Negi, Y. S. (2016). Converting paper mill sludge into neutral lipids by oleaginous yeast *Cryptococcus vishniaccii* for biodiesel production. *Bioresource Technology, 213*, 96–102.

Domenichiello, A. F., Kitson, A. P., & Bazinet, R. P. (2014). Is docosahexaenoic acid synthesis from a-linolenic acid sufficient to supply the adult brain. *Progress in Lipid Research, 59*, 54–66.

Du, H., Liao, X., Gao, Z., Li, Y., Lei, Y., Chen, W., Chen, L., Fan, X., Zhang, K., & Chen, S. (2019). Effects of methanol on carotenoids as well as biomass and fatty acid biosynthesis in *Schizochytrium limacinum* B4D1. *Applied and Environmental Microbiology, 85*, e01243–e01219.

Dyerberg, J., Leaf, A., & Galli, C. (1995). ISSFAL board statement: Recommendations for the essential fatty acid requirement for infant formulas. *Journal of the American College of Nutrition, 14*, 21314.

Fernandez, F., Reis, A., Wijffels, R. H., Barbosa, M., Verdelho, V., & Llams, B. (2021). The role of microalgae in the bioeconomy. *New Biotechnology, 61*, 99–107.

Furlan, V., Maus, V., Batista, I., & Bandarra, N. M. (2017). Production of docosahexaenoic acid by Aurantiochytrium sp. ATCC PRA-276. *Brazilian Journal of Microbiology, 48*, 359–365.

Geng, L., Chen, S., Sun, X., Hu, X., Ji, X., Huang, H., & Ren, L. (2019). Fermentation performance and metabolomic analysis of an engineered high-yield PUFA-producing strain of Schizochytrium sp. *Bioprocess and biosystems engineering, 42*, 71–81.

Gong, Y., Wan, X., Jiang, M., Hu, C., Hu, H., & Huang, F. (2014). Metabolic engineering of microorganisms to produce omega-3 very long-chain polyunsaturated fatty acids. *Progress in Lipid Research, 56*, 19–35.

Guiheneuf, F., Mimouni, V., Ulmann, L., & Tremblin, G. (2009). Combined effects of irradiance level and carbon source on fatty acid and lipid class composition in the microalga *Pavlova lutheri* commonly used in mariculture. *Journal of Experimental Marine Biology and Ecology, 369*(2), 136–143.

Hamilton, M. L., Warwick, J., Terry, A., Allen, M. J., Napier, J. A., & Sayanova, O. (2015). Towards the industrial production of omega-3 long chain polyunsaturated fatty acids from a genetically modified diatom *Phaeodactylum tricornutum. PLoS One, 10*, e0144054.

Hamilton, M. L., Powers, S., Napier, J. A., & Sayanova, O. (2016). Heterotrophic production of omega-3 long-chain polyunsaturated fatty acids by trophically converted marine diatom *Phaeodactylum tricornutum. Marine Drugs, 14*(3), 53.

Hibbelna, J. R., & Gowb, R. V. (2015). Omega-3 fatty acid and nutrient deficits in adverse neurodevelopment and childhood behaviors. *Child and Adolescent Psychiatric Clinics of North America, 23*, 555–590.

Higashiyama, K., Fujikawa, S., Park, E. Y., & Shimizu, S. (2002). Production of arachidonic acid by *Mortierella* fungi. *Biotechnology and Bioprocess Engineering, 7*(5), 252–262.

Horrocks, L. A., & Yeo, Y. K. (1999). Health benefits of docosahexaenoic acid (DHA). *Pharmacological Research, 40*(3), 211–225.

Hossain, N., & Mahlia, T. M. (2019). Progress in physicochemical parameters of microalgae cultivation for biofuel production. *Critical Reviews in Biotechnology, 39*, 835–859.

Hu, H., & Gao, K. (2003). Optimization of growth and fatty acid composition of a unicellular marine picoplankton, *Nannochloropsis* sp., with enriched carbon sources. *Biotechnology Letters, 25*(5), 421–425.

Huang, Y. S., Pereira, S. L., & Leonard, A. E. (2004). Enzymes for transgenic biosynthesis of long-chain polyunsaturated fatty acids. *Biochimie, 86*(11), 793–798.

Jareonkitmongkol, S., Shimizu, S., & Yamada, H. (1993). Production of an eicosapentaenoic acid-containing oil by a Δ12 desaturase-defective mutant of *Mortierella alpina* 1S-4. *Journal of the American Oil Chemists' Society, 70*(2), 119–123.

Jiang, Y., & Chen, F. (2000a). Effects of medium glucose concentration and pH on docosahexaenoic acid content of heterotrophic *Crypthecodinium cohnii. Process Biochemistry, 35*(10), 1205–1209.

Jiang, Y., & Chen, F. (2000b). Effects of temperature and temperature shift on docosahexaenoic acid production by the marine microalge *Crypthecodinium cohnii. Journal of the American Oil Chemists' Society, 77*(6), 613–617.

Khoo, K. S., Chew, K. W., Yew, G. Y., Leong, W. H., Chai, Y. H., Show, P. L., & Chen, W. H. (2020). Recent advances in downstream processing of microalgae lipid recovery for biofuel production. *Bioresource Technology, 304*, 122996.

Kiy, T. (2005). Production of docosahexaenoic acid by the marine microalga, *Ulkenia* sp. In *Single cell oil*. AOCS Press (pp. 36–52).

Kobayashi, T., Sakaguchi, K., Matsuda, T., Abe, E., Hama, Y., Hayashi, M., & Ito, M. (2011). Increase of eicosapentaenoic acid in thraustochytrids through thraustochytrid ubiquitin promoter-driven expression of a fatty acid Δ5 desaturase gene. *Applied and Environmental Microbiology, 77*(11), 3870–3876.

Kralovec, J. A., Zhang, S., Zhang, W., & Barrow, J. C. (2011). A review of the process in enzymatic concentration and microencapsulation of omega-3 rich oil from fish and microbial sources. *Food Chemistry, 131*, 639–644.

Kyle, D. J., Reeb, S. E., & Sicotte, V. J. (1995). *U.S. patent no. 5,407,957*. U.S. Patent and Trademark Office.

Larsson, G. (2011). Cultivation technology Compendium.

Li, Y., Qin, J. G., Ball, A. S., & Moore, R. B. (2009). *Perspectives on marine phytoplankton as sources of nutrition and bioenergy 187-203*. Marine phytoplankton. Nova Science Pub Inc, New York.

Lima, R. A., Andrade, R. F., Ribeaux, D. R., Santos, P. N., Albuquerque, C. D., & Campos-Takaki, G. M. (2015). Production of very long chain polyunsaturated omega 3 and omega 6 fatty acids by Candida glabrata strains. *International Journal of Current Microbiology and Applied Sciences, 4*(8), 244–252.

Liu, Y., Koh, C. M., & Ji, L. (2018). *US patent no. 10,081,821*. US Patent and Trademark Office.

Mendes, A., Lopes da Silva, T., & Reis, A. (2007). DHA concentration and purification from the marine heterotrophic microalga *Crypthecodinium cohnii* CCMP 316 by winterization and urea complexation. *Food Technology and Biotechnology, 45*(1), 38–44.

Mendes, A., Reis, A., Vasconcelos, R., Guerra, P., & Lopes da Silva, T. (2009). *Crypthecodinium cohnii* with emphasis on DHA production: A review. *Journal of Applied Phycology, 21*(2), 199–214.

Mironov, A. A., Nemashkalov, V. A., Stepanova, N. N., Kamzolova, S. V., Rymowicz, W., & Morgunov, I. G. (2018). The effect of pH and temperature on arachidonic acid production by glycerol-grown *Mortierella alpina* NRRL-A-10995. *Fermentation, 4*(1), 17.

Nagy, K., & Tiuca, I. (2017). *Importance of fatty acids in physiopathology of human body*. Corpus ID: 54513514.

Ochsenreither, K., Gluck, C., Stressler, T., Fischer, L., & Syldatk, C. (2016). Production strategies and applications of microbial single cell oils. *Frontiers in Microbiology, 7*, 1539.

Patel, A., Arora, N., Sartaj, K., Pruthi, V., & Pruthi, P. A. (2016). Sustainable biodiesel production from oleaginous yeasts utilizing hydrolysates of various non-edible lignocellulosic biomasses. *Renewable and Sustainable Energy Reviews, 62*, 836–855.

Patil, V., Kallqvist, T., Olsen, E., Vogt, G., & Gislerod, H. R. (2007). Fatty acid composition of 12 microalgae for possible use in aquaculture feed. *Aquaculture International, 15*(1), 1–9.

Pleissner, D., & Eriksen, N. T. (2012). Effects of phosphorous, nitrogen, and carbon limitation on biomass composition in batch and continuous flow cultures of the heterotrophic dinoflagellate *Crypthecodinium cohnii. Biotechnology and Bioengineering, 109*(8), 2005–2016.

Qiu, X., Hong, H., & MacKenzie, S. L. (2001). Identification of a Δ4 fatty acid desaturase from Thraustochytrium sp. involved in the biosynthesis of docosahexanoic acid by heterologous expression in *Saccharomyces cerevisiae* and Brassica juncea. *Journal of Biological Chemistry, 276*(34), 31561–31566.

Rathod, R., Kale, A., & Joshi, S. (2016). Novel insights into the effect of vitamin B12 and omega-3 fatty acids on brain function. *Journal of Biomedical Science, 23*(1), 1–7.

Ratledge, C. (2004). Fatty acid biosynthesis in microorganisms being used for single cell oil production. *Biochimie, 86*(11), 807–815.

Ratledge, C., Bowater, MDV., Taylor, PN. (1997). Correlation of ATP: Citrate lyase activity with lipid accumulation in developing seeds of Brassica napus L. *Lipids, 32*, 7–12.

Ren, L. J., Ji, X. J., Huang, H., Qu, L., Feng, Y., Tong, Q. Q., & Ouyang, P. K. (2010). Development of a stepwise aeration control strategy for efficient docosahexaenoic acid production by *Schizochytrium* sp. *Applied Microbiology and Biotechnology, 87*(5), 1649–1656.

Rubio-Rodriguez, N., Beltran, S., Jaime, I., Sara, M., Sanz, M. T., & Carballido, J. R. (2010). Production of omega-3 polyunsaturated fatty acid concentrates: A review. *Innovative Food Science & Emerging Technologies, 11*(1), 1–12.

Saenge, C., Cheirsilp, B., Suksaroge, T. T., & Bourtoom, T. (2011). Potential use of oleaginous red yeast *Rhodotorula glutinis* for the bioconversion of crude glycerol from biodiesel plant to lipids and carotenoids. *Process Biochemistry, 46*, 210–218.

Sajjadi, B., Chen, W. Y., Raman, A., & Ibrahim, S. (2018). Microalgae lipid and biomass for biofuel production: A comprehensive review on lipid enhancement strategies and their effects on fatty acid composition. *Renewable and Sustainable Energy Reviews, 97*, 200–232.

Sengmee, D., Cheirsilp, B., Suksaroge, T. T., & Prasertsan, P. (2017). Biophotolysis-based hydrogen and lipid production by oleaginous microalgae using crude glycerol as exogenous carbon source. *International Journal of Hydrogen Energy, 42*, 1970–1976.

Seto, A., Wang, H. L., & Hesseltine, C. W. (1984). Culture conditions affect eicosapentaenoic acid content of *Chlorella minutissima. Journal of the American Oil Chemists' Society, 61*(5), 892–894.

Shene, C., Leyton, A., Esparza, Y., Flores, L., Quilodran, B., Hinzpeter, I., & Rubilar, M. (2010). Microbial oils and fatty acids: Effect of carbon source on docosahexaenoic acid (c22: 6 n3, DHA) production by thraustochytrid strains. *Journal of Soil Science and Plant Nutrition, 10*, 207–216.

Shimiziu, S., Kawashima, H., Shinmen, Y., Akimoto, K., & Yamada, H. (1988). Production of eicosapentaenoic acid by Mortierella fungi. *Journal of the American Oil Chemists' Society, 65*(9), 1455–1459.

Sijtsma, L., & De Swaaf, M. E. (2004). Biotechnological production and applications of the ω-3 polyunsaturated fatty acid docosahexaenoic acid. *Applied Microbiology and Biotechnology, 64*(2), 146–153.

Singh, A., & Ward, O. P. (1996). Production of high yields of docosahex-aenoc acid by Thraustochytrium roseau ATCC 28210. *J Ind Microbiol Biotechnol, 16*, 370–373.

Swaaf, D., De Rijk, T. C., Eggink, G., & Sijtsma, L. (1999). Optimisation of docosahexaenoic acid production in batch cultivations by Crypthecodinium cohnii. *Progress in Industrial Microbiology, 35*, 185–192.

Swaaf, D., Sijtsma, L., & Pronk, J. T. (2003). High-cell-density fed-batch cultivation of the docosahexaenoic acid producing marine alga *Crypthecodinium cohnii. Biotechnology and Bioengineering, 81*(6), 666–672.

Tuttle, R. C., & Loeblich, A. R. (1975). An optimal growth medium for the dinoflagellate *Crypthecodinium cohnii. Phycologia, 14*(1), 1–8.

Vadivelan, G., & Venkateswaran, G. (2014). Production and enhancement of omega-3 fatty acid from Mortierella alpina CFR-GV15: Its food and therapeutic application. *BioMed Research International, 2014*, 1–9.

Van Wagenen, J., Miller, T. W., Hobbs, S., Hook, P., Crowe, B., & Huesemann, M. (2012). Effects of light and temperature on fatty acid production in *Nannochloropsis* salina. *Energies, 5*(3), 731–740.

Wang, K., Sun, T., Cui, J., Liu, L., Bi, Y., Pei, G., Chen, L., & Zhang, W. (2018). Screening of chemical modulators for lipid accumulation in *Schizochytrium* sp. S31. *Bioresource Technology, 260*, 124–129.

Ward, O. P., & Singh, A. (2005). Omega-3/6 fatty acids: Alternative sources of production. *Process Biochemistry, 40*(12), 3627–3652.

Wen, Z., & Chen, F. (2010). Production of eicosapentaenoic acid using heterotrophically grown microalgae. In *Single cell oils* (pp. 151–177).

Wynn, J. P., Behrens, P., Sundararajan, H., & Apt, K. (2005). *Single Cell Oils*. AOCS Publishing.

Xue, Z., Sharpe, P. L., Hong, S. P., Yadav, N. S., Xie, D., Short, D. R., & Zhu, Q. (2013). Production of omega-3 eicosapentaenoic acid by metabolic engineering of *Yarrowia lipolytica. Nature Biotechnology, 31*(8), 734–740.

Yazawa, K. (1996). Production of eicosapentaenoic acid from marine bacteria. *Lipids, 31*, S297–S300.

Yazawa, Z., Araki, K., Okazaki, N., Watanabe, K., Ishikawa, C., Inoue, A., & Kondo, K. (1988). Production of eicosapentaenoic acid by marine bacteria. *The Journal of Biochemistry, 103*(1), 5–7.

Yongmanitchai, W., & Ward, O. P. (1991). Screening of algae for potential alternative sources of eicosapentaenoic acid. *Phytochemistry, 30*(9), 2963–2967.

Yuan, S. F., & Alper, H. S. (2019). Metabolic engineering of microbial cell factories for production of nutraceuticals. *Microbial Cell Factories, 18*(1), 1–11.

Zhang, K., Chen, L., Liu, J., Gao, F., He, R., Chen, W., Guo, W., Chen, S., & Li, D. (2017). Effects of butanol on high value product production in *Schizochytrium limacinum* B4D1. *Enzyme and Microbial Technology, 102*, 9–15.

# Chapter 8
# Lactic Acid Bacteria as a Source of Functional Ingredients

**Maryam Fatima, Sultan Mehmood Ghani, Samia Khanum, Abdel Rahman Mohammad Said Al-Tawaha, Arun Karnwal, Abhijit Dey, Abdel Razzaq M. Al Tawaha, and Iftikhar Ali**

**Abstract** The formation of fermented foods relies heavily on lactic acid bacteria. They are employed as starters in various dairy, meat, vegetable, and beverage fermentations because of their metabolic capabilities. Their metabolic by-products enhance food's nutritional content, organoleptic qualities, and microbiological safety. Therefore, we examine the present situation of these items in this chapter. Lactic acid bacteria make antimicrobial substances like bacteriocins and organic acids that stop many pathogenic microorganisms from growing. The biofortification of vitamins by lactic acid bacteria reduces deficiencies and increases food value. Exopolysaccharides produced by LAB provide the dual functions of enhancing food texture and serving as a component of functional foods. They also enhance the flavor of fermented foods through aroma chemicals produced by metabolizing citrate and amino acids. People are apprehensive about processed foods and artifi-

M. Fatima · S. M. Ghani
National Institute of Food Science and Technology, University of Agriculture Faisalabad, Faisalabad, Pakistan

S. Khanum
Department of Botany, University of the Punjab, Lahore, Pakistan

A. R. Mohammad Said Al-Tawaha (✉)
Department of Biological Sciences, Al Hussein Bin Talal University, Ma'an, Jordan
e-mail: Abdel-al-tawaha@ahu.edu.jo

A. Karnwal
Department of Microbiology, Lovely Professional University, Phagwara, Punjab, India

A. Dey
Department of Life Sciences, Presidency University, Kolkata, India

A. R. M. Al Tawaha
Department of Crop Science, Faculty of Agriculture, Universiti Putra Malaysia, Serdang, Selangor, Malaysia

I. Ali
School of Life Science, Chinese University of Hong Kong, Hong Kong, China

Centre for Plant Science and Biodiversity, University of Swat, Charbagh, Pakistan



A. Karnwal, A. R. Mohammad Said Al-Tawaha (eds.), *Food Microbial Sustainability*, https://doi.org/10.1007/978-981-99-4784-3_8

cial preservatives. The use of LAB in goods or processing is recognized as a natural method of food preservation and health promotion.

**Keywords** Exopolysaccharides · Lactic acid bacteria · Microbiological safety · Fermentation · Functional food

## 8.1 Introduction

The production of dietary supplements that can alter the composition of the gut microbial population is the main mechanism through which new functions of food, such as the prevention of various diseases, are highlighted by developments in the food industry. Also, the demand for additive-free foodstuffs among many customers has assisted in developing and marketing a range of distinctive commodities by the food business, known as functional foods (Oliveira et al., 2018). Functional foods have an appropriate nutritional impact and positively influence specific biological functions, providing therapeutic benefits. Probiotics, prebiotics, and symbiotics are significant components of functional foods (Wong et al., 2022). In addition, the safety, shelf life, nutritional content, taste, and general quality of fermented goods can all be enhanced by lactic acid bacteria (LAB), frequently employed as starter cultures for the wide fermentation range of foods. Therefore, finding and using strains that deliver health-improving chemicals is interesting. These microbes may be found on the mucosal surfaces of humans, animals, and fish and in other places where carbohydrates are abundant. They are a typical microbiota component that naturally occupies the gastrointestinal tract and comprises a range of bacterial strains in both human and animal bodies (Gaspar et al., 2013). LAB adapts to several environments. They are present in milk, meat, vegetables, grains, and humans' vaginal microbiota and digestive systems. Their intestinal colonization requires strict nutritional requirements. Fermentable sugars and other crucial elements ensure and expedite their metabolism. This creates various metabolites with culinary and medicinal applications (Mazzoli et al., 2014).

In order to make lactic acid and energy, lactic acid bacteria must first break down a wide variety of carbohydrates and related compounds. *Oenococcus*, *Tetragenococcus*, *Aerococcus*, *Lactobacillus*, *Carnobacterium*, *Streptococcus*, *Enterococcus*, *Vagococcus*, and *Weissella* are all examples of common lactic acid bacteria of the Lactobacillus order. Lactic acid bacteria are Gram-positive, non-spore-forming anaerobes without a respiratory chain and catalase that can survive in an aerobic environment (Papagianni, 2012). Bacteria that produce lactic acid fall into one of two categories based on their hexose metabolic pathways: homofermentative bacteria, which produce lactic acid as their principal product, and heterofermentative bacteria, which produce lactic acid in addition to other products, including acetic acid, carbon dioxide, or aroma compounds. Lactic acid and several other valuable chemicals are produced as by-products of their metabolic process. In addition, they produce substances that are beneficial to human health, including as vitamin D, ethanol, polysaccharides, and antibiotics. In addition to it, they produce biofuels such as ethanol and butanol (Nuraida, 2015).

Although certain members of the genus Pediococcus deteriorate the food, which leads to spoilage, many strains of lactic acid bacteria are among the most significant microbes utilized in the food sectors. Lactic acid bacteria are utilized to preserve food and to alter the organoleptic qualities of food, such as taste and texture (Korhonen, 2010). Furthermore, LAB plays a significant part in producing chemicals, drugs, or other valuable goods in the industrial sector. In addition, lactic acid production using biotechnology has recently been reported as providing a remedy for the polluted environment caused by the petroleum industry (Hamdan & Sonomoto, 2011). This chapter will cover contemporary uses of LAB as a source of exopolysaccharides, starter cultures, vitamins, enzymes, and probiotics, particularly those that can meet the consumer demand for functional foods concerning health benefits.

## 8.2 Lactic Acid Bacteria as a Source of Probiotics

The word "probiotics" derives from the Greek "probios," meaning "for life," according to its origin. According to Parker, probiotics are microbes and chemicals that help the gut microbial equilibrium. Fuller said that a probiotic is a live microorganism-feeding ingredient that benefits the host by maintaining its gut microflora (Florou-Paneri et al., 2013). Probiotics have been utilized for over a hundred years to cure many infections, but usage has significantly reduced ever since the development of antibiotics. Because probiotics may colonize the microbiota in the gut and exert their positive impact without the need for continued medical intervention, they are so popular. Because of the rise in antibiotic resistance and the growing need to cut treatment costs, probiotics are increasingly being considered as an alternative to antibiotics (Iqbal et al., 2014). The initial probiotics that were readily accessible generally included one type of microorganism, often from the Saccharomyces or Lactobacillus genus. Further studies demonstrated the benefit of using such probiotics in preventing infectious diarrhea and post-antibiotic diarrhea, such as colitis caused by *Clostridium difficile* (Goldenberg et al., 2017). The quantity and type of microbes in probiotic formulations increased, ranging from $10^8$ to more than $10^{10}$ microorganisms. Most probiotic strains have been initially created for their ability to withstand low stomach pH, which led to several variations with unidentified physiological traits. Correlations are challenging due to the wide range of microbial associations, which presents the appearance of drug class effects and results in insufficient probiotic prescription medications (Ohkusa et al., 2019). Most identified Lactobacillus species, including *L. acidophilus*, *L. rhamnosus*, *L. reuteri*, *L. casei*, *L. plantarum*, *L. delbrueckii*, and *L. helveticus*, are regarded to be commercial probiotics.

## 8.2.1 *Mechanism of Action of Probiotics*

Probiotics affect the host in a variety of ways. The primary ways that probiotic bacteria work to strengthen the gastrointestinal tract's mucosal defenses include the following:

### 8.2.1.1 Disease Prevention

Probiotic strains inhibit the pathogenic microorganisms because they have to compete for the limited substrates needed for fermentation. Furthermore, enhancing the mucosal barrier function and releasing gut-protective chemicals inhibit the adhesion of harmful bacteria to the host cell. Organic acids, hydrogen peroxide, and bacteriocins secreted by probiotics also fight bacteria. Many strains treat lactose intolerance, acute diarrhea from bacterial and viral infections like rotavirus infections in children, Clostridium difficile gastroenteritis, diarrhea in tube-fed patients, chemo or radiotherapy-induced diarrhea, and irritable bowel diarrhea (Hemaiswarya et al., 2013).

### 8.2.1.2 Enhancement of Systemic and Mucosal Host Immunity

Immuno-modulation is achieved by controlling the actin cytoskeleton and protein phosphorylation, which increases mucus secretion. Probiotic bacteria and pathogenic bacteria fight for epithelial complex formation, preventing strains from colonization (O'Shea et al., 2012). Probiotic bacteria interact with the gut epithelial cells to improve mucosal barrier function. Probiotics may assist the host in several conditions, including Type 1 diabetes, by improving mucosal barrier function. External microbes can enter the gut wall by translocation through the epithelial layer.

Native intestinal bacteria, such as lactobacilli, can pass through mucosal layer and sustain for many days in the spleen or other organs that promote phagocytic activity. The intestinal mucus layer's physical reaction to lactobacilli given by mouth and its thickness is significant in the immunological response (Quinto et al., 2014). The chemicals synthesized with immuno-modulatory and anti-inflammatory activities can stimulate immune cells by the gastrointestinal microflora and regulates the immune system. The probiotic bacteria interactions with epithelial cells, monocytes/macrophages, and lymphocytes provide these immune-modulation properties. The modulatory immune system's response is one of the primary ways of action probiotics. B and T cells are essential for the immune system's adaptive response because they attach to specific antigens (D'Amelio & Sassi, 2018). The innate system, in comparison, reacts to typical molecular patterns that occur in most infections, known as pathogen-associated molecular patterns. Pattern recognition receptors attach the pathogen-associated molecular pattern and provide primary infection prevention. Trans-membrane proteins produce pattern recognition receptors in B-cells, natural killer cells, macrophages, fibroblasts, epithelial cells, and endothelial cells (Gómez-Llorente et al., 2010).

#### 8.2.1.3 Probiotics on Non-gastrointestinal Problems

Probiotics have benefits besides treating gastrointestinal disorders, such as anti-allergic action. Probiotics are utilized to cure allergic diseases such as atopic dermatitis, rhinitis, and food allergies (Hemaiswarya et al., 2013). Atopic illness is a symptom of food hypersensitivity caused by intestinal bacteria digesting food antigens in the gastrointestinal tract. Probiotics can alter potential antigens' structural characteristics, lessen their immunogenicity, decrease intestinal permeability, and diminish the production of pro-inflammatory cytokines, which are high in individuals with a range of allergy illnesses. By the activation of anti-inflammatory mediators or enhanced secretion that aids in the elimination of antigens from the intestinal mucosa, they significantly influence the immune system's response (Plaza-Diaz et al., 2019).

### *8.2.2 Probiotics and Health*

Today's consumers know the connection between lifestyle, diet, and good health, fueling interest in products that can improve health and overall nutrition. The number of health advantages attributed to functional food is expanding, and probiotics are among the food groups with the quickest growth rates for which scientific studies have shown therapeutic proof. Probiotics have a variety of medicinal uses, including the prevention of urogenital illnesses, relief from constipation, prevention of diarrhea, lowering of hypercholesterolemia, prevention of colon and bladder cancer, prophylaxis of osteoporosis, and reduction of food allergies (Saeed & Salam, 2013).

Probiotics are live microorganisms that are used to promote health nowadays in both humans and animals. Various food and beverages processed by fermentation are currently provided, making up around one-third of all human nutrition globally. Probiotics can be added to various food products (Dekumpitiya et al., 2016). Due to its widespread consumption, yogurt is regarded as the best source of probiotics. Due to growing knowledge and health-promoting qualities of functional foods containing probiotics and prebiotics, their utilization is increased in the food and beverage industry. They reduce the risk illness while improving intestinal health (O'Sullivan et al., 2020). All probiotic products differ in their nutritional and therapeutic properties according to several factors, including the strain's genetic makeup, the number of probiotics utilized, the product's intended usage, and its shelf stability. The chosen probiotic strain's production, effect, and host-benefits are significant factors. A probiotic food must contain $10^6$ CFU/g of probiotic microorganism to provide a health benefit. $10^7$–$10^9$ CFU/mg/day is the recommended dose for human intake (Islam, 2016).

It is well-established that the specific strain used to produce a particular product has a bearing on the results that one might anticipate from consuming probiotic foods. Lactobacillus, Escherichia coli, Bifidobacterium, Enterococcus, and Leuconostoc are essential genera for acquiring helpful probiotic strains. Saccharomyces, Pediococcus, and Leuconostoc, are also important (Singh & Singh, 2014). Saliva samples typically reveal the presence of lactobacillus strains of the *L. paracasei*, *L. plantarum*, and *L. rhamnosus* species. Bifidobacterium species that occur both in the oral cavity and breast milk are considered completely harmless. Different variables such as diet, environmental issues, incompatibilities, diseases, and treatment methods significantly impact the gut microbiota's formation, stability and function (Zommiti et al., 2020).

## 8.3 Lactic Acid Bacteria Starter Culture

### *8.3.1 Starter Cultures and Functional Starter Cultures in Fermented Foods*

Human diets are enriched by fermented foods because they provide and preserve many nutrients in a broad range of flavors, fragrances, and textures (Franz et al., 2014). It is believed that local microbes found in raw materials, containers, utensils, and the surroundings act as functional and non-functional microorganisms to ferment most of the world's fermented foods. These bacteria change the biochemical components of raw materials, enhancing flavor, digestibility, and fragrance while adding nutritious and pharmaceutical benefits to select fermented foods generally accepted by customers (Tamang et al., 2015). A starter culture is a microbiological mixture of a significant number of microbes of the same or different type that is added to raw materials to speed up and direct the fermenting process to generate a fermented meal.

Over the last 20 years, research on fermented products has shown that the native microbiota or the starter culture promotes probiotic activities, the bioavailability of macro and micronutrients, and the development of anti-oxidative, pro, and antimicrobial substances and other beneficial components during fermentation. These health-improving elements enhance customer knowledge of the value of consuming ancient foods to boost health and prevent diseases (Marco et al., 2017). Traditional fermented foods are mostly produced by the local LAB from the resources at hand, which continues the fermentation without additional artificial starting culture. The total fermentation process is controlled by the precise inclusion of starter cultures in raw materials, which largely minimizes the danger of fermentation failure while enhancing the quality of the finished product. In addition, many physiological features of LAB, such as antioxidant properties by Lactobacillus and Bifidobacteria, have been identified (Mishra et al., 2015).

Lactic acid bacteria are often utilized as starting cultures in fermented foods due to their potential to improve nutritional value, flavor, microbiological quality, and shelf life. Lactic acid starts acidification quickly and efficiently by producing organic acids from carbs. Lactic acid is most prevalent, followed by acetic acid. LAB may also create ethanol, bacteriocins, fragrance compounds, exopolysaccharides, and enzymes (Alvarez-Sieiro et al., 2016). Traditional methods of producing fermented meals and drinks relied on the bacteria already in the raw ingredients. Subsequently, food businesses favored directly injecting a defined starting culture into the food material (Bratulić et al., 2021). Many fermented food producers rely on commercial starter cultures, often supplied as frozen and freeze-dried concentrate or lyophilized formulations, to inject the food item directly. Functional starting cultures are being studied for fermenting food and drink. Each culture improves fermentation and product quality and safety. Health benefits abound. Carefully choosing beginning cultures may exclude D-lactic acid, a racemate of lactic acid (DL), and biogenic amines (Boeck et al., 2022).

### *8.3.2 Role of LAB as a Functional Starter Culture in Fermented Dairy and Non-dairy Products*

Lactic acid bacteria are frequently the most common microorganisms used in the food industry because most Lactic acid bacterial species achieve food certification (generally recognized as safe). Lactic acid bacteria cause the fermentation of the carbon sources present in natural foods to lactic acid, simultaneously reducing the pH that plays a crucial role in improving the textural, safety, and quality characteristics. Moreover, they promote the nutritional content, flavor, and texture of these foods such as sourdough bread, dairy products (buttermilk, cheese, yogurt, and fermented milk), and fermented beverages. In addition, lactic acid bacteria are recognized for producing probiotic-like enzymes, sugar polymers, vitamins, antibacterial and aromatic chemicals, and sugar polymers (Landete, 2017). LAB's characteristics related to safety include its origin, hemolytic potential, metabolic functions, non-pathogenicity, toxins generation, adverse impact in human trials, and epidemiological surveillance of adverse consumers.

Similarly, functional characteristics can be linked to GI tract maintenance, survival in low and high pH environments, bile salt tolerance, antibiotic resistance, hydrophobic interactions, immuno-modulation, antagonism, and anti-mutagenic qualities (Rul et al., 2012). Technological factors also include the potential for development at varying pH, temperature, NaCl threshold, and their capacity for acidification and metabolism. NaCl addition, which has a salinizing effect, is a standard practice in most fermented dairy foods and impacts starting cultures' growth (Ruiz Rodríguez et al., 2019).

The primary role of starter cultures in milk synthesis is the breakdown of peptides, which allows the coagulant to create smaller peptides and amino acids. Moreover, starter cultures break down casein and transform amino acids into various flavoring substances. The starter cultures release the intracellular proteases into the cheese matrix via cell disruption for flavor production and to regulate bitterness in cheese maturation (Yepez & Tenea, 2015). Several traditional fermented meals or different ingredients consumed by starting cultures to make fermented food have been reported to include lactic acid bacteria with diverse characteristics. Depending on the kind of raw material, a variety of LAB species are used to generate the required fermented products. Depending on the type of cheese being made, different species of *Lactococcus lactis* subsp. *lactis*, *Leuconostoc mesenteroides* subsp. *cremoris*, *Lactobacillus (Lb.) delbrueckii* subsp. *lactis*, *Lb. helveticus*, and *Lb. casei* are commonly found in dairy products like cheese. The two most common Lactic acid bacteria that ferment citrate in dairy starts are *Leuconostoc* sp. and *L. diacetylactis* (Tamime & Thomas, 2018). Short-chain fatty acids, ethanol, hydrogen peroxide, bacteriocins, bacteriocins-like inhibiting substances, and antifungal compounds are only a few of the non-specific antimicrobial chemicals that lactic acid bacteria create. Due to this functional feature, the significance of employing LAB as a substitute for synthetic additives and chemicals in food bio-preservation has increased. Bacteriocins, produced in situ using a bacterio-cinogenic starting culture or ex-situ as food additives, are the most promising antibacterial chemicals created by lactic acid bacteria (Mokoena, 2017).

Amylolytic lactic acid bacteria (ALAB) efficiently convert starch into lactic acid by combining fermentation and saccharification in one stage. Probiotics or hypoallergenic children's foods may also be fermented using ALAB strains and cereal-based foods and beverages (Petrova et al., 2013). Extracellular amylases that produce ALAB are employed as the starter in sourdough to enhance bread's textural properties and shelf stability. During glycolysis, lipolysis, and proteolytic reactions, cheese-ripening procedures have used proteases and peptidases from lactic acid bacteria to produce a variety of sensory attributes (Patel et al., 2013). The peptidases that *Lc. lactis* subsp. *cremoris* produces specifically improved sensory characteristics of cheese. Casein and whey hydrolysate are alternatively produced in the dairy industry using non-coagulant proteases (Feijoo-Siota et al., 2014).

## 8.4 Lactic Acid Bacteria as a Source of Antimicrobial Agents

### *8.4.1 The Microbial Ecosystem*

The interaction between microorganisms in the microbial ecosystem, which is perfectly balanced, impacts other microorganisms in the population. The host–pathogen interactions in the ecological system of microorganisms include predation, commensalism, synergism, parasitism, resistance, and food competition (Ye et al.,

2021). LAB seed cultures have been utilized in meat and dairy for decades. Under strict circumstances, homo- and heterofermentative bacteria are employed to make yogurt, sour cream, butter, cheese, fermented drinks, and other delicacies. A group of microorganisms called "mixed cultures" in food often interact with one another to increase their metabolic activity and create desired results on the safety and quality of the product (Ibrahim et al., 2021).

Many extrinsic factors can influence microbial proliferation. Several microbes interact symbiotically with one another. For instance, one kind of microbe may create chemical molecules that serve as major primary sources of energy for another form of bacterium to metabolize. This interaction is typically referred to as symbiosis. Food matrices with two or more microorganisms typically include this type of interaction. Another instance of microbe-microbe interaction is when two or more microbial species interact and inhibit one another's development ability (Prosser et al., 2007). Microbes do not exist in pure culture. Microbial associations are mostly studied using mixed cultures. A traditional illustration employs mixed microbial isolates to research mutualism and competition in populations. Moreover, the ability of synthetic mixed cultures to perform specified tasks, such as producing biofuel and bioremediation, was confirmed. These cultures are now suitable for industrial applications (Hibbing et al., 2010).

### *8.4.2 Effect of LAB on Microbial Growth*

Maintaining food items' safety and stability is the main goal of shelf life extension, and this may often be done by limiting the development of harmful bacteria and rotting microorganisms. One or two antimicrobial agents, such as nisin, might be used synergistically toward the target microorganisms to combat and prevent the proliferation of pathogenic microbes. The activity of these antimicrobial compounds does not adversely affect the nutrition and sensory attributes of foods while maintaining their physiochemical composition (Hibbing et al., 2010). When added to food, LAB are advantageous because they can, suppress the development of dangerous enteric pathogens, remove toxic food components from the intestines, enhance the immune-modulatory mechanism, increase the immune system efficiency, and improve the peristalsis response of the GI tract (Agriopoulou et al., 2020).

The probiotic qualities of LAB must include antagonistic action against foods and gastrointestinal pathogen so the antimicrobial activity becomes the desired trait for choosing strains. Lactic acid bacteria are essential for fermentation, conservation, and storage because they have antibacterial capabilities that can fight fungi and several Gram-positive and Gram-negative bacteria (Awaisheh & Ibrahim, 2009). The antimicrobial chemicals the strains generate are what give LAB strains their antibacterial characteristics. These antibiotics may be categorized into three groups: lactic acid and butyric acetic acids are examples of organic acids. Other minor molecules include diacetyl, hydrogen peroxide, acetaldehyde, acetone, reuterin, and reutericyclin (Sieuwerts et al., 2018).

#### 8.4.2.1 Bacteriocins

Gram-positive and gram-negative bacteria create bacteriocins, important protein-based compounds with bactericidal or bacteriostatic activity. These compounds compete with gram-positive bacteria mostly within a single biological niche or nutrition pool (Mani-López et al., 2018; Zacharof & Lovitt, 2012). Bacteria use chemicals called bacteriocins as a defence against germs that could be harmful (Juturu & Wu, 2018). Certain bacteriocins may also have the ability to fight off viruses and fungi. While bacteriocins are created during the lagged phase and categorized as primary metabolites, they differ from antibiotics produced at the end of bacterial activity and categorized as secondary metabolites (Sharma et al., 2019). In addition, bacteriocins can be divided into two groups based on the range of their inhibitory effects. The first type is also known as narrow-spectrum bacteriocins, and the second category, broad-spectrum bacteriocins, exhibits their inhibitory effect on bacteria belonging to several species.

A bacteriocin is regarded as perfect when just minimal quantities are needed for it to work, when it has a broad spectrum of activity against several pathogenic bacteria, when it does not harm the product to which it is applied, and when its production does not come at a high cost (O'Connor et al., 2015). Bacteriocins are removed by the proteolytic enzymes of the digestive system, and lactic acid bacteria are odorless and colorless. Therefore, they do not influence the sensory qualities and physicochemical properties of food (Perez et al., 2014). The qualities above make lactic acid bacteria-produced bacteriocins the preferable choice for usage in food. There are primarily two types of bacteriocins:

- Lactoccocin is one example of a class I bacteriocin that works largely by preventing the production of peptidoglycans.
- Nisin is an example of a class II bacteriocin that creates holes in the cytoplasmic membrane to destabilize it (Negash & Tsehai, 2020).

#### 8.4.2.2 Organic Acids

Several organic acids that lactic acid bacteria generate have general antibacterial properties. For instance, certain species of Acetobacter aceti, Propionibacterium, and Lactobacillus produce organic acids such as acetic acid, propionic acid, and lactic acid. Formate and succinate are two common organic acids that LAB can generate. Organic acids are appealing since they prevent the development of yeast, molds, and Gram-positive and Gram-negative bacteria in various food items. Moreover, organic acids are often thought to be safe for human usage. The ability of organic acids to kill microorganisms has been linked to their fragmented molecules, which deprotonate when they enter biological membranes (Table 8.1).

**Table 8.1** Organic acid and their food applications

| Organic acids | Prominent LAB Producer | Food pathogen application | References |
|---|---|---|---|
| Lactic acid | *Lactobacillus delbrueckii* subsp. *bulgaricus* | *Pseudomonas* spp. | Ayivi et al. (2020) |
| Formic acid | *Lactococcus lactis* subsp. *cremoris* | *Esherichia coli*, *Listeria* spp., *Salmonella* spp., *Clostridium perfringens* | Özcelik et al. (2016), Ricke et al. (2020) |
| Succinic acid | *Lactococcus lactis* subsp. *lactis* | *Salmonella* spp. | Özcelik et al. (2016), Ricke et al. (2020) |
| Malic acid | *Limosilactobacillus reuteri* | *Staphylococcus* | Ghoul and Mitri (2016) |
| Propionic acid | *Lactococcus lactis* subsp. *lacti* | *Campylobacter* spp. | Özcelik et al. (2016), Peh et al. (2020) |
| Acetic acid | *Lactobacillus acidophilus* | *Pseudomonas* spp. | Sallam (2007) |
| Butyric Acid | *Lactobacillus acidophilus* | *Salmonella* spp. | Özcelik et al. (2016) |

### 8.4.2.3 Small Molecules Having Antimicrobial Properties

LAB generates a variety of small compounds with antimicrobial properties, including diacetyl, hydrogen peroxide, and reuterin. Hydrogen peroxide is a potent antibacterial agent that works in conjunction with heat. Deactivating necessary enzymes leads to a change in catalytic activity, which produces the antimicrobial mechanism of action. The dicarbonyl group of diacetyls reacts with arginine in the enzymes, which results in their inactivation (Hertzberger et al., 2014). Without intracellular catalase, pseudo catalase, or peroxidase, lactic acid bacteria produce hydrogen peroxide under aerobic circumstances. Hydrogen peroxide possesses bacteriostatic activity, and activating the lactoperoxidase thiocyanate system in raw milk enhances its antimicrobial action. Lactic acid bacteria prevent harmful bacteria from adhering to their host tissue by enhancing the gastrointestinal mucosal barrier function. The impact of the microbial flora due to the creation of antibiotic substances seems to be another consequence of lactic acid bacteria (Cizeikiene et al., 2013). The antibacterial properties of LAB on foodborne pathogens have been documented in several types of research. Lactic acid bacteria also prevent the fungus (molds) that cause food to deteriorate and produce mycotoxin. In addition to producing organic acids with distinctive fungi-static and fungicidal qualities that inhibit fungi and yeast, including Aspergillus, lactic acid bacteria are equipped with bacteriocin-like compounds (Agriopoulou et al., 2020).

## 8.5 Lactic Acid Bacteria as a Vitamin Source

Vitamins are organic substances required for numerous vital metabolic processes that sustain life. In contrast to other nutrients, vitamins serve as enzyme cofactors, participate in oxidation-reduction metabolic reactions, or operate as hormones. They do not, however, fulfill structural roles or serve as an energy source (Combs & McClung, 2016). Furthermore, humans cannot synthesize most vitamins. Thus, they must be received exogenously from diet and other sources like gut bacteria. Due to their biological effects, some vitamins, including folate, have even been referred to as functional food elements.

Even though a wide range of foods contains vitamins, the global epidemic of vitamin deficiencies is primarily caused by poor eating habits. Because of this, foods with a lot of natural vitamin forms could be made using microorganisms that make vitamins instead of the artificial vitamins that are usually added to foods to make them healthier. LeBlanc et al. (2013) say that biofortified foods are a cheaper way to get vitamins than fortification and help people meet the RDI. Consumer perceptions toward food consumption changed due to lifestyle changes. People nowadays prefer meals that provide additional health benefits along with basic nutrients. In this regard, lactic acid bacteria produce vitamins to improve the food's value and increase consumers' perception of functional foods. Different processes like fermentation can replace the chemical synthesis of vitamins like riboflavin and are both cost-effective and beneficial for the environment (Thakur et al., 2016a, b). Microbial synthesis seems environmentally friendly and produces high-quality end products using renewable resources (Zhu et al., 2020).

### *8.5.1 Riboflavin-Producing Lactic Acid Bacteria*

Vitamin B2, often known as riboflavin, is a water-soluble vitamin produced by numerous microbes and plants. Riboflavin is the precursor of the enzymes and coenzymes like flavin mononucleotide (FMN) and flavin adenine dinucleotide (FAD) that serve as coenzymes in the electron transport chain during oxidation–reduction reactions (Aili et al., 2013). All aerobic life forms require riboflavin as a critical component for proper cellular growth and functioning. Even though riboflavin is present in many foods, inadequate intake may cause ariboflavinosis (Watson et al., 2017). Eye disorders, CVDs, preeclampsia, anemia, liver and skin diseases, and disruptions in brain glucose metabolism are linked to riboflavin deficiency. Depending on sex, age, and physiological state, different amounts of riboflavin are recommended for people. As the body cannot effectively conserve riboflavin, normal individuals must intake around 0.9–1.6 mg of riboflavin daily (LeBlanc et al., 2015).

As mentioned above, the utilization of riboflavin-producing microorganisms has become more significant in recent years as a biofortification technique and the fortification of foods using chemically synthesized vitamins. The nutritional content of meals is increased when certain LAB strains are used during fermentation because they can produce riboflavin. Additionally, riboflavin quantities in certain products may occasionally differ due to processing techniques and the microorganisms' behavior in food preparation. It has been shown that riboflavin-producing LAB may increase the levels of this vitamin in food matrices such as milk, soymilk, whey, and pseudocereals. These lactic acid bacteria have been isolated from several ecological niches (Thakur et al., 2016a, b). In LAB, species- or strain-specific features represent the genetic code for riboflavin production. Although an interrupted rib operon is occasionally seen in some strains, comparative genome research reveals that many of the sequenced members of LAB can manufacture riboflavin (Capozzi et al., 2012). These findings demonstrated that the vitamins produced were bioavailable and comparable to synthetic vitamins, suggesting they may help prevent vitamin deficiencies. In a different investigation, riboflavin synthesis genes were examined using a PCR-based approach to test up to 60 lactobacilli for the capacity to produce too much riboflavin. Riboflavin synthesis-related genes were strain-specific among the lactobacilli examined across several species (Liu et al., 2020).

### 8.5.2 Folate-Producing Lactic Acid Bacteria

Folic acid and similar substances with the same biological function are folates or vitamin B9/B11. This vitamin plays a vital role in many essential biological processes, including DNA synthesis and methylation, amino acids, nucleotides, and other vitamins (Nazki et al., 2014). Even though many foods contain folates, dietary consumption may not sufficiently meet daily needs. Food fortification through synthetic chemicals can address this issue, but it has certain consequences. Certain bacteriocins may also have the ability to fight off viruses and fungi. While bacteriocins are created during the lagged phase and categorized as primary metabolites, they differ from antibiotics produced at the end of bacterial activity and categorized as secondary metabolites (Sharma et al., 2019). In addition, bacteriocins can be divided into two groups based on the range of their inhibitory effects. The first type is also known as narrow-spectrum bacteriocins, and the second category, broad-spectrum bacteriocins, exhibit their inhibitory effect on bacteria belonging to several species.

Previous research suggested that *Lact. Delbrueckii* subsp. *Bulgaricus* requires folates for growth, whereas *Streptococcus thermophilus* produces folate. Nevertheless, it was shown that some strains of *Lact. Delbrueckii* subsp. *Bulgaricus*, isolated from Argentina's fermented milk products, could grow on a folate free culture medium and generate significant quantities of folate. It was proven that *Lact. Bulgaricus* could produce both intracellular and extracellular folates. This strain produced yogurt that was naturally bio-enriched with folates and had a four times higher vitamin content than unfermented milk and two times higher than natural yogurt after being inoculated with *Streptococcus thermophilus* (Laiño et al., 2013).

### *8.5.3 Vitamin K-Producing Lactic Acid Bacteria*

Blood coagulation, tissue calcification, atherosclerotic plaque, bone health, and renal function are all impacted by vitamin K. Green plants contain vitamin K as phylloquinone (vitamin K1). At the same time, some gut bacteria, such as LAB and particular strains of the genera Lactococcus, *Enterococcus*, *Leuconostoc*, and *Streptococcus,* generate menaquinone (K2) as well. Clinical conditions include cerebral hemorrhage in newborns and potential bone breakage due to osteoporosis, which has been linked to vitamin K insufficiency. Menaquinone produced by LAB could help complement human vitamin K needs (Binda & Ouwehand, 2019).

Bacteria involved in the fermentation process are responsible for producing long-chain vitamin K2 and the specified structure of menaquinones. These food-grade bacteria might be used to biotechnologically produce long-chain menaquinones for fortifying fermented foods. Bacillus subtilis produces MK-7, and some strains are utilized to produce fermented soybean foods (Walther et al., 2013). In the electron transport chain, lactic acid bacteria need vitamin K2 in their cytoplasmic membranes (Kurosu & Begari, 2010). LAB are facultative anaerobes; however, certain lactococci and pediococci respire with heme (or menaquinone plus heme) supplementation. *L. lactis* generates menaquinone in static fermentation, aerobic fermentation, and aerobic respiration (Brooijmans et al., 2009).

## 8.6 Enzymes Derived from Lactic Acid Bacteria as a Potential Source

Lactic acid bacteria make a lot of enzymes that can change how the food is made, how it is processed, and how it tastes. These bacteria make different digestive enzymes that work together with the digestive system to reduce problems with not being able to absorb food. Mtshali (2007) says that these bacteria can be used to make enzymes that could help with fermentation. Pediococcus and Lactococcus species are the LAB frequently connected with fermented foods. The most stable enzymes generated by LABs, amylases in particular, may be employed in sourdough technology to improve bread texture naturally. Moreover, LAB enhances the flavor and fragrance of fermented meals. The sensory quality of cheese was enhanced by specific peptidases generated by *Lactococcus lactis* subsp. *Cremoris*. Moreover, proteolysis and lipolysis also improve the aroma of most cheese variants (González et al., 2010).

Enzymes are crucial to the production of wine. Apart from grape aromas and alcoholic fermentation, the taste and fragrance of wine are produced mainly by the lactic acid by the activity of their enzymes. Following alcoholic fermentation, these bacteria proliferate in wine throughout malolactic fermentation, which enhances the flavor and taste of wine through a variety of secondary alterations (Florou-Paneri et al., 2013).

## 8.7 Exopolysaccharide Production by Lactic Acid Bacteria

LAB can produce glycosidic polymers that can secrete certain compounds called exopolysaccharides. Lactic acid bacteria of the genera Enterococcus, Lactobacillus, Lactococcus, Leuconostoc, Pediococcus, Oenococcus, Streptococcus, and Weissella have been found to produce EPS (exopolysaccharides) (Amari et al., 2013; Rühmkorf et al., 2013). The properties of the EPS generated by LAB can vary in four different ways:

1. Composition (different types of monosaccharide units linked by various linkages).
2. Structure (varying levels and types of branching).
3. Molecular mass.
4. Overall structural configuration.

The concentration of the polymers and their interactions with food and these properties determine the function of EPS in food. Various lactic acid bacteria produce EPS, but at elevated prices for bacterial polysaccharides compared to polysaccharides derived from algal, vegetable, and animal sources, they are not used in isolated form by the food industry. However, plant and seaweed polysaccharides lack the necessary reliability due to their inconsistent uniformity and limited production volumes. However, EPS synthesized in the laboratory has significant industrial applications as thickeners, stabilizers, emulsifiers, and gelling, thickening, and water-binding agents (Zannini et al., 2016).

Certain EPS have positive effects on health, such as lowering cholesterol and having anticancer properties. In addition, they inhibit pathogen biofilm development, and LAB-produced glucans have prebiotic and immune-modulating properties. As a result, EPS inclusion in food may potentially benefit human health. Nevertheless, for this to happen, it will be necessary to create processes allowing the bacteria to grow in inexpensive culture media, improve the yield and recovery of EPS from culture supernatants, and promote their in situ production. Notararigo et al. (2013) say that LAB-synthesized EPS is used in and helps make dairy and cereal-based products. In addition to their technological advantages, some EPS made from LAB is said to have positive physiological impacts on human health. These advantages can be seen in minimal quantities. Because of the higher viscosity of EPS, it may stay in the gastrointestinal system longer, which is advantageous for the temporary colonization of probiotics. Short-chain fatty acids are produced in the gut by the colonic microflora's breakdown, providing an additional advantage. There is a chance that certain fatty acids can help prevent colon cancer. Moreover, EPS produced in the laboratory seems to have anti-tumor, anti-ulcer, immunomodulating, and cholesterol-lowering properties (Zarour et al., 2017).

## 8.8 Conclusion

Lactic acid bacteria and their by-products are a big part of what makes traditional fermented foods and drinks healthy and safe. Moreover, strains of lactic acid bacteria can produce metabolites like EPS, vitamins, and bacteriocins through the metabolic engineering of lactic acid bacteria.

Lactic acid bacteria hold great potential as sources for new products and uses, including exceptionally functional foods that may meet the rising consumer demand for natural goods and functional meals. The ability of lactic acid bacteria to produce antimicrobial compounds and other value-added goods is well known. Unquestionably, the value of human existence has been enhanced by these probiotics.

These can be included in the diets of both people and animals, focusing on improving human health. Even though new things have happened, research on LAB and their functional parts is still in its early stages and has not yet reached its full potential.

## References

Agriopoulou, S., Stamatelopoulou, E., Sachadyn-Król, M., & Varzakas, T. (2020). Lactic acid bacteria as antibacterial agents to extend the shelf life of fresh and minimally processed fruits and vegetables: Quality and safety aspects. *Microorganisms, 8*(6), 952.

Aili, A., Hasim, A., Kelimu, A., Guo, X., Mamtimin, B., Abudula, A., & Upur, H. (2013). Association of the plasma and tissue riboflavin levels with C20orf54 expression in cervical lesions and its relationship to HPV16 infection. *PLoS One, 8*(11), e79937.

Alvarez-Sieiro, P., Montalbán-López, M., Mu, D., & Kuipers, O. P. (2016). Bacteriocins of lactic acid bacteria: Extending the family. *Applied Microbiology and Biotechnology, 100*, 2939–2951.

Amari, M., Arango, L. F. G., Gabriel, V., Robert, H., Morel, S., Moulis, C., et al. (2013). Characterization of a novel dextransucrase from Weissella confusa isolated from sourdough. *Applied Microbiology and Biotechnology, 97*(12), 5413–5422. https://doi.org/10.1007/s00253-012-4447-8

Awaisheh, S. S., & Ibrahim, S. A. (2009). Screening of antibacterial activity of lactic acid bacteria against different pathogens found in vacuum-packaged meat products. *Foodborne Pathogens and Disease, 6*(9), 1125–1132.

Ayivi, R. D., Gyawali, R., Krastanov, A., Aljaloud, S. O., Worku, M., Tahergorabi, R., et al. (2020). Lactic acid bacteria: Food safety and human health applications. *Dairy, 1*(3), 202–232.

Binda, S., & Ouwehand, A. C. (2019). Lactic acid bacteria for fermented dairy products. In *Lactic acid bacteria* (pp. 175–198). CRC Press.

Boeck, T., Ispiryan, L., Hoehnel, A., Sahin, A. W., Coffey, A., Zannini, E., & Arendt, E. K. (2022). Lentil-based yogurt alternatives fermented with multifunctional strains of lactic acid bacteria—Techno-functional, microbiological, and sensory characteristics. *Foods, 11*(14), 2013.

Bratulić, M., Mikuš, T., Cvrtila, Ž., Cenci-Goga, B. T., Grispoldi, L., Pavunc, A. L., et al. (2021). Quality of traditionally produced Istrian sausage and identification of autochthonous lactic acid bacteria strains as potential functional starter cultures. *European Food Research and Technology, 247*(11), 2847–2860.

Brooijmans, R., Smit, B., Santos, F., Van Riel, J., de Vos, W. M., & Hugenholtz, J. (2009). Heme and menaquinone induced electron transport in lactic acid bacteria. *Microbial Cell Factories, 8*(1), 1–11.

Capozzi, V., Russo, P., Dueñas, M. T., López, P., & Spano, G. (2012). Lactic acid bacteria producing B-group vitamins: A great potential for functional cereals products. *Applied Microbiology and Biotechnology, 96*, 1383–1394.

Cizeikiene, D., Juodeikiene, G., Paskevicius, A., & Bartkiene, E. (2013). Antimicrobial activity of lactic acid bacteria against pathogenic and spoilage microorganism isolated from food and their control in wheat bread. *Food Control, 31*(2), 539–545.

Combs, G. F., Jr., & McClung, J. P. (2016). *The vitamins: Fundamental aspects in nutrition and health*. Academic Press.

D'Amelio, P., & Sassi, F. (2018). Gut microbiota, immune system, and bone. *Calcified Tissue International, 102*, 415–425.

Dekumpitiya, N., Gamlakshe, D., Abeygunawardena, S., & Jayaratne, D. (2016). Identification of the microbial consortium in Sri Lankan buffalo milk curd and their growth in the presence of prebiotics. *Journal of Food Science and Technology Nepal, 9*, 20–30.

Feijoo-Siota, L., Blasco, L., Luis Rodriguez-Rama, J., Barros-Velázquez, J., de Miguel, T., Sánchez-Pérez, A., & Villa, T. G. (2014). Recent patents on microbial proteases for the dairy industry. *Recent Advances in DNA & Gene Sequences (Formerly Recent Patents on DNA & Gene Sequences), 8*(1), 44–55.

Florou-Paneri, P., Christaki, E., & Bonos, E. (2013). Lactic acid bacteria as source of functional ingredients. In *Lactic acid bacteria-R & D for food, health and livestock purposes*. IntechOpen.

Franz, C. M., Huch, M., Mathara, J. M., Abriouel, H., Benomar, N., Reid, G., et al. (2014). African fermented foods and probiotics. *International Journal of Food Microbiology, 190*, 84–96.

Gaspar, P., Carvalho, A. L., Vinga, S., Santos, H., & Neves, A. R. (2013). From physiology to systems metabolic engineering for the production of biochemicals by lactic acid bacteria. *Biotechnology Advances, 31*(6), 764–788.

Ghoul, M., & Mitri, S. (2016). The ecology and evolution of microbial competition. *Trends in Microbiology, 24*(10), 833–845.

Goldenberg, J. Z., Yap, C., Lytvyn, L., Lo, C. K. F., Beardsley, J., Mertz, D., & Johnston, B. C. (2017). Probiotics for the prevention of Clostridium difficile-associated diarrhea in adults and children. *Cochrane Database of Systematic Reviews*, (12), CD006095.

Gómez-Llorente, C., Munoz, S., & Gil, A. (2010). Role of toll-like receptors in the development of immunotolerance mediated by probiotics. *Proceedings of the Nutrition Society, 69*(3), 381–389.

González, L., Sacristán, N., Arenas, R., Fresno, J. M., & Tornadijo, M. E. (2010). Enzymatic activity of lactic acid bacteria (with antimicrobial properties) isolated from a traditional Spanish cheese. *Food Microbiology, 27*(5), 592–597.

Hamdan, A. M., & Sonomoto, K. (2011). Production of optically pure lactic acid for bioplastics. In *Lactic acid bacteria and bifidobacteria: Current progress in advanced research*. Caister Academic Press.

Hemaiswarya, S., Raja, R., Ravikumar, R., & Carvalho, I. S. (2013). Mechanism of action of probiotics. *Brazilian Archives of Biology and Technology, 56*, 113–119.

Hertzberger, R., Arents, J., Dekker, H. L., Pridmore, R. D., Gysler, C., Kleerebezem, M., & de Mattos, M. J. T. (2014). H2O2 production in species of the Lactobacillus acidophilus group: A central role for a novel NADH-dependent flavin reductase. *Applied and Environmental Microbiology, 80*(7), 2229–2239.

Hibbing, M. E., Fuqua, C., Parsek, M. R., & Peterson, S. B. (2010). Bacterial competition: Surviving and thriving in the microbial jungle. *Nature Reviews Microbiology, 8*(1), 15–25.

Ibrahim, S. A., Ayivi, R. D., Zimmerman, T., Siddiqui, S. A., Altemimi, A. B., Fidan, H., et al. (2021). Lactic acid bacteria as antimicrobial agents: Food safety and microbial food spoilage prevention. *Foods, 10*(12), 3131.

Iqbal, M. Z., Qadir, M. I., Hussain, T., Janbaz, K. H., Khan, Y. H., & Ahmad, B. (2014). Probiotics and their beneficial effects against various diseases. *Pakistan Journal of Pharmaceutical Sciences, 27*(2), 405.

Islam, S. U. (2016). Clinical uses of probiotics. *Medicine, 95*(5), e2658.

Juturu, V., & Wu, J. C. (2018). Microbial production of bacteriocins: Latest research development and applications. *Biotechnology Advances, 36*(8), 2187–2200.

Korhonen, J. (2010). *Antibiotic resistance of lactic acid bacteria*. Itä-Suomen yliopisto.

Kurosu, M., & Begari, E. (2010). Vitamin K2 in electron transport system: Are enzymes involved in vitamin K2 biosynthesis promising drug targets? *Molecules, 15*(3), 1531–1553.

Laiño, J. E., del Valle, M. J., de Giori, G. S., & LeBlanc, J. G. J. (2013). Development of a high folate concentration yogurt naturally bio-enriched using selected lactic acid bacteria. *LWT - Food Science and Technology, 54*(1), 1–5.

Landete, J. M. (2017). A review of food-grade vectors in lactic acid bacteria: From the laboratory to their application. *Critical Reviews in Biotechnology, 37*(3), 296–308.

LeBlanc, J. G., Milani, C., De Giori, G. S., Sesma, F., Van Sinderen, D., & Ventura, M. (2013). Bacteria as vitamin suppliers to their host: A gut microbiota perspective. *Current Opinion in Biotechnology, 24*(2), 160–168.

LeBlanc, J. G., Laiño, J. E., del Valle, M. J., de Giori, G. S., Sesma, F., & Taranto, M. P. (2015). B-group vitamins production by probiotic lactic acid bacteria. In *Biotechnology of lactic acid bacteria: Novel applications* (pp. 279–296). John Wiley & Sons, Ltd..

Liu, S., Hu, W., Wang, Z., & Chen, T. (2020). Production of riboflavin and related cofactors by biotechnological processes. *Microbial Cell Factories, 19*(1), 1–16.

Mani-López, E., Palou, E., & López-Malo, A. (2018). Biopreservatives as agents to prevent food spoilage. In *Microbial contamination and food degradation* (pp. 235–270). Elsevier.

Marco, M. L., Heeney, D., Binda, S., Cifelli, C. J., Cotter, P. D., Foligné, B., et al. (2017). Health benefits of fermented foods: Microbiota and beyond. *Current Opinion in Biotechnology, 44*, 94–102.

Mazzoli, R., Bosco, F., Mizrahi, I., Bayer, E. A., & Pessione, E. (2014). Towards lactic acid bacteria-based biorefineries. *Biotechnology Advances, 32*(7), 1216–1236.

Mishra, V., Shah, C., Mokashe, N., Chavan, R., Yadav, H., & Prajapati, J. (2015). Probiotics as potential antioxidants: A systematic review. *Journal of Agricultural and Food Chemistry, 63*(14), 3615–3626.

Mokoena, M. P. (2017). Lactic acid bacteria and their bacteriocins: Classification, biosynthesis and applications against uropathogens: A mini-review. *Molecules, 22*(8), 1255.

Mtshali, P. S. (2007). *Screening and characterization of wine related enzymes produced by wine associated lactic acid Bacteria*. University of Stellenbosch.

Nazki, F. H., Sameer, A. S., & Ganaie, B. A. (2014). Folate: Metabolism, genes, polymorphisms and the associated diseases. *Gene, 533*(1), 11–20.

Negash, A. W., & Tsehai, B. A. (2020). Current applications of bacteriocin. *International Journal of Microbiology, 2020*, 4374891.

Notararigo, S., Nácher-Vázquez, M., Ibarburu, I., Werning, M. L., de Palencia, P. F., Dueñas, M. T., et al. (2013). Comparative analysis of production and purification of homo-and hetero-polysaccharides produced by lactic acid bacteria. *Carbohydrate Polymers, 93*(1), 57–64.

Nuraida, L. (2015). A review: Health promoting lactic acid bacteria in traditional Indonesian fermented foods. *Food Science and Human Wellness, 4*(2), 47–55. https://doi.org/10.1016/j.fshw.2015.06.001

O'Connor, P. M., Ross, R. P., Hill, C., & Cotter, P. D. (2015). Antimicrobial antagonists against food pathogens: A bacteriocin perspective. *Current Opinion in Food Science, 2*, 51–57.

O'Shea, E., Cotter, P. D., Stanton, C., Ross, R. P., & Hill, C. (2012). Production of bioactive substances by intestinal bacteria as a basis for explaining probiotic mechanisms: Bacteriocins and conjugated linoleic acid. *International Journal of Food Microbiology, 152*(3), 189–205.

O'Sullivan, J. N., Rea, M. C., Hill, C., & Ross, R. P. (2020). Protecting the outside: Biological tools to manipulate the skin microbiota. *FEMS Microbiology Ecology, 96*(6), fiaa085.

Ohkusa, T., Koido, S., Nishikawa, Y., & Sato, N. (2019). Gut microbiota and chronic constipation: A review and update. *Frontiers in Medicine, 6*, 19.

Oliveira, A., Amaro, A. L., & Pintado, M. (2018). Impact of food matrix components on nutritional and functional properties of fruit-based products. *Current Opinion in Food Science, 22*, 153–159.

Özcelik, S., Kuley, E., & Özogul, F. (2016). Formation of lactic, acetic, succinic, propionic, formic and butyric acid by lactic acid bacteria. *LWT - Food Science and Technology, 73*, 536–542.

Papagianni, M. (2012). Metabolic engineering of lactic acid bacteria for the production of industrially important compounds. *Computational and Structural Biotechnology Journal, 3*(4), e201210003.

Patel, A., Shah, N., & Prajapati, J. (2013). Biosynthesis of vitamins and enzymes in fermented foods by lactic acid bacteria and related genera-A promising approach. *Croatian Journal of Food Science and Technology, 5*(2), 85–91.

Peh, E., Kittler, S., Reich, F., & Kehrenberg, C. (2020). Antimicrobial activity of organic acids against Campylobacter spp. and development of combinations—A synergistic effect? *PLoS One, 15*(9), e0239312.

Perez, R. H., Zendo, T., & Sonomoto, K. (2014). Novel bacteriocins from lactic acid bacteria (LAB): Various structures and applications. *Microbial Cell Factories, 13*(1), 1–13.

Petrova, P., Petrov, K., & Stoyancheva, G. (2013). Starch-modifying enzymes of lactic acid bacteria–Structures, properties, and applications. *Starch-stärke, 65*(1–2), 34–47.

Plaza-Diaz, J., Ruiz-Ojeda, F. J., Gil-Campos, M., & Gil, A. (2019). Mechanisms of action of probiotics. *Advances in Nutrition, 10*(Suppl 1), S49–S66.

Prosser, J. I., Bohannan, B. J., Curtis, T. P., Ellis, R. J., Firestone, M. K., Freckleton, R. P., et al. (2007). The role of ecological theory in microbial ecology. *Nature Reviews Microbiology, 5*(5), 384–392.

Quinto, E. J., Jiménez, P., Caro, I., Tejero, J., Mateo, J., & Girbés, T. (2014). Probiotic lactic acid bacteria: A review. *Food and Nutrition Sciences, 5*(18), 1765.

Ricke, S. C., Dittoe, D. K., & Richardson, K. E. (2020). Formic acid as an antimicrobial for poultry production: A review. *Frontiers in Veterinary Science, 7*, 563.

Rühmkorf, C., Bork, C., Mischnick, P., Rübsam, H., Becker, T., & Vogel, R. F. (2013). Identification of Lactobacillus curvatus TMW 1.624 dextransucrase and comparative characterization with Lactobacillus reuteri TMW 1.106 and Lactobacillus animalis TMW 1.971 dextransucrases. *Food Microbiology, 34*(1), 52–61.

Ruiz Rodríguez, L. G., Mohamed, F., Bleckwedel, J., Medina, R., De Vuyst, L., Hebert, E. M., & Mozzi, F. (2019). Diversity and functional properties of lactic acid bacteria isolated from wild fruits and flowers present in Northern Argentina. *Frontiers in Microbiology, 10*, 1091.

Rul, F., Zagorec, M., & Champomier-Vergès, M.-C. (2012). Lactic acid bacteria in fermented foods. In *Proteomics in foods: Principles and applications* (pp. 261–283). Springer.

Saeed, A. H., & Salam, A. I. (2013). Current limitations and challenges with lactic acid bacteria: A review. *Food and Nutrition Sciences, 4*, 73.

Sallam, K. I. (2007). Antimicrobial and antioxidant effects of sodium acetate, sodium lactate, and sodium citrate in refrigerated sliced salmon. *Food Control, 18*(5), 566–575.

Sharma, V., Ranveer, R., Jain, N., & Aseri, G. (2019). Bacteriocins: Production, different strategies of purification and applications. *International Journal of Research in Pharmaceutical Sciences, 10*, 1808–1817.

Sieuwerts, S., Bron, P. A., & Smid, E. J. (2018). Mutually stimulating interactions between lactic acid bacteria and Saccharomyces cerevisiae in sourdough fermentation. *LWT - Food Science and Technology, 90*, 201–206.

Singh, H., & Singh, H. (2014). Probiotics–An emerging concept. *International Journal of Scientific and Research Publications, 4*(6), 1–3.

Tamang, J. P., Thapa, N., Tamang, B., Rai, A., & Chettri, R. (2015). Microorganisms in fermented foods and beverages. In *Health benefits of fermented foods and beverages* (pp. 1–110). Routledge.

Tamime, A. Y., & Thomas, L. V. (2018). *Probiotic dairy products*. John Wiley & Sons.

Thakur, K., Lule, V. K., Rajni, C., Kumar, N., Mandal, S., Anand, S., et al. (2016a). Riboflavin producing probiotic Lactobacilli as a biotechnological strategy to obtain riboflavin-enriched fermented foods. *Journal of Pure and Applied Microbiology, 10*, 161–166.

Thakur, K., Tomar, S. K., & De, S. (2016b). Lactic acid bacteria as a cell factory for riboflavin production. *Microbial Biotechnology, 9*(4), 441–451.

Walther, B., Karl, J. P., Booth, S. L., & Boyaval, P. (2013). Menaquinones, bacteria, and the food supply: The relevance of dairy and fermented food products to vitamin K requirements. *Advances in Nutrition, 4*(4), 463–473.

Watson, R. R., Collier, R. J., & Preedy, V. R. (2017). *Nutrients in dairy and their implications for health and disease*. Academic Press.

Wong, W.-Y., Chan, B. D., Leung, T.-W., Chen, M., & Tai, W. C.-S. (2022). Beneficial and anti-inflammatory effects of formulated prebiotics, probiotics, and synbiotics in normal and acute colitis mice. *Journal of Functional Foods, 88*, 104871.

Ye, P., Wang, J., Liu, M., Li, P., & Gu, Q. (2021). Purification and characterization of a novel bacteriocin from Lactobacillus paracasei ZFM54. *LWT - Food Science and Technology, 143*, 111125.

Yepez, L., & Tenea, G. N. (2015). Genetic diversity of lactic acid bacteria strains towards their potential probiotic application. *Romanian Biotechnology Letters, 20*(2), 10191–10199.

Zacharof, M.-P., & Lovitt, R. (2012). Bacteriocins produced by lactic acid bacteria a review article. *APCBEE Procedia, 2*, 50–56.

Zannini, E., Waters, D. M., Coffey, A., & Arendt, E. K. (2016). Production, properties, and industrial food application of lactic acid bacteria-derived exopolysaccharides. *Applied Microbiology and Biotechnology, 100*(3), 1121–1135. https://doi.org/10.1007/s00253-015-7172-2

Zarour, K., Vieco, N., Pérez-Ramos, A., Nácher-Vázquez, M., Mohedano, M. L., & López, P. (2017). Food ingredients synthesized by lactic acid bacteria. In *Microbial production of food ingredients and additives* (pp. 89–124). Elsevier.

Zhu, Y.-Y., Thakur, K., Feng, J.-Y., Cai, J.-S., Zhang, J.-G., Hu, F., et al. (2020). Riboflavin-overproducing lactobacilli for the enrichment of fermented soymilk: Insights into improved nutritional and functional attributes. *Applied Microbiology and Biotechnology, 104*, 5759–5772.

Zommiti, M., Feuilloley, M. G., & Connil, N. (2020). Update of probiotics in human world: A nonstop source of benefactions till the end of time. *Microorganisms, 8*(12), 1907.

# Chapter 9
# Applications of Microbial Enzymes in the Food Industry

**Akanksha Shukla, Madhuri Girdhar, and Anand Mohan**

**Abstract** The application of enzymes improves the quality of processed foods, which enhances their flavor and nutritional content. As enzymes have a significant part in the processing of food as well as in assisting to reduce energy consumption, improve product biodegradability, and replace hazardous, chemical-based procedures with safer, more environmentally friendly ones. They are employed in the biological process as activators to strengthen most chemical processes necessary to maintain and sustain life in organisms. Microorganisms are regarded as favorable sources for industrial enzymes due to their simple availability and rapid multiplication rate. Several companies, including those in the food and dairy sectors, have recently increased their usage of microbial enzymes to speed up processing and use less energy. The microbial enzymes utilized in the food industry are tannase, α-amylase, inulinase, Xylanase, Phytase, transglutaminase, and lipase. Microbial enzymes have been widely employed in the food business to improve food diversity, quality, and variety. The production of large quantities of regulated enzymes from microorganisms is always favored over other sources since it is quicker and less expensive. This chapter includes various microbial enzymes used in the food industry and their biotechnological application.

**Keywords** Microbial enzymes · Tannase · Xylanase · Transglutaminase · Phytase · Inulinase · α-amylase · Lipase

## 9.1 Introduction

For a long time, the global economy has placed the food business among its top objectives. The food business continues to emphasize health and nutrition. The agri-food sector has made significant strides in resolving the global food problem, and cutting-edge technologies are at the core of the whole food chain, from raw

A. Shukla · M. Girdhar · A. Mohan (✉)
School of Bioengineering and Biosciences, Lovely Professional University, Phagwara, India



A. Karnwal, A. R. Mohammad Said Al-Tawaha (eds.), *Food Microbial Sustainability*, https://doi.org/10.1007/978-981-99-4784-3_9

materials and ingredient procurement to enhancing the quality of food processing and packaging by improving quality assurance of processed food (Liu et al., 2021; Khatami et al., 2021). Food is crucial to life's survival and continued existence. Microorganisms swallow food with the help of macromolecules such as carbohydrates, lipids, nitrogenous chemicals, vitamins, and minerals. Organisms at various evolutionary stages eat food in diverse ways. Higher eukaryotes, such as humans, in contrast, hand, consume a more sophisticated variety of food, such as dairy products, fruits, vegetables, grains, meats, and pulses. The abuse and improper management of this essential resource, food, at any point in its life cycle raise substantial social, economic, and environmental concerns. The food industry is continuously confronted with ceasing the shake-out contamination caused by microbiological in the food chain and alleviating the financial losses put forward by microbial spoilage. Therefore, food safety is still a major problem in the present world (Sharma et al., 2020a, b). It is crucial to ensure the safety of microbiological dangers when preparing food. Microorganisms are present, move through, and continue to proliferate when food is processed, with mixed evolutionary traits and uncertainty. To accurately forecast and estimate the risk of microbiological risks in food processing, a specific risk assessment is required to quantify yield in the food processing industry (Chen et al., 2021). Enzymes and microorganisms are commonly used in our society's food production industry. Due to their specificity for substrates and end products, mild reaction conditions, low by-product production, and high yield, these are widely used in various industries (Ravindran & Jaiswal, 2016). These microorganisms are the principal source of industrial enzymes; fungi and yeast account for 50%, bacteria for 35%, and plants or animals account for the remaining 15% of microbial enzyme production (Garg et al., 2016). Exopolysaccharides produced by microorganisms are inherently biodegradable and less hazardous than manufactured polymers. Although polysaccharide made by plants (pectin and starch) and seaweeds (alginate and agar) has been used for decades in the food industry, cosmetics products, medical field, agricultural fields, and research labs, in recent times, the characteristics and composition of microbial exopolysaccharide have gained more attention from the scientific community (Rana & Upadhyay, 2020). They are crucial components in many goods and manufacturing procedures. New application domains have been established and pioneered as new technology innovations to compete with the present world; microorganisms and other enzymes are widely utilized to enhance flavor and aroma, leading to tremendous financial advantages for food enterprises (Chen et al., 2022). Because of their numerous benefits, microbial enzymes have traditionally been the preferred choice. It is further estimated that the food industry enzymes sector comprises the largest part of the worldwide industrial enzymes market and is anticipated to increase from around $1.5 billion in 2016 to $1.9 billion in 2021 (Fernández-Lucas et al., 2017). These benefits include rapid growth of microorganisms, which results in a broad range of catalytic efficiency; stock availability due to the absence of seasonal changes; and simplicity of genetic manipulation, which allows for rapid multiplication and metabolite of enzymes as well as improved characteristics of the enzymatic performance. Sorbitol, mannitol, erythritol, and maltitol are commonly used sugar alcohols that are significantly used in the food and pharmaceutical industries. Mannitol and erythritol are to be used for diabetic food products due to their lesser calorie content (Zhuang et al., 2021). Liquefaction and saccharification are the two phases that make up the starch processing process. When starch is liquefied, α-amylase cleaves β-1,4 glycosidic linkages, resulting in dextrins with comparatively low polymerization degrees. Starch is frequently transformed into glucose, malt oligosaccharides, and dextrins with varying degrees of polymerization for industrial uses. To alter the starch molecules by the process of hydrolyzing and depolymerization of complex structure, debranching enzymes (EC 3.2.1.10, EC 3.2.1.41, and EC 3.2.1.41/142) work in conjunction with degrading starch enzymes (EC 3.2.1.1, 1,4-D-glucan glucanohydrolase), and glucoamylase (EC 3.2.1.3, glucan 1,4-D-glucosidase) (Xia et al., 2021). Due to their functions and prowess in creating food and beverage goods, the enzymes utilized in the food industry are diverse (Mehta et al., 2021). Enzymes are broken down into basic categories based on the types of processes they catalyze in the form of tannase, α-amylase, inulinase, lipase, xylanase, phytase, and transglutaminase. These are utilized for biotransformation of products and are labeled as natural process (Verma et al., 2022). Wide-ranging usage of microbial enzymes may be found in industrial processes and goods, primarily in the food sector. With the aid of these enzymes, various sugar syrups, including glucose, fructose, and inverted sugar syrups, prebiotics, including galactooligosaccharides and fructooligosaccharides, and isomaltulose—an intriguing sugar substitute for sucrose that enhances the sensory qualities of juices and wines while lowering lactose in milk can be produced by microbial enzymes used in food industry (Contesini et al., 2013). Oligosaccharides (OS) are several classes of simple carbohydrate and soluble fibers made up of monomers with varied levels of polymerization (DP) that range from 3 to 10 units. Due to their relatively low production costs, other OS groups like post-harvest derived xylo-oligosaccharides (XOS) and pectic oligosaccharides are additionally considered. At present, fructooligosaccharides (FOS) and galactooligosaccharides (GOS) are among the most commonly available functional OS in the food and beverages industry (Catenza & Donkor, 2021).

## 9.2 Requirement of Microbial Enzyme in the Food Industry

In solving the problem of food safety, there is a requirement for innovative technologies to stop the spread of food contamination. Compared to cellulose obtained from plants, bacterial cellulose, a pure exocellular polymer put together with the help of microbes, exhibits several superior qualities, including a high capacity for retaining water, a large surface area, rheological characteristics, and biocompatibility. Due to its fluid suspension quality thickening, water-retention, and sustaining the bulky fluid competency, BC is a viable lesser calorie intake and augment ingredient used for the production of innovative, enhanced functional meals in a variety of forms, such as powder gelatinous or shred foams, which allow its application in the food sector (Lin et al., 2020). Consumers are clued up for the environmental effects caused by food processing and the enhanced nutritional value, safety, and sensory attributes of the produced food items. By combining many methods to produce the required high-quality product with a longer shelf life, hurdle technology can potentially reduce the depletion of the ozone layer. An effective diffusion mechanism must be included in the design of an industrial-scale ozone treatment process for grains. Ozone treatment is less known to consumers than other methods. A specific product with a manageable processing cost can solve the problem (Sivaranjani et al., 2021). Thus, using microbial enzymes in food processing is widely accepted worldwide. Food safety and quality during microbial action are severe problems for organizations and employees in the food supply business. With improved social connectivity of epidemics of foodborne diseases, quality assurance has received immense attention from the public. Expanding consumer awareness of the hazards associated with increased global commerce has led to adopting several strict safety and quality criteria (Gupta et al., 2021; Udayakumar et al., 2021). As to overcome the issues related to food safety, the agri-food industry has developed and adopted several sustainable technologies ideal for farming and food processing. Thus, there is a requirement to fulfill green technology and sustainability standards; the food sector has reduced energy usage, waste production, and environmental footprints while enhancing by-product recycling. The criteria for green technology are satisfied by novel food processing techniques such as pulsed electric field, high-pressure processing, powerful pulse lighting, ohmic heating, ultrasound, irradiation, supercritical fluid extraction, nonthermal plasma, ozone, electrolysis, or a combination of these techniques. These techniques have more effect on the environment and food quality (López-Pedrouso et al., 2019). Food safety and hygiene are influential to agencies working with food supply and related personnel. With enhanced social platform coverage of the flare-up of foodborne diseases, quality assurance has received unprecedented attention. The amelioration of global commerce has enhanced consumers' awareness of the hazards involved, leading to the acquisition of various strict quality and safety regulations and standards. Therefore, there is a requirement for techniques that require antimicrobial activity to enhance the quality and assurance of food processing standards.

The linear portion of starch is principally affected by the actions of β-amylases, α-amylases, and α-glucosidase, which hydrolyze β-1,4 glucoside linkages that may be broken by α-1,6 bonds. The remaining component such as "limit dextrin," must be further degraded by debranching the α-1,6 branches (Reshmy et al., 2021). Thus, microorganisms that stick to the contact surfaces of food products might be a significant possibility of contamination, resulting in substantial hygiene issues and financial losses from food spoiling. To evaluate the shift in the indigenous microflora from amenities to processed food products over its shelf life, a food material containing inadequate beginning microbial proportion was used that increases yield in the food industry (Delhalle et al., 2020). These increments initially give rise to the disintegration of food particles and subsequent enzyme actions that permit the bonding of unbound phenolic compounds and boost the activity of antioxidants in processed food. For the world's ever-rising population, improving these beneficial components and boosting the intake of fermentable food are essential due to enzymatic activity (Adebo & Medina-Meza, 2020). As well as it is considered as the food business is more likely to embrace enzymatic alteration of food ingredients than the more typical chemical techniques. As time flows, various enzyme traits with unique abilities in processing food are developed and are in the urge of developing various more strain to be used in the food industry. Table 9.1 illustrates various microbial enzyme used in the food industry and their unique class number.

## 9.3 Different Enzymes Used in Food Industries

Numerous microorganisms produce numerous industrial enzymes from molasses, including lipase, protease, lactase, inulinase, and phytase. In the food, pharmaceuticals, chemical, and textile sectors, a few of these enzymes are crucially used in these sectors. Some pure enzyme compositions have been utilized extensively in food preservation, food processing, and the synthesis of food additives, particularly as safe catalysts in the food processing industry. Figure 9.1 shows the various microbial enzymes utilized in the food industry along with their family and the function they performed to catalyze various linkages between various compounds.

**Table 9.1** Source of production of various microbial enzyme

| Microbial enzymes | Class number | Source of enzyme | References |
|---|---|---|---|
| Tannase | EC 3.1.1.20 | *Aspergillus* sp., *Candida* sp., *Mucor* sp., *Bacillus velezensis* | Lekshmi et al. (2020) |
| α-Amylase | EC 3.2.1.1 | *Rhizomucor miehei* | Wang et al. (2020) |
| Inulinase | EC 3.2.1.7 | *Aspergillus* sp., *Bacillus* sp., *Penicillium* sp., *Cryptococcus* sp., *Clostridium* sp., *Pseudomonas* sp., *Arthrobacter* sp., *Candida* sp., *Staphylococcus* sp., *Xanthomonas* sp., *Kluyveromyces* sp., *Pichia* sp., and *Sporotrichum* sp. | Mohan et al. (2018), Singh et al. (2020) |
| Lipase | EC 3.1.1.3 | Bacterial Lipase-*Pseudomonas* and *Bacillus* sp., Fungi Lipase-*Aspergillus* sp., *Candida* sp., *Mucor* sp., *Rhizopus* sp. | Bharathi and Rajalakshmi (2019) |
| Xylanase | EC 3.2.1.8 | *Trichoderma* sp., *Aspergillus* sp., *Aureobasidium pullulans* NRRL Y-2311-1 | Juturu and Wu (2014) |
| Phytase | EC 3.1.3.26<br>EC 3.1.3.8 | *Bacillus stipitis* ATCC 6653, *Aspergillus tubingensis* | Sharma et al. (2020a, b) |
| Transglutaminase | EC 2.3.2.13 | *Chlamydomonas reinhardtii*, *Dunaliella salina* | Aloisi et al. (2016) |

### 9.3.1 Tannase

Tannase is an adaptable intracellular or extracellular enzyme that belongs to the esterase superfamily. As a result of hydrolyzing the ester and depside linkages in tannins, it primarily catalyzes the synthesis of glucose, gallic acid, and different galloyl esters (Lekshmi et al., 2020). The ester linkages found in the hydrolyzable tannins and gallic acid esters are hydrolyzed by a substance known as tannase, also known as tannin acyl hydrolase (EC, 3.1.1.20). It is produced commercially by microorganisms in submerged cultures (SmC), where the action is primarily exhibited in intracellular form and hence involves significant manufacturing expenditures (Lekshmi et al., 2021). The creation of quick tea or acorn liquor and the manufacturing of gallic acid, a crucial intermediate in the creation of the antibacterial medication trimethoprim, utilized in both the pharmaceutical and food industries, is now the tannase's most significant commercial uses. Gallic acid serves as a substrate for propyl gallate's enzymatic or chemical synthesis, a powerful antioxidant. Additionally, tannase is a clarifying agent used in various wines, fruit juices, and beverages with coffee flavors (Belmares et al., 2004). Various species, including fungi, yeast, and bacteria, are used for biosynthesizing inducible enzyme tannase. The main tannic acid hydrolysis product is gallic acid, which is used in cosmetics and the manufacturing of many antioxidants (Chávez González et al., 2016). The majority of research on these sources has been focused on fungus tannase. However, the genetic complexity and sluggish growth rate of tannase prevent the use of fungal isolates for the massive manufacturing of tannase. A potential source for the synthesis of commercially effective tannase may be bacterial strains that can withstand high pH and temperature (Lekshmi et al., 2020).

### 9.3.2 α-Amylase

Endoamylases like α-amylase (EC 3.2.1.1) primarily break down α-1,4-D-glucan linkages in substrates like starch and glycogen to create α-anomeric mono- or oligosaccharides counterparts. Thus, β-1,4 glycosidic bonds in polysaccharides are hydrolyzed by the enzyme α-amylase, which is utilized to break down the food's starch. The α-amylase family, commonly known as the GH13 family, includes enzymes that function as isomerases, transglycosydases, and hydrolases (Farias et al., 2021). Although fungal α-amylases are more effective in the hydrolysis of starch conversions and generation of more maltodextrins that has a greater volume of maltose syrup in them and are of significant economic relevance, the bulk of α-amylases is derived chiefly from Bacillus species (Wang et al., 2020). The maltose syrup obtained from amylase has several uses in the food business. Like molasses, an end product of the sugar industry, that also includes significant levels of cellulose and hemicellulose.

**Fig. 9.1** Various microbial enzymes

Similarly, sugar beet pulp is mainly utilized as an affordable carbon source for generating microorganisms that add value to chemicals (Zhang et al., 2021a, b). Amylases are a crucial industrial enzyme class that accounts for around 25% of the global enzyme market. Many industrial sectors, including the biorefinery, food, pharmaceutical, textile, and paper industries, employ α-amylases to address the needs of particular applications, and work is continuously being done to produce novel α-amylases in the food industry (Maske et al., 2021). It is noted that starch is partially digested in the small intestine after ingestion by α-amylases made in the pancreas, but other elements, such as the ratio of amylose and amylopectin in the starch granules, the crystalline size, and the method of cooking, might affect how well starch is absorbed and utilized in the food industry (Maske et al., 2021). It was estimated that thermostable amylases aid in lowering cooling costs, enhancing substrate solubility, lowering the risk of microbial contamination, and providing resistance to denaturing chemicals mainly utilized in industries (Bhatt et al., 2020). They are commonly utilized as flavor enhancers and antistaling agents in the baking industry to intensify the bread standard. α-amylases are mixed into the dough before baking to steer starch into small oligosaccharides, which are then quickly baked (Raveendran et al., 2018).

### *9.3.3 Inulinase*

The glycoside hydrolases GH 32 family includes inulinase, an inulin-catalyzing enzyme. Glycosylhydrolases are a group of enzymes that include microbial inulinases (GH 32 family). The probable sources of inulinase are bacteria, fungi, and yeasts. Due to its numerous uses in the current biotechnological period, including the generation of fructooligosaccharides, high fructose corn syrup, bioethanol, organic acids, single-cell oil, 2,3-butanediol, single-cell proteins, etc., inulinase is attracting much interest (Mohan et al., 2018). It is widely noted that due to better interactions between inulin and inulinase for the hydrolysis of poly fructans, researchers worldwide are utilizing inulinase. The knowledge of inulinases' structural structure, which is sparse in the literature, is crucial to comprehending these interactions. Many microorganisms generate inulinases, which hydrolyze the β-(2 → 1) fructosyl bond (fungi, bacteria, and yeast). The ultimate fructose constituent of inulin, which is held together by hydrolyzing β-fructofuranosidic bonds, is released by exo-inulinase (EC 3.8.1.80; D-fructan fructohydrolase or β-D-fructokinase). As a fermentable substrate, high fructose syrup (up to 95%) produced by inulinase activity can be used. This method is superior to those processes as cellulose and starch need enzyme consortia to produce feedstock glucose (Rawat et al., 2017). With the aid of this enzyme, it is possible to produce fructose and fructooligosaccharides, which seem to be vital components of the food and pharmaceutical industries. It has become an industrial food enzyme that has received much interest lately (Ramapriya et al., 2018).

### 9.3.4 Lipase

Serine hydrolases, also referred to as the lipase enzyme and members of the triacylglycerol ester hydrolase family, are widely distributed as EC 3.1.1.3. They are associated with the carboxylic ester hydrolase class, triacylglycerol acyl hydrolases, or lipases, are a group of hydrolytic enzymes that catalyze the hydrolysis of insoluble triacylglycerol to glycerol, acylglycerols, and free fatty acids (Chandra et al., 2020). Long-chain triacyl-glycerols, a subtype of lipases with minimal water solubility, are used to catalyze reactions at the lipid-water interface (Bharathi & Rajalakshmi, 2019). The oxyanion hole, or "pocket," found in lipases is another crucial component. It aids in stabilizing the highly reactive intermediate formed during ester-bond hydrolysis. Two residues make up the oxyanion hole, one of which always comes after the nucleophilic residue, and the other can move about (Borrelli & Trono, 2015). Lipase is obtained from various species of fungi and bacterial strains. These include Pseudomonas sp., such as *P. aeruginosa*, *P. fluorescence*, *P. fragi*, *P. alcaligenes*, and some from *Bacillus* sp., while the species of fungi include *Trichoderma*, *Penicillium*, and *Aspergillus niger* (Freitas et al., 2007; Yao et al., 2021). In several lipases, the location of a movable component known as the lid gains access to the enzyme's active site. The "lid," a movable domain with one, two, or three helices or a loop area that is a common feature of most lipases, is what gives them their name. Lacking the lid domain, the lipases from the bacteria *Bacillus subtilis*, *Bacillus pumilus*, and *Bacillus licheniformis*, as well as one from the pancreas of a guinea pig, have lower molecular weight weights and do not exhibit interfacial activation (Secundo et al., 2006). Microbial lipases were estimated to be worth 425 million in 2018, and it is anticipated that the market will increase by 6.8% annually to USD 590 million by 2023 (Almeida et al., 2020). The exceptional adaptability of lipases as a green catalyst, which is acknowledged as the most significant category of biocatalysts in the field of biotechnology, enables their use in a variety of sectors, including the manufacture of food, detergent, pharmaceuticals, leather, textiles, cosmetics, biodiesel, and paper. Lipases are utilized in food to change the flavor by creating an esters bond of short-chain fatty acids and alcohol, which are well-recognized for chemicals associated with taste and aroma in the food industry (Salgado et al., 2022). They have appealing characteristics for industrial use, such as simple catalytic property modification that may be sculpted by altering the genetic structure or reaction conditions, high specificity, and strong enantioselectivity, which enables use in organic synthesis. Thus, a significant emphasis on the synthesis of esters (RCOOR′) in the form of glycerides, the main constituent of fats and oils, has enhanced the interest in natural products in the food industry (De Souza et al., 2020).

### 9.3.5 Xylanase

The enzyme xylanase has been identified as a member of the glycoside hydrolase family. Xylanase is an enzyme that catalyzes the breakdown of xylan, a complex mixture of monosaccharides joined by ester and glycosidic linkages to form a substrate. As a renewable polysaccharide, xylan ranks second in abundance on earth. A group of glycosyl hydrolases called exo-xylanases are crucial in the hydrolysis of xylan into xylose. They are members of the eighth glycosyl hydrolases (GH) family, and their molecular structures contain the recognizable ($\alpha/\alpha$) 6-barrel fold. Exo-xylanases typically have a molecular mass of around 40 kDa and a catalytic domain (CD) responsible for carrying out enzyme activities. A glycosyl hydrolase (GH), the CD of this family of enzymes, hydrolyzes glycosidic bonds found in xylo-oligomers.

In most cases, the production of xylanase on a large scale was carried out through microbial biosynthesis using *Aspergillus* sp. and *Trichoderma* sp., fungi like *Aureobasidium pullulans* produce high moiety xylanase (Yegin, 2017; Juturu & Wu, 2014). The xylanase enzyme is a crucial biocatalyst that is advantageous for various applications together with the treatment of the pulp to enhance the paper standard, improvement of bread quality, treatment of lignocellulose waste, manufacture of xylose sugar, and yield of biological fuels (Kaushal et al., 2021). This is due to the ability of enzymes to break down the β-1,4-glycoside bond in the xylan polymer. To clarify fruit juice, xylanase has most recently been used in the food business. Since the commercial product needs to be devoid of extra polysaccharides that enhance the long storage life of juice, turbidity brought on by colloidal polysaccharides found in freshly squeezed fruit juice creates a problem for the fruit juice business (Mohan et al., 2020).

### 9.3.6 Phytase

Phytases, a microbial enzyme in the food industry, assist in phytate compound breakdown. Phytate is considered the primary source of phosphate storage among plants and is utilized as a naturally occurring compound. Enzyme technology has made a substantial jump from the perspective of biotechnology by greatly expanding the commercial applications of Phytase (Sharma et al., 2020a). Three kinds of phytases can be distinguished in accord with their mechanism of catalytic action and are classified into three groups Histidine acid phosphatases, or acid phosphatases, are members of Group 1 (EC 3.1.3.2), β-propeller phytases are members of Group 2 (EC 3.1.3.8), and purple acid phosphatases or cysteine phytases are members of Group 3 (EC 3.1.3.2). Histidine acid phosphatases are a group of enzymes that include phytases from fungi and *E. coli*. These enzymes share the catalytic dipeptide, ten cysteine residues, and active site sequence (RHGXRXP). Bacillus phytases comprise the bulk of Group 2 enzymes, namely β-propeller phytases (Rao et al., 2009).

Yeasts synthesize phytases under the control of external phosphate concentrations and other variables, including carbon source, temperature, pH, and the presence of metal cations. If an enzyme is secreted, it may do so intracellularly, periplasmatically, directly into the culture media, or as a cell wall-bound form (Suryani et al., 2021). Phytase is engaged in extensive processes of dephytinization involved in industrial activity, which removes glycinin and B-conglycinin to produce soy-based goods, as well as in the food and beverage processing industries, breweries, and bakery products to enhance the texture and quality of bread (Handa et al., 2020). Protein, carbohydrate, and fat digestibility are all decreased by phytase enzymes. Proteins become less soluble when they form a compound with phytase, which helps them to withstand proteolysis. Interactions with the amylase enzyme, phytic acid, and polyphenols may influence starch digestion. It was estimated that the amount of phytic acid consumed and zinc bioavailability have an antagonistic connection (Vohra & Satyanarayana, 2003). Table 9.2 indicates the various microbial enzymes used in the food industry, with the main emphasis on their application in various industries. The amount of wheat grain removed during the milling process determines the chemical makeup of the flour used to make bread. A loaf of whole wheat bread is becoming more popular because of its nutritional advantages, such as its high fiber content. High levels of iron and phytate, which come from the bran, are present in whole-grain flour (Gontia-Mishra & Tiwari, 2013).

### *9.3.7 Transglutaminase (TGase)*

The calcium-dependent enzyme transglutaminase (EC 2.3.2.13) catalyzes the acyl transfer process involving γ-carboxamide moieties of residues of glutamine proteins, peptides, and other primary amines. The functional characteristics of proteins in food systems have frequently been altered using microbial transglutaminase (MTGase). TGase was utilized to catalyze the acyl transfer reaction between primary amines as the acyl acceptors and γ-carboxyl amide of glutaminyl residues as the acyl donors (Akbari et al., 2021). The acyl transfer process involving free amino groups and protein-bound Gln residues is facilitated by TGase, which are $Ca^{2+}$-dependent enzymes. Several fundamental amino groups may function as the donor of the amine group, but the γ-carboxamide group of protein-bound Gln is the only donor of the acyl substrate of TGase (Ando et al., 1989). The transglutaminases (TGs) biocatalysis results in the transamination of aliphatic amines to glutamine residues, eventually leading to protein polymerization and the generation of new covalent isopeptide bonds (crosslinks) involving lysine and glutamine, peptide bonded residues. In both normal and pathological situations, TGs cause the post-translational alteration of proteins via a catalytic triad of cysteine/histidine/aspartic acid (Cys/His/Asp) (Savoca et al., 2018). The usage of transglutaminase in research

**Table 9.2** Various microbial enzymes application in different industries

| Microbial enzymes | Different industry | Application | References |
|---|---|---|---|
| Xylanase | Brewing industry | To improve filtration | Juturu and Wu (2014), Raveendran et al. (2018) |
| | Baking Industry | Improves texture, the shelf life of bread, and loaf volume | |
| Lipase | Dairy industry | Enhancing flavoring agents in cheese, milk, and butter, hydrolyze the milk fat, modify the length of the fatty acid chain | Mehta et al. (2021), Chandra et al. (2020) |
| | Meat industry | Reduces the fat present in the meat and fish, development of flavor | |
| | Bakery industry | Prolongation of shelf life of bakery products | |
| Transglutaminase | Dairy industry | Formation of protein films that improve properties of dairy products | Kieliszek and Misiewicz (2014), Romeih and Walker (2017), Akbari et al. (2021) |
| | Meat industry | Improves the quality of meat by enhancing nutritive value with lower fat content | |
| | Bakery industry | Improve texture, flour, and volume of bread, improve pore size, water adsorption, and elasticity in a dough | |
| α-Amylase | Bakery industry | Improve bread quality by acting as an antistaling agent; enhancement of flavor helps in the conversion of starch to smaller dextrins | Zhang et al. (2021a, b), Wang et al. (2020) |
| | Starch industry | For liquefaction of starch that efficiently converts starch to glucose and fructose syrup | |
| | Juice industry | Clarification of fruit juice | |
| Phytase | | Enhance the bioactivity of essential elements Mg, Ca, Fe, Zn | Vohra and Satyanarayana (2003), Handa et al. (2020) |
| | Bakery industry | Reduces the phytate content in dough and lessens the fermentation time | |
| | Cereal grains | Produces soluble non-starch carbohydrates, insoluble fibers, oil fraction, and high-value proteins in grains | |
| Tannase | Beverages | Clarifying agents in coffee-flavored beverages improve aromatic compounds | Belmares et al. (2004) |
| | Food industry | Gallic acid is utilized as a substrate for the synthesis of food preservatives in the forms of gallates and pyrogallol | |

on the food industry and applications in medical assistance is increasing due to the recent discovery, opening the door to practical implementation. The inclusion of MTGase in dairy products changes milk proteins' technical and functional characteristics, such as their ability to gelate, withstand heat, emulsify, contain water, and foam with consistency (Romeih & Walker, 2017).

Similarly, Transglutaminase is utilized in the baking industry to improve the flour quality, the volume and texture of bread, and the texture of cooked pasta (Kieliszek & Misiewicz, 2014). From a health perspective, TG can lessen allergies, regulate energy intake from meals, and serve as a facilitator in the healing of wounds. In addition to all of these advantages, research has shown that the activity of transglutaminase (mTG) in food items may result in the development of autoantigen in the celiac disease (CD) population. Hence, the enzymes may deamidate gluten and so resemble native tissue transglutaminase, microbial transglutaminase-crosslinked gluten may be harmful to CD (tTG) (Amirdivani et al., 2018).

## 9.4 Biotechnological Applications of Different Microbial Enzymes

Every microbial enzyme has various biotechnological applications for the production of various bioproducts. These natural preservatives are used more frequently in food than synthetic ones, and their significance as preservatives is expanding so widely and facilitating the food industry. These natural preservatives are divided into three subcategories, which include antimicrobial, antibrowning, and antioxidant agents. The antimicrobial agents include various microbial enzymes that have biotechnological applications in the food industry (Gokoglu, 2019). The synthesis of extracellular metabolites in the form of xylitol, citric acid, ethanol, erythritol, exopolysaccharides, and biosurfactants is connected to the biotechnological usage of Candida yeasts. These materials are employed in the food processing, cosmetics, and pharmaceutical industries (Kieliszek et al., 2017). It has been demonstrated that the actions of several enzymes, including phosphatase, trypsin, amylase, and tyrosinase, are inhibited by phytic acid and inositol Penta-phosphate (Vohra & Satyanarayana, 2003). Microbial enzymes are employed in food processing as biological catalysts for various processes, including baking, coagulation, ripening, cell rupture, brewing, hydrolysis, and molecular structural change. In many food products, enzymes enhance the quality, longevity, freshness, appearance, usability, nutritional content, and aroma and form desired structures. Enzymes are occasionally employed to purify juices or to speed up the breakdown of the cell walls of oil crops to increase extraction output (Ermis, 2017). From inulin and inulin-rich sources of inulinases that may be used to produce high fructose syrup (HFS) and fructooligosaccharides (FOS) (Singh et al., 2020). Globally, there is a high demand for HFS as a value-added flavor booster. It is employed in various sectors (beverages, food, etc.). It has a GRAS (Generally Regarded as Safe) designation due to its high nutritional value and technological superiority (Gahlawat et al., 2017), while FOS, an important family of prebiotics with remarkable influences on health effects, fructooligosaccharides are low-calorie, indigestible, beneficial nutritional constituents. FOSs are among the several prebiotics, such as galactooligosaccharides (GOSs), Xylo-oligosaccharides (XOSs), and fructooligosaccharides (FOSs), that are in high demand globally because of their multiple bioactive components. In the human colon, FOSs promotes the formation of beneficial microflora, including Bifidobacterium and Lactobacilli (Singh & Singh, 2010). Similarly, microbial lipases are a safe and effective green tool utilized in the food industry. Lipases are considered crucial microbial enzymes, to improve the aroma, flavor, and texture of food. The utilization of lipases has increased due to developments in protein engineering and immobilization methods. Thus, microbial lipases are considered viable biocatalysts for industry use (Salgado et al., 2022). The primary components of food, fragrance, and flavor, enhance the properties of organoleptic in food and increase consumer access to food. Similarly, MTGase has a variety of possible usage in foods that are primarily protein-based in structure. One of MTGase's most promising uses is in the meat sector, notably in reformed meat-based products. MTGase makes items harder and opens up new potential for manufacturing low-fat processed meat with excellent sensory qualities compared to its conventional counterpart (Romeih & Walker, 2017). This broad family of TGase as protein-remodeling enzymes has sparked attention due to the many biotechnological uses. In the last 50 years, the food industry has used both human and microbial TG polymerizing activity to improve the quality, texture, and nutritional content of foods in addition to their appearance and commercial viability (Savoca et al., 2018). Furthermore, another microbial enzyme, lipase, has various biotechnological advancements in the food industry with the catalyzed reaction that replaces the traditional method utilized in the food industry. It is used in the refining of oil by eliminating free fatty acids. Lower-quality oils can be changed to create products with a high value-added (dos Santos Aguilar & Sato, 2018). It is also used in the dairy industry to hydrolyze milk fat which could significantly enhance the taste and aroma of dairy products and raise their nutritional content. Lipase can also be utilized to improve the quality of baked foods; it is also utilized to elongate the shelf life of baked foods and whiten darker loaves in the bread (Yao et al., 2021). Significant improvements aimed at creating high-quality, secure, and nutrient-rich bread products with prolonged shelf lives are projected to result from integrating bio-protection techniques for active packaging or processing technology (Dong & Karboune, 2021). Lipases are extensively utilized in the dairy, bakery, wine, fruit juice, and beer sectors in the beverage and food sector. Despite having several uses in numerous sectors, lipase only accounts for less than 10% of the worldwide industrial enzyme market. Commercial lipases are mainly engaged in the process related to meals full of fat containing ingredients that generate flavors in dairy products. By the way of their influence on the milk lipids to produce free fatty acids followed by hydrolysis process, which can improve the distinctive flavor of cheese (Raveendran et al., 2018). Thus, natural preservatives are used more frequently in food than that of synthetic ones, their significance as preservatives is expanding so widely and facilitating food industry

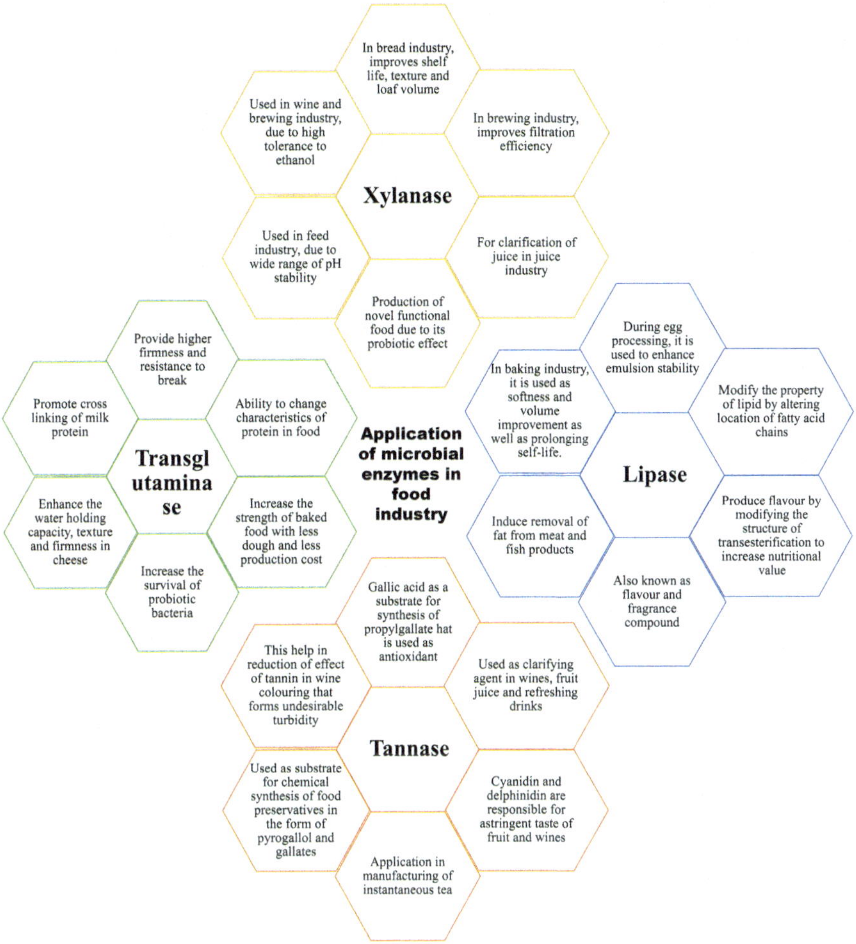


**Fig. 9.2** Application of microbial enzymes in the food industry

(Figueroa et al., 2021). Figure 9.2 illustrates biotechnological application of microbial enzymes in food industry.

## 9.5 Conclusion

Depending on the microbial enzyme and reaction conditions, a wide range of food ingredients may be obtained with maximum nutritional value, flavor, and aroma. The oligo-saccharide market is already sizable and is still growing quickly. The production of healthy meals presents an extraordinary chance to enhance the health and well-being of consumers. There is various possibilities to discover larger industrial applications for the many microbial-origin enzymes that have not yet been fully explored. Various microbial enzymes accomplish innumerable biochemical, physiological, and regulatory processes. Thus, due to their ecologically benign nature of enzymes that show effective control process, unreasonable yield, affordable cost of the refining process, and process safety, enzyme-based techniques are now preferred to chemical ones in many industries.

## References


Adebo, O. A., & Medina-Meza, I. G. (2020). Impact of fermentation on the phenolic compounds and antioxidant activity of whole cereal grains: A mini review. *Molecules, 25*(4), 1–19. https://doi.org/10.3390/molecules25040927

Akbari, M., Razavi, S. H., & Kieliszek, M. (2021). Recent advances in microbial transglutaminase biosynthesis and its application in the food industry. *Trends in Food Science and Technology, 110*, 458–469. https://doi.org/10.1016/j.tifs.2021.02.036

Almeida, J. M., Alnoch, R. C., Souza, E. M., Mitchell, D. A., & Krieger, N. (2020). Metagenomics: Is it a powerful tool to obtain lipases for application in biocatalysis. *Biochimica et Biophysica Acta, Proteins and Proteomics, 1868*(2), 140320. https://doi.org/10.1016/j.bbapap.2019.140320

Aloisi, I., Cai, G., Serafini-Fracassini, D., & Del Duca, S. (2016). Transglutaminase as polyamine mediator in plant growth and differentiation. *Amino Acids, 48*(10), 2467–2478. https://doi.org/10.1007/s00726-016-2235-y

Amirdivani, S., Khorshidian, N., Fidelis, M., Granato, D., Koushki, M. R., Mohammadi, M., Khoshtinat, K., & Mortazavian, A. M. (2018). Effects of transglutaminase on health properties of food products. *Current Opinion in Food Science, 22*, 74–80. https://doi.org/10.1016/j.cofs.2018.01.008

Ando, H., Adachi, M., Umeda, K., Matsuura, A., Nonaka, M., Uchio, R., Tanaka, H., & Motoki, M. (1989). Purification and characteristics of a novel transglutaminase derived from microorganisms. *Agricultural and Biological Chemistry, 53*(10), 2613–2617. https://doi.org/10.1080/00021369.1989.10869735

Belmares, R., Contreras-Esquivel, J. C., Rodríguez-Herrera, R., Coronel, A. R., & Aguilar, C. N. (2004). Microbial production of tannase: An enzyme with potential use in food industry. *LWT - Food Science and Technology, 37*(8), 857–864. https://doi.org/10.1016/j.lwt.2004.04.002

Bharathi, D., & Rajalakshmi, G. (2019). Microbial lipases: An overview of screening, production and purification. *Biocatalysis and Agricultural Biotechnology, 22*, 101368. https://doi.org/10.1016/j.bcab.2019.101368

Bhatt, K., Lal, S., Srinivasan, R., & Joshi, B. (2020). Bioconversion of agriculture wastes to produce α-amylase from bacillus velezensis KB 2216: Purification and characterization. *Biocatalysis and Agricultural Biotechnology, 28*, 101703. https://doi.org/10.1016/j.bcab.2020.101703

Borrelli, G. M., & Trono, D. (2015). Recombinant lipases and phospholipases and their use as biocatalysts for industrial applications. *International Journal of Molecular Sciences, 16*(9), 20774–20840. https://doi.org/10.3390/ijms160920774

Catenza, K. F., & Donkor, K. K. (2021). Recent approaches for the quantitative analysis of functional oligosaccharides used in the food industry: A review. *Food Chemistry, 355*, 129416. https://doi.org/10.1016/j.foodchem.2021.129416

Chandra, P., Enespa, R. S., & Arora, P. K. (2020). Microbial lipases and their industrial applications: A comprehensive review. *Microbial Cell Factories, 19*, 169. https://doi.org/10.1186/s12934-020-01428-8

Chávez González, M. L., Buenrostro-Figueroa, J., Rodríguez Durán, L. V., Zárate, P. A., Rodríguez, R., Rodríguez-Jasso, R. M., Ruiz, H. A., & Aguilar, C. N. (2016). Tannases. In *Current developments in biotechnology and bioengineering: Production, isolation and purification of industrial products* (pp. 471–489). Elsevier. https://doi.org/10.1016/B978-0-444-63662-1.00020-8

Chen, Q., Zhao, Z., Wang, X., Xiong, K., & Shi, C. (2021). A dynamic risk assessment approach based on stochastic hybrid system: Application to microbial hazards in food processing. *Microbial Risk Analysis, 19*, 100163. https://doi.org/10.1016/j.mran.2021.100163

Chen, Z., Zhang, C., Huan, D., Chen, C., Xue, Q., & Yongjin, H. (2022). Effect of starter cultures on dynamics succession of microbial communities, physiochemical parameters, enzyme activities, taste and volatile flovor compounds during sufu fermentation. *Food Chemistry Advances, 1*, 100057. https://doi.org/10.1016/j.focha.2022.100057

Contesini, F. J., de Alencar Figueira, J., Kawaguti, H. Y., de Barros Fernandes, P. C., de Oliveira Carvalho, P., da Graça Nascimento, M., & Sato, H. H. (2013). Potential applications of carbohydrases immobilization in the food industry. *International Journal of Molecular Sciences, 14*(1), 1335–1369. https://doi.org/10.3390/ijms14011335

De Souza, W., Reis, M., Deda, R., Ferreira, M., & Benedito, E. (2020). Immobilization of commercial lipase onto different supports: Characterization and application in esterification reaction. *International Journal of Engineering Research and Science, 1*, 15–24.

Delhalle, L., Taminiau, B., Fastrez, S., Fall, A., Ballesteros, M., Burteau, S., & Daube, G. (2020). Evaluation of enzymatic cleaning on food processing installations and food products bacterial microflora. *Frontiers in Microbiology, 11*, 1827. https://doi.org/10.3389/fmicb.2020.01827

Dong, Y. N., & Karboune, S. (2021). A review of bread qualities and current strategies for bread bioprotection: Flavor, sensory, rheological, and textural attributes. *Comprehensive Reviews in Food Science and Food Safety, 20*(2), 1937–1981. https://doi.org/10.1111/1541-4337.12717

Dos Santos Aguilar, J. G., & Sato, H. H. (2018). Microbial proteases: Production and application in obtaining protein hydrolysates. *Food Research International, 103*, 253–262. https://doi.org/10.1016/j.foodres.2017.10.044

Ermis, E. (2017). Halal status of enzymes used in food industry. *Trends in Food Science and Technology, 64*, 69–73. https://doi.org/10.1016/j.tifs.2017.04.008

Farias, T. C., Kawaguti, H. Y., & Koblitz, M. G. B. (2021). Microbial amylolytic enzymes in foods: Technological importance of the bacillus genus. *Biocatalysis and Agricultural Biotechnology, 35*, 102054. https://doi.org/10.1016/j.bcab.2021.102054

Fernández-Lucas, J., Castañeda, D., & Hormigo, D. (2017). New trends for a classical enzyme: Papain, a biotechnological success story in the food industry. *Trends in Food Science and Technology, 68*, 91–101. https://doi.org/10.1016/j.tifs.2017.08.017

Figueroa, J. G., Borrás-Linares, I., Del Pino-García, R., Curiel, J. A., Lozano-Sánchez, J., & Segura-Carretero, A. (2021). Functional ingredient from avocado peel: Microwave-assisted extraction, characterization and potential applications for the food industry. *Food Chemistry, 352*, 129300. https://doi.org/10.1016/j.foodchem.2021.129300

Freitas, L., Bueno, T., Perez, V. H., Santos, J. C., & De Castro, H. F. (2007). Enzymatic hydrolysis of soybean oil using lipase from different sources to yield concentrated of polyunsaturated fatty acids. *World Journal of Microbiology and Biotechnology, 23*(12), 1725–1731. https://doi.org/10.1007/s11274-007-9421-8

Gahlawat, S. K., Salar, R. K., Siwach, P., Duhan, J. S., Kumar, S., & Kaur, P. (2017). *Plant biotechnology: Recent advancements and developments* (pp. 1–390). Springer. https://doi.org/10.1007/978-981-10-4732-9

Garg, G., Singh, A., Kaur, A., Singh, R., Kaur, J., & Mahajan, R. (2016). Microbial pectinases: An ecofriendly tool of nature for industries. *3 Biotech, 6*(1), 1–13. https://doi.org/10.1007/s13205-016-0371-4

Gokoglu, N. (2019). Novel natural food preservatives and applications in seafood preservation: A review. *Journal of the Science of Food and Agriculture, 99*(5), 2068–2077. https://doi.org/10.1002/jsfa.9416

Gontia-Mishra, I., & Tiwari, S. (2013). Molecular characterization and comparative phylogenetic analysis of phytases from fungi with their prospective applications. *Food Technology and Biotechnology, 51*(3), 313–326.

Gupta, R., Raza, N., Bhardwaj, S. K., Vikrant, K., Kim, K. H., & Bhardwaj, N. (2021). Advances in nanomaterial-based electrochemical biosensors for the detection of microbial toxins, pathogenic bacteria in food matrices. *Journal of Hazardous Materials, 401*, 123379. https://doi.org/10.1016/j.jhazmat.2020.123379

Handa, V., Sharma, D., Kaur, A., & Arya, S. K. (2020). Biotechnological applications of microbial phytase and phytic acid in food and feed industries. *Biocatalysis and Agricultural Biotechnology, 25*, 101600. https://doi.org/10.1016/j.bcab.2020.101600

Juturu, V., & Wu, J. C. (2014). Microbial exo-xylanases: A mini review. *Applied Biochemistry and Biotechnology, 174*(1), 81–92. https://doi.org/10.1007/s12010-014-1042-8

Kaushal, J., Khatri, M., Singh, G., & Arya, S. K. (2021). A Multifaceted enzyme conspicuous in fruit juice clarification: An elaborate review on xylanase. *International Journal of Biological Macromolecules, 193*, 1350–1361. https://doi.org/10.1016/J.IJBIOMAC.2021.10.194

Khatami, K., Atasoy, M., Ludtke, M., Baresel, C., Eyice, Ö., & Cetecioglu, Z. (2021). Bioconversion of food waste to volatile fatty acids: Impact of microbial community, pH and retention time. *Chemosphere, 275*, 129981. https://doi.org/10.1016/j.chemosphere.2021.129981

Kieliszek, M., & Misiewicz, A. (2014). Microbial transglutaminase and its application in the food industry. A review. *Folia Microbiologica, 59*(3), 241–250. https://doi.org/10.1007/s12223-013-0287-x

Kieliszek, M., Kot, A. M., Bzducha-Wróbel, A., BŁażejak, S., Gientka, I., & Kurcz, A. (2017). Biotechnological use of candida yeasts in the food industry: A review. *Fungal Biology Reviews, 31*(4), 185–198. https://doi.org/10.1016/j.fbr.2017.06.001

Lekshmi, R., Arif Nisha, S., Kaleeswaran, B., & Alfarhan, A. H. (2020). Pomegranate peel is a low-cost substrate for the production of tannase by bacillus velezensis TA3 under solid state fermentation. *Journal of King Saud University - Science, 32*(3), 1831–1837. https://doi.org/10.1016/j.jksus.2020.01.022

Lekshmi, R., Arif Nisha, S., Thirumalai Vasan, P., & Kaleeswaran, B. (2021). A Comprehensive review on tannase: Microbes associated production of tannase exploiting tannin rich agro-industrial wastes with special reference to its potential environmental and industrial applications. *Environmental Research, 201*, 111625. https://doi.org/10.1016/j.envres.2021.111625

Lin, D., Liu, Z., Shen, R., Chen, S., & Yang, X. (2020). Bacterial cellulose in food industry: Current research and future prospects. *International Journal of Biological Macromolecules, 158*, 1007–1019. https://doi.org/10.1016/j.ijbiomac.2020.04.230

Liu, J. M., Chen, L., Jensen, P. R., & Solem, C. (2021). Food grade microbial synthesis of the butter aroma compound butanedione using engineered and non-engineered lactococcus lactis. *Metabolic Engineering, 67*, 443–452. https://doi.org/10.1016/j.ymben.2021.08.006

López-Pedrouso, M., Díaz-Reinoso, B., Lorenzo, J. M., Cravotto, G., Barba, F. J., Moure, A., Domínguez, H., & Franco, D. (2019). Green technologies for food processing: Principal considerations. In *Innovative thermal and non-thermal processing, bioaccessibility and bioavailability of nutrients and bioactive compounds*. Elsevier. https://doi.org/10.1016/B978-0-12-814174-8.00003-2

Maske, B. L., de Melo Pereira, G. V., Vale, A. d. S., de Carvalho Neto, D. P., Karp, S. G., Viesser, J. A., Lindner, J. D. D., Pagnoncelli, M. G., Soccol, V. T., & Soccol, C. R. (2021). A review on enzyme-producing lactobacilli associated with the human digestive process: From metabolism to application. *Enzyme and Microbial Technology, 149*, 109836. https://doi.org/10.1016/j.enzmictec.2021.109836

Mehta, A., Guleria, S., Sharma, R., & Gupta, R. (2021). The lipases and their applications with emphasis on food industry. In *Microbial biotechnology in food and health*. Elsevier. https://doi.org/10.1016/b978-0-12-819813-1.00006-2

Mohan, A., Flora, B., & Girdhar, M. (2018). Inulinase: An important microbial enzyme in food industry. In *Microbial bioprospecting for sustainable development* (pp. 237–248). Springer. https://doi.org/10.1007/978-981-13-0053-0_12

Mohan, A., Kaur, S., & Girdhar, M. (2020). Hydrolytic xylanase production from fungi-An assessment for their production, properties and computional analysis of their gene sequences. In *New and future developments in microbial biotech and bioengineering* (pp. 79–89). Elsevier. https://doi.org/10.1016/B978-0-12-821008-6.00008-6

Ramapriya, R., Thirumurugan, A., Sathishkumar, T., & Manimaran, D. R. (2018). Partial purification and characterization of exoinulinase produced from Bacillus sp. *Journal, Genetic Engineering & Biotechnology, 16*(2), 363–367. https://doi.org/10.1016/j.jgeb.2018.03.001

Rana, S., & Upadhyay, L. S. B. (2020). Microbial exopolysaccharides: Synthesis pathways, types and their commercial applications. *International Journal of Biological Macromolecules, 157*, 577–583. https://doi.org/10.1016/j.ijbiomac.2020.04.084

Rao, D. E. C. S., Rao, K. V., Reddy, T. P., & Reddy, V. D. (2009). Molecular characterization, physicochemical properties, known and potential applications of phytases: An overview. *Critical Reviews in Biotechnology, 29*(2), 182–198. https://doi.org/10.1080/07388550902919571

Raveendran, S., Parameswaran, B., Ummalyma, S. B., Abraham, A., Mathew, A. K., Madhavan, A., Rebello, S., & Pandey, A. (2018). Applications of microbial enzymes in food industry. *Food Technology and Biotechnology, 56*(1), 16–30. https://doi.org/10.17113/ftb.56.01.18.5491

Ravindran, R., & Jaiswal, A. K. (2016). Microbial enzyme production using lignocellulosic food industry wastes as feedstock: A review. *Bioengineering, 3*(4), 30. https://doi.org/10.3390/bioengineering3040030

Rawat, H. K., Soni, H., Treichel, H., & Kango, N. (2017). Biotechnological potential of microbial inulinases: Recent perspective. *Critical Reviews in Food Science and Nutrition, 57*(18), 3818–3829. https://doi.org/10.1080/10408398.2016.1147419

Reshmy, R., Philip, E., Madhavan, A., Sindhu, R., Pugazhendhi, A., Binod, P., Sirohi, R., Awasthi, M. K., Tarafdar, A., & Pandey, A. (2021). Advanced biomaterials for sustainable applications in the food industry: Updates and challenges. *Environmental Pollution, 283*, 117071. https://doi.org/10.1016/j.envpol.2021.117071

Romeih, E., & Walker, G. (2017). Recent advances on microbial transglutaminase and dairy application. *Trends in Food Science and Technology, 62*, 133–140. https://doi.org/10.1016/j.tifs.2017.02.015

Salgado, C. A., dos Santos, C. I. A., & Vanetti, M. C. D. (2022). Microbial lipases: Propitious biocatalysts for the food industry. *Food Bioscience, 45*, 101509. https://doi.org/10.1016/J.FBIO.2021.101509

Savoca, M. P., Tonoli, E., Atobatele, A. G., & Verderio, E. A. M. (2018). Biocatalysis by transglutaminases: A review of biotechnological applications. *Micromachines (Basel), 9*(11), 562. https://doi.org/10.3390/mi9110562

Secundo, F., Carrea, G., Tarabiono, C., Gatti-Lafranconi, P., Brocca, S., Lotti, M., Jaeger, K. E., Puls, M., & Eggert, T. (2006). The lid is a structural and functional determinant of lipase activity and selectivity. *Journal of Molecular Catalysis B: Enzymatic, 39*(1–4), 166–170. https://doi.org/10.1016/j.molcatb.2006.01.018

Sharma, A., Ahluwalia, O., Tripathi, A. D., Singh, G., & Arya, S. K. (2020a). Phytases and their pharmaceutical applications: Mini-review. *Biocatalysis and Agricultural Biotechnology, 23*, 101439. https://doi.org/10.1016/j.bcab.2019.101439

Sharma, P., Gaur, V. K., Kim, S. H., & Pandey, A. (2020b). Microbial strategies for bio-transforming food waste into resources. *Bioresource Technology, 299*, 122580. https://doi.org/10.1016/j.biortech.2019.122580

Singh, R. S., & Singh, R. P. (2010). Production of fructooligosaccharides from inulin by endoinulinases and their prebiotic potential. *Food Technology and Biotechnology, 48*(4), 435–450.

Singh, R. S., Singh, T., Hassan, M., & Kennedy, J. F. (2020). Updates on inulinases: Structural aspects and biotechnological applications. *International Journal of Biological Macromolecules, 164*, 193–210. https://doi.org/10.1016/j.ijbiomac.2020.07.078

Sivaranjani, S., Arun Prasath, V., Pandiselvam, R., Kothakota, A., & Khaneghah, A. M. (2021). Recent advances in applications of ozone in the cereal industry. *LWT - Food Science and Technology, 146*, 111412. https://doi.org/10.1016/j.lwt.2021.111412

Suryani, A. E., Anggraeni, A. S., Istiqomah, L., Damayanti, E., & Karimy, M. F. (2021). Isolation and identification of phytate-degrading yeast from traditional fermented food. *Biodiversitas, 22*(2), 866–873. https://doi.org/10.13057/biodiv/d220241

Udayakumar, G. P., Muthusamy, S., Bharathi Selvaganesh, N., Sivarajasekar, K. R., Banat, F., Sivamani, S., Sivakumar, N., Hosseini-Bandegharaei, A., & Show, P. L. (2021). Biopolymers and composites: Properties, characterization and their applications in food, medical and pharmaceutical industries. *Journal of Environmental Chemical Engineering, 9*(4), 105322. https://doi.org/10.1016/j.jece.2021.105322

Verma, D. K., Gddoa, S. T., Niamah, A. K., Thakur, M., Shah, N., Singh, S., Baranwal, D., Patel, A. R., Utama, G. L., & Aguilar, C. N. (2022). Recent trends in microbialflavour compounds: A review on chemistry, synthesis mechanism and their application in food. *Saudi Journal of Biological Science, 29*, 1565–1579.

Vohra, A., & Satyanarayana, T. (2003). Phytases: Microbial sources, production, purification, and potential biotechnological applications. *Critical Reviews in Biotechnology, 23*(1), 29–60. https://doi.org/10.1080/713609297

Wang, Y. C., Hu, H. F., Ma, J. W., Yan, Q. J., Liu, H. J., & Jiang, Z. Q. (2020). A novel high maltose-forming α-amylase from rhizomucor miehei and its application in the food industry. *Food Chemistry, 305*, 125447. https://doi.org/10.1016/j.foodchem.2019.125447

Xia, W., Zhang, K., Lingqia, S., & Jing, W. (2021). Microbial starch debranching enzymes: Developments and applications. *Biotechnology Advances, 50*, 107786. https://doi.org/10.1016/j.biotechadv.2021.107786

Yao, W., Liu, K., Liu, H., Jiang, Y., Wang, R., Wang, W., & Wang, T. (2021). A valuable product of microbial cell factories: Microbial lipase. *Frontiers in Microbiology, 12*, 1–16. https://doi.org/10.3389/fmicb.2021.743377

Yegin, S. (2017). Single-step purification and characterization of an extreme halophilic, ethanol tolerant and acidophilic xylanase from aureobasidium pullulans NRRL Y-2311-1 with application potential in the food industry. *Food Chemistry, 221*, 67–75. https://doi.org/10.1016/j.foodchem.2016.10.003

Zhang, H., Fengfeng, W., Dan, X., Ali, B., Jiali, Q., & Xueming, X. (2021a). Endogenous alpha-amylase alters the pasting properties of starch during starch separation by proteases. *Journal of Cereal Science, 101*, 103311. https://doi.org/10.1016/j.jcs.2021.103311

Zhang, S., Wang, J., & Jiang, H. (2021b). Microbial production of value-added bioproducts and enzymes from molasses, a by-product of sugar industry. *Food Chemistry, 346*, 128860. https://doi.org/10.1016/j.foodchem.2020.128860

Zhuang, N., Jieyu Ma, L., Yang, R. X., Qian, X., Chen, M., & Zhang, S. (2021). Rapid determination of sucrose and glucose in microbial fermentation and fruit juice samples using engineered multi-enzyme biosensing microchip. *Microchemical Journal, 164*(30), 106075. https://doi.org/10.1016/j.microc.2021.106075

# Chapter 10
# Microbial Enzymes in Food Industries: Enhancing Quality and Sustainability

**Surender Jangra and Sarita Srivastava**

**Abstract** Even before the discovery of enzymes, they have been used by our ancestors in the production of various food products, and even today they are used routinely in the production of various food products in both households and food industries such as baking, brewing, beverages, juice, dairy, oil refinery, food packaging, etc. Enzymes exhibit various properties such as eco-friendly in nature, work under normal pressure, temperature, and pH, consume less energy, cost-effective, do not produce any greenhouse gases, show high activity and turnover number, high biodegradability, and unsurpassed specificity. Furthermore, enzymes produce specific products with no or minimal byproducts or wastes. Because of these properties they are preferred over inorganic catalysts or chemicals in food and other industries. Enzymes are widely distributed and are present in all living organisms. But in food industries, microbes are given priority for the isolation of enzymes as microbes are easily available, grow at faster speed, and can be manipulated through genetic engineering to produce improved or novel enzymes that can sustain and work optimally at different industrial conditions. Alternatively, enzymes can be extracted from microorganisms that can grow at extreme conditions. Food industries are largest market for enzymes where approx. 55% of all industrial enzymes are used. They are known to enhance the flavor, aroma, palatability, and overall quality of the products produced. In this chapter the authors will discuss the applications of some microbial enzymes used in food industries for various purposes.

**Keywords** Enzyme · Food industry · Food products · Microbes · Apoenzyme

S. Jangra (✉) · S. Srivastava
Department of Chemistry and Biochemistry, School of Basic Sciences and Research, Sharda University, Greater Noida, India



A. Karnwal, A. R. Mohammad Said Al-Tawaha (eds.), *Food Microbial Sustainability*, https://doi.org/10.1007/978-981-99-4784-3_10

## 10.1 Introduction

Enzymes are biocatalysts that enhance the speed of chemical reactions by lowering down the activation energy. All enzymes are proteinaceous in nature except ribozymes (catalytic RNA). They are macromolecules and their molecular weight varies from 10 to 2000 kDa (Okpara, 2022). Some enzymes require cofactors; some require coenzymes while some require both cofactors and coenzymes for their biological activity. Enzymes with their cofactors or/and coenzymes part are active and are called holoenzymes, whereas enzymes without their cofactors or/and coenzyme part are inactive and are called apoenzymes. Active site is made up of less than ten amino acids and because of its shape and charge properties it enables the enzymes to bind to their substrates specifically, making a particular enzyme unique from other enzymes (Robinson, 2015). Active sites can be present on enzyme's surface, in cleft or pocket or buried deeply within enzyme globular structure (Pravda et al., 2014). On the basis of specificities of enzymes, Enzyme Commission (EC) has classified, named, and numbered over 3000 known enzymes (Patel et al., 2017). Enzymes are broadly classified into six categories based on the reactions catalyzed by them, viz. oxidoreductase, transferase, hydrolase, lyase, isomerase, and ligase. In 2018, a new and seventh class of enzymes named translocases has been introduced by enzyme commission. Translocases catalyze the movement of ions and molecules across the membrane, for example, carnitine-acylcarnitine translocase, ADP/ATP translocase, etc. (https://iubmb.qmul.ac.uk/enzyme/). Enzymes of every category have some role in food industries but enzymes of hydrolases category have more frequent role. Functions of enzymes depend upon their tertiary or quaternary structures which are ultimately determined by their primary structures. A small change in the primary structure of enzyme can modify the functions of enzymes. This property can be exploited to produce improved or new biocatalysts with desired features through enzyme or protein engineering (Quin & Schmidt-Dannert, 2011).

Enzymes exhibit the various properties such as eco-friendly in nature, work under normal pressure, temperature, and pH, consume less energy, cost-effective, do not produce any greenhouse gases, show high activity and turnover number, high biodegradability, and unsurpassed specificity (exhibit group specificity, absolute specificity, and stereospecificity). Moreover, enzymes produce specific products with no or minimal byproducts or wastes (biodegradable and non-toxic), consequently, reduce the need for complex downstream processes. This all will be helpful in the development of sustainable food system having positive impact on the environment. Because of these properties they are preferred over inorganic catalysts or chemicals in food and other industries; therefore, industrial enzymes market has grown from 0.31 billion USD in 1960 to 6 billion USD in 2020, which is expected to grow to 9 billion USD by 2027 (Okpara, 2022). Consequently, to fulfill the industrial demand of enzymes, many enzyme producing companies such as DSM, DuPontTM, Novozymes, Advanced Enzyme Technologies, Megazyme, Biocatalysts Ltd., MetGen, and many others came into existence. But among the over 3000 known enzymes only 5% enzymes have been used in industries (Okpara, 2022).

## 10.2 Sources of Enzymes

Enzymes in food industries can be taken from animals, plants, and microorganisms (fungi and bacteria). In industries, microbes are given priority for the isolation of enzymes due to following reasons: (1) Microbes are easily accessible; (2) They grow at faster speed; (3) They are easy to grow in cheap growth media; (4) Most enzymes (plants, animals, and microbial sources) work optimally at temperature 37 °C, pH 7, and in the absence of inhibitors. But sometimes industrial production conditions are not optimum for the working of enzymes. Temperature, pressure, salinity, pH, or presence of inhibitors during the production process could affect the enzyme activity. In this case microbes can be manipulated through genetic engineering to produce improved or novel enzymes that can sustain and work optimally at different industrial conditions. Alternatively, enzymes can be extracted from microorganisms that can grow at extreme conditions such as extreme temperature, salinity, or pH level; (5) Production of microbial enzymes through fermentation process is cost-effective; (6) Microbial enzymes can be produced in bulk amount in minimal time and space; and (7) Microbial enzymes are free from harmful phenolic compounds as in case of plant derived enzymes and endogenous enzyme inhibitors and protease as in case of animal derived enzymes; and (8) Microbial enzymes are more stable (Patel et al., 2017; Singh et al., 2019; Okpara, 2022; Littlechild, 2015). In industries more than 100 enzymes are used, out of which more than 50% enzymes have been taken from microorganisms. *Aspergillus niger*, *A. oryzae*, *Mucor, Saccharomyces cerevisiae*, *Serratia, Bacillus subtilis*, *B. licheniformis, B. amyloliquefaciens*, *Lactobacillus casei*, *L. acidophilus*, *L. delbrueckii*, *Rhizopus oryzae, Corynebacterium glutamicum* are some common microbial species that are employed in the generation of many commercial enzymes. Enzymes such as lipase, amylase, protease, rennet, pectinase, invertase, cellulase, glucose oxidase, raffinose, catalase, lactase, etc. are obtained from microbial sources and are used in food production and processing industries (Singh et al., 2019).

## 10.3 Need for Microbial Enzymes in the Food Industries

Biological agents have been used by our ancestors in processing of various food products, viz. tenderization of meat by papaya leaves, making of wine, beer, cheese, curd, soy sauce, vinegar, etc. Traditionally acid hydrolysis method has been used in food industries to hydrolyze starch for the preparation of glucose syrup. Acids such as hydrochloric acid, sulfuric acid, formic acid, nitric acid, and trifluoroacetic acid have been employed for acid hydrolysis of starch. This method is simple and cheap but has few drawbacks such as relatively low yield, high process temperature (120–150 °C), formation of undesirable products, viz. furfural generated from pentose sugar and hydroxymethylfurfural generated from hexose sugar and use of acid can cause serious environmental issues (Azmi et al., 2017). In the late 1990s this traditional method of starch hydrolysis was replaced by enzymatic method where enzymes, viz. amylase and glucoamylase (amyloglucosidase) have been used (Crabb & Shetty, 1999; Azmi et al., 2017; Haq et al., 2010). Thereafter, demand of enzymes in food industries for production and manufacturing of products has increased rapidly. Furthermore, development of different technologies such as protein engineering, enzyme immobilization, and in 1980s recombinant DNA technology have made a huge impact on the applications and uses of enzymes in food industries (Liu et al., 2013). Of all industries, food industries are largest market for enzymes where approx. 55% of industrial enzymes are used (Guerrand, 2017). In 2020 global food enzymes market was approx. 2.3 billion USD which was valued approx. 40% of all the industrial enzyme market and by 2026 it is expected to grow to 3.3 billion USD (Research and Markets, 2021).

## 10.4 Microbial Enzymes: Sources and Applications in Food Industries

Even before the discovery of enzymes, they have been used by our ancestors in the production of food many products, viz. meat tenderization by papaya leaves, preparation of soy sauce, beer, bread, curd, cheese, wine and vinegar, and even today they are used routinely in the preparation of many products in both households and industries such as dairy, baking, brewing, beverages, juice, oil refinery, food packaging, etc. They enhance the flavor, palatability, and improve the quality of the product. In food industries enzymes' main role is the processing of the food products.

### *10.4.1 α-Amylase (EC 3.2.1.1)*

Amylose and amylopectin are the two main components of starch. Amylose is linear and made up of glucose units that attached to each other by α-1,4 glycosidic bond, whereas amylopectin is branched and made up of glucose units that are attached to each other by α-1,4 and α-1,6-glycosidic bonds. α-Amylase (endoamylase) breaks the α-1,4 glycosidic bond of the starch and releases glucose, maltose, maltotriose, and short chain dextrins.

#### 10.4.1.1 Sources

Three types of amylases have been found in nature named as α-amylase, β-amylase, and γ-amylase. α-Amylases can be obtained from archaea, fungi, bacteria, plants, animals, and humans, β-amylases can be obtained from plants and microbes, and γ-amylases can be obtained from animals and plants (Azzopardi et al., 2016). Among microbial sources *Bacillus* spp. are exploited most for the commercial production of α-amylase. *B. amyloliquefaciens, B. licheniformis, B. stearothermophilus, Anoxybacillus* sp. AH1, *Geobacillus thermoleovorans, Bacillus* sp. BCC 01-50, and *Bacillus subtilis* JS-2004 are used widely for the isolation of α-amylase at industrial scale (Elyasi Far et al., 2020; Okpara, 2022). Thermostable α-amylase is extracted from *Bacillus* spp. such as *B. licheniformis, B. subtilis, B. amyloliquefaciens,* and *B. stearothermophilus* (Prakash & Jaiswal, 2010). In the processing of starch some steps such as saccharification, gelatinization, and liquefaction are done at high temperature; therefore, thermostable α-amylases are used in such harsh conditions. Salt tolerant α-amylase can be taken from halophilic bacteria such as *Halobacillus* sp., *Halomonas meridian, Bacillus dipsosauri, Chromohalobacter* sp., and *Haloarcula hispanica*. Salt tolerant enzyme is used in food processes where salt concentration is high (Singh et al., 2019). *Aspergillus* spp. such as *A. oryzae, A. terreus* NCFT4249.10, *A. awamori, A. fumigatus* KIBGE-$IB_{33}$, *A. niger* WLB42 are some examples of α-amylase producing strains (Okpara, 2022). α-Amylases of *Aspergillus* spp. are less thermostable as compared to α-amylases from *Bacillus* spp. (Okpara, 2022). Few *Penicillium* sp. are also used for the production of α-amylase.

#### 10.4.1.2 Applications

In food industries α-amylases are widely used in baking, brewing, starch liquefaction, preparation of livestock feed, fruit juice, starch syrups, and digestive aids. In early nineteenth century acid hydrolysis method was developed for the production of glucose syrup from starch and till 1990s diluted acid hydrolysis method was employed commercially to produce glucose syrup from starch. In late 1990s enzymatic hydrolysis method was developed (Crabb & Shetty, 1999; Haq et al., 2010; Azmi et al., 2017). Commercial production of glucose syrup from starch is a two-step process. First step is the liquefaction where short chain dextrins are produced by α-amylase of *B. stearothermophilus, B. licheniformis, or B. amyloliquefaciens.* Second step is saccharification where starch hydrolysate is converted to high concentration glucose syrup by using exo-glucoamylases which is obtained from *A. niger* (Haq et al., 2010). Glucose syrup can be converted to fructose syrup by using glucose isomerase. Fructose syrup is employed as sweetener and added to many processed products such as soft drink, fruit drink, yogurts, and breads (Parker et al., 2010).

During bread making α-amylase enzyme is used to hydrolyze the starch in flour into fermentable sugars. Later these sugars are used to produce $CO_2$ by yeast fermentation. Production of $CO_2$ causes the leavening or rising of the dough. Staling (reduced moisture and crispiness in bread) and starch crystallization cause the huge economic loss to the baking industries. Addition of α-amylase along with other enzymes and chemicals reduces the starch crystallization and staling of the bread.

Moreover, shelf-life, aroma, taste, softness, and quality of bread are improved (Singh et al., 2019; Okpara, 2022). During baking enzymes should be added within the prescribed range as overdoses of enzymes can result in some undesirable changes in the bread. For example, excess use of α-amylase can cause the stickiness in the dough due to production of maltodextrin (Van Der Maarel et al., 2002).

Production of alcohols or biofuels from starch is a demand of today's scenario. Conventional methods for starch liquefaction (acid hydrolysis) require strong chemicals such as caustic soda, lime, sulfuric acid, etc. and 120–150 °C temperature. Maintenance of temperature and pH in conventional methods is a challenge and adds additional cost to the industries (Robertson et al., 2006; Singh et al., 2019). Therefore, in industries enzymatic methods for starch liquefaction are preferred. Simple and fermentable sugars can be obtained from starch by the action of α-amylase. Later these sugars are fermented into alcohols or biofuels by using yeast. Yeast (GRAS category, high productivity, and better tolerance) is commonly used in the industries for the production of alcohol or bioethanol. Fermentation of raw starch into ethanol can be improved by the use of α-amylase in combination with glucoamylase. α-Amylase isolated from *Streptococcus bovis* and glucoamylase isolated from *Rhizopus oryzae* form the most effective combination (Singh et al., 2019). α-Amylase is also used as clarifying agents during the production of beer (Okpara, 2022).

Ingredients present in the feed are not fully digested and absorbed by livestock. Consequently, ingredients of the feed without being utilized fully are passed on into the feces of animals. Therefore, feed industries are adding some enzymes, viz. α-amylase, xylanase, phytase, and protease into the feed of animals. α-Amylase breaks down the starch into glucose units which results in more digestibility of carbohydrates by animals. This results in improvement of body weight of animals, feed conversion ratio, and milk production (Silva et al., 2006; Sidkey et al., 2011; Jegannathan & Nielsen, 2013). In juice industries use of α-amylase is prescribed for improving the juice extraction and its yield. It also acts as clarifying agent in the generation of juice (Vaillant et al., 2001; Sivaramakrishnan et al., 2006).

### *10.4.2 Protease (EC 3.4)*

These are proteolytic enzymes which break down the peptide bonds in the proteins resulting in generation of amino acids and smaller peptides. Allergens in the food are mostly proteins. As proteases degrade the proteins, they can be used to reduce the allergenic properties of food products. On the basis of site of cleavage proteases are categorized into endoproteases and exoproteases (aminopeptidases and carboxypeptidases) and on the basis of mechanism of catalysis proteases are categorized into six classes, viz. metalloproteases, glutamic proteases, aspartic proteases, threonine proteases, serine proteases, and cysteine proteases. Glutamic proteases are yet to be discovered in mammals. Aspartic proteases, glutamic proteases, and metalloproteases use water molecules as nucleophile to break peptide bond, while threonine proteases, serine proteases, and cysteine proteases use amino acids (threonine, serine, and cysteine, respectively) present in their active sites as nucleophile to break peptide bond (López-Otín & Bond, 2008). On the basis of pH optima proteases are categorized into acidic proteases, neutral proteases, and alkaline proteases.

#### 10.4.2.1 Sources

Proteases are found in animals, plants, archea, fungi, and bacteria. In food industries, fungi and bacteria are considered preferred sources of proteases. Of the total global enzymes sales microbial proteases contribute to approx. 40% (Singh et al., 2019). Fungi such as *A. usamii*, *A. niger*, *A. flavus*, *A. fumigatus*, *A. oryzae* DRDFS13MN726447, etc. and bacteria such as *B. subtilis* SMDFS 2B MN715837, *B. licheniformis*, *Chryseobacterium* sp., etc. are exploited for the production of proteases in food industries (Okpara, 2022). *Bacillus* produces the neutral and alkaline proteases, while *Pseudomonas* produces the alkaline protease. Neutral proteases of bacteria are heat sensitive and show activity in the pH range of 5–8. Because of their moderate activities neutral proteases cause the little bitterness in hydrolyzed food proteins when compared to animal proteases; therefore, they are preferred in the food industries. Different strains of *Pseudomonas aeruginosa* are used for the extraction of variety of proteases. Fungi produce more varieties of proteases as compared to bacteria, for example, acid proteases, neutral proteases, and alkaline proteases are produced by *A. oryzae* (Singh et al., 2019). Common proteases produced by plants are papain, bromelain, keratinases, and ficin.

#### 10.4.2.2 Applications

Proteases are used in baking, dairy, food processing, brewing, meat, and animal feed industries. Peptides generated due to partial hydrolysis of proteins by proteases may have some biological functions and can be used in food as functional component. In dairy industries proteases are mainly used for making of cheese. In making of cheese proteases (rennin or rennet) are used to hydrolyze the peptide bond between $Phe^{105}$ and $Met^{106}$ in kappa casein to produce para-k-casein and glycomacropeptide during milk coagulation. Organoleptic and rheological properties of cheese are also improved by proteases. Four types of proteases are used for coagulation/curdling of milk, viz. animal rennet, vegetable rennet, microbial proteases, and genetically engineered chymosin. Chymosin is preferred in dairy industries for making cheese because of its high specificity for casein (Singh et al., 2019; Okpara, 2022). Recently, Mamo et al. (2020) reported that organoleptic properties of cheese were improved by milk clotting protease originated from *A. oryzae* DRDFS13MN726447.

Proteases extracted from *A. oryzae* have been used in the baking industry since 1952 (Lyons, 1982). More the gluten content in the dough more it will be difficult to stretch the dough, consequently, bread will be tough and chewy. Chemical like sodium bisulfite can be used to weaken the gluten structure but it affects the nutritional properties of dough. Therefore, in bakery industries proteases are used to weaken the structure of gluten. Moreover, use of proteases does not affect the nutritional properties of the dough (Okpara, 2022). In bakery industries, gluten-free products are prepared using proteases for individuals who are intolerant to gluten. Acid proteases extracted from *A. usamii* have been used to improve the rheological properties of wheat gluten used for baking (Deng et al., 2016).

Proteins in the alcoholic beverages (e.g. beer) cause the formation of haze; therefore, microbial proteases are used in brewing industries to prevent haze formation. Moreover, proteases cause the breakdown of proteins of the wort into smaller peptides and/or amino acids. This leads to improvement of fermentation process and the overall quality of beer or whiskey (Singh et al., 2019; Okpara, 2022).

Connective tissues, for example, ligaments, tendons, silver skin, and muscle fibers are present in the raw meat. It causes the meat hard to chew, consequently, meat palatability decreases. Use of papaya leaves (source of protease) for meat tenderization has been a practice of long time. In today's scenario, meat industries are using various proteases extracted from microorganisms for meat tenderization, e.g. aspartic protease from *A. oryzae*, collagenase from *Clostridium histolyticum*, caldolysin from *Thermus* strain, thermophile protease from *Bacillus*. Meat tenderization causes the improvement of rheological properties of meat (Bekhit et al., 2014).

Aspartame or Nutrasweet is a non-saccharide based artificial sweetener that is made up of only two amino acids, viz. L-aspartic acid and L-phenylalanine. As diabetes is growing globally, demand of aspartame and other zero-calorie sweeteners is increasing exponentially. Therefore, 3000–6000 metric tons of aspartame is produced every year by industries. Because aspartame can only be produced by amino acids in L-configuration and maintenance of stereospecificity chemically adds additional cost to the industries. Therefore, synthesis of aspartame by enzymes is recommended as enzymes are stereospecific in nature (Singh et al., 2019). In animal feed industries proteases are added for the breakdown of proteins present in the animal feed. It improves the digestibility and utilization of proteins by animals and reduces the excretion of nitrogen content from the livestock (Okpara, 2022).

### *10.4.3 Lipase (EC 3.1.1.3)*

Lipases causes the breakdown of triglycerides into its components, viz. free fatty acids and glycerol. Lipases aid in proper digestion of fats and lipids in animals including humans.

#### 10.4.3.1 Sources

Lipases are widely distributed enzymes and can be found in all living organisms. In humans and animals, lipases are produced by pancreas and stomach and can be found in blood, pancreatic secretions, intestinal juices, gastric juices, and adipose tissue. In plants lipases are found in lipid storing tissue such as seeds. High lipase content is found in seeds of plants that belong to Euphorbiaceae, Ranunculaceae and Papaveraceae families. Barley, corn, and cotton seeds are used frequently for extraction of lipase enzyme (Singh et al., 2019) In industries microbes are given priority for the isolation of enzymes as microbes are easily available, microbial enzymes have high stability and are cheap. For industrial applications lipases are extracted from various species of fungi such as *Rhizomucor miehei*, *A. niger* F0215, *A. repens*, *A. oryzae*, *Penicillium camemberti*, *Candida rugosa*, *Mucor javanicus*, *Thermomyces lanuginosus*, and various species of bacteria such as *Bacillus subtilis* LP2, *Staphylococcus aureus*, *S. caprae* NCU S6, *B. cepacia*, *B. megaterium*, *P. aeruginosa* JCM5962(T), *Burkholderia glumae*, *P. alcaligenes*, *P. mendocina*, *S. hyicus* , *S. simulans* PMRS35, *Rhodothermus marinus*, *Serratia marcescens* (Okpara, 2022).

#### 10.4.3.2 Applications

In baking industries lipase is used either along with or in place of traditional emulsifiers to release emulsifying lipids in situ through hydrolysis of wheat lipids. They are employed to enhance the flavor of baked products through production of esters of short chain fatty acids. Along with other enzymes used in the baking industries lipases are used to enhance the loaf volume, shelf-life, texture, and softness of baked products (Guerrand, 2017). Egg white is needed for the production of many baked products. Lipid present (0.02%) in egg white can deteriorate the quality of the dough. Therefore, lipases are needed to decrease the lipid content of the egg white through hydrolysis. This leads to improvement of quality of baked products. In some baked products lipases are added as preservative too (Okpara, 2022).

In dairy industries, lipases from various sources are used in the making of varieties of the cheese, viz. Italian cheese, Cheddar, Feta, Manchego, Rumi, Ras, Domiati, blue cheese, etc. Production of free fatty acids through action of lipases on milk fat leads to the generation of cheese and related dairy products. Moreover, production of free fatty acids promotes cheese ripening, increases flavor, taste, texture, and softness of the cheese (Aravindan et al., 2007; Okpara, 2022).

In alcoholic industries lipases are employed for the breakdown of lipids present in barley seeds to produce free fatty acids. Lipid hydrolysis enhances the aroma of alcoholic beverages, for example, sake. For improving the alcoholic content and aroma in apple wine, patented lipase extracted from *R. delemar* or *Candida* sp. have been used for fermentation by Japanese company Tanabe Seiyaku (Singh et al., 2019). In meat industries, lipases are employed to produce lean meat by removal of excess fat. Flavor of meat products can be enhanced by the use of lipases. Lipases of micrococcaceae and lactobacilli are used for the improvement of flavor and ripening of dry sausages (Singh et al., 2019).

In tea industries lipases are employed to improve the flavor and aroma of tea. For example, lipases isolated from *Rhizomucor miehei* have been found to improve the production of flavored compounds that contributes to the development of flavor and aroma in the tea (Ramarethinam et al., 2002).

In oil industries lipases are employed as degumming agents to eliminate the phosphatides from crude vegetable oils. This process enhances the production and quality of refined oils. Treatment of crude vegetable oils by lipases also enhances the organoleptic properties of refined oils through production of flavor esters (Okpara, 2022). A novel lipase called An-lipase isolated from *A. niger* F0125, and was used to produce butyl butyrate, ethyl lactate and ethyl caprylate flavour esters in crude soybean oil (Cong et al., 2019). Lipase treatment also enhances the content of ω-3 polyunsaturated fatty acids in some oils such as salmon and sardine oil (Okada & Morrissey, 2007; Kahveci & Xu, 2011).

### *10.4.4 Rennet*

Rennet is a mixture of enzymes, viz. lipase, pepsin, and chymosin. It is produced in the stomach of weaning ruminants for the proper digestion of mother's milk. Casein makes up the 80% of total milk protein and is found in milk in the form of micelle. Three types of casein are found in the milk, viz. alpha, beta, and kappa casein. Of total casein alpha casein content is highest and accounts approx. 55%, beta casein accounts approx. 30%, while kappa casein is present in least amount and accounts approx. 15%. But depending upon the size of micelle proportion of each type of casein varies. Rennet or rennin hydrolyzes the peptide bond of kappa casein between $Phe^{105}$ and $Met^{106}$ and releases macropeptide with *N*-acetyl neuraminic acid (glycomacropeptide) and para-k-casein (Morr, 1975; Michael Eskin & Douglas Goff, 2013).

#### 10.4.4.1 Sources

Rennet can be obtained from microorganisms, plants, and mammals. Initially, rennet obtained from mammals was used exclusively for making all kinds of cheese. But with increase in the demand of cheese and need to derive rennet from non-animal sources has pushed industrialist to get rennet from alternative sources. Plants derived proteases can also be used as milk coagulants. Kosher and halal cheeses can be produced using plants derived rennet. However, all kosher cheese are being produced using microbial derived rennet. Globally, currently approx. 30% of the cheese is produced using rennet obtained from microorganisms. Production of rennet from microorganisms is advantageous over rennet derived from animals. Microbial rennet is having low cost, can be produced in bulk, has unlimited availability, and as in case of animal rennet, chances of disease transmission are low. Commercially, rennet can be obtained from microorganisms, viz. *R. miehei*, *R. pusillus*, *A. oryzae*, *Irpex lactis*. Rennet obtained from microorganisms and plants have high proteolytic activity when compared to animal's rennet. High proteolytic activity of rennet, consequently, decreases the yield as well as taste of cheese and increases the cheese bitterness due to generation of bitter peptides during cheese ripening process. Consequently, people have started producing animal rennet (calf chymosin) using genetic engineering. Calf rennet or chymosin have been produced in genetically modified organisms such as fungi or yeasts (Singh et al., 2019).

#### 10.4.4.2 Applications

Major application of rennet in dairy industries is the production of cheese. Globally, hundreds of varieties of cheese are produced through fermentation. Type of cheese depends upon many factors, viz. type of milk, diet of animal, fat content of milk, type of microorganism used (rennet source), and processing and aging conditions of cheese. It is good source of proteins, minerals, and vitamins.

### *10.4.5 Catalase (EC 1.11.1.6)*

Catalase is one of the three antioxidant enzymes working in human body. The function of catalase is to remove or detoxify free radicals generated by metabolic reactions. It is a tetrameric enzyme and made up of 4 identical subunits of molecular weight 60 kDa each. Turnover number of catalase is highest of all known enzymes. In 1 s, catalase can degrade 2.8 million $H_2O_2$ molecules to produce $H_2O$ and $O_2$.

#### 10.4.5.1 Sources

All aerobic organisms like plants, animals, and microbes are equipped with catalase. Mostly microorganisms are exploited for the generation of catalase commercially. *A. niger*, *Pyrobaculum calidifontis*, *Micrococcus luteus*, *Rhizobium radiobacter* 2-1, *Ureibacillus thermosphaericus* FZSF03, *Bacteroides fragilis*, *Bacillus maroccanus*, *Enterococcus faecalis* are some examples of microbes exploited for the production of catalase at industrial scale (Okpara, 2022).

#### 10.4.5.2 Applications

Catalases are employed in dairy industries to eliminate peroxides from milk to avoid milk rancidity (Abada, 2018). For production of wine with lower alcohol contents catalase and glucose oxidase can be employed in combination. Combined activity of these enzymes decreases the alcohol content to 2% within 30 h but aroma profile and organoleptic properties of the wine affected significantly. Catalase and glucose oxidase in combination are also used to eliminate $O_2$ from wine to enhance its shelf-life (Röcker et al., 2016). Packaged food items are susceptible to spoilage due to oxidation of food. Therefore, catalases are needed for food packaging to avoid oxidation of food products. This treatment increases the shelf-life of packaged food items (Okpara, 2022). Along with other enzymes catalases are also used in small amount during production of cheese and processing of eggs.

### *10.4.6 Cellulase (EC 3.2.1.4)*

Cellulose is linear and made up of glucose units that are attached to each other by β-1,4 glycosidic bond. Cellulase breaks glycosidic bond in cellulose and releases the oligosaccharides, cellobiose, and glucose.

#### 10.4.6.1 Sources

For commercial production of cellulase microorganisms such as fungi and bacteria are preferred. Fungi such as *A. niger* and *Trichoderma reesei* and bacteria such as *B. subtilis* ABDR01, *Paenibacillus* spp., *Streptomyces* sp. strain J2 and *B. licheniformis* are exploited most for the production of cellulase at industrial scale (Okpara, 2022).

#### 10.4.6.2 Applications

During extraction of juice from fruits and vegetables cellulose and hemicellulose present in the fruits and vegetables cause the turbidity or cloudiness in the juice. Ultimately, quality of juices gets reduced. Therefore, cellulases along with other enzymes, viz. pectinases and hemicellulases are used for softening of fruits in juice industries. These enzymes hydrolyze the cellulose, hemicellulose, and pectin in the raw fruit and vegetables. This ultimately increases the extraction, clarification, stabilization, and yield of juices (Bhat, 2000). Cellulase, hemicellulase, and pectinase are also used in the wine industries for improving extraction, skin maceration, clarification, and overall quality of the wine.

Cellulase is also employed in the processing purees of vegetables and fruits. Cellulase treatment increases the yield and decreases the viscosity of purees. Animal feed is made up of complex polysaccharides such as lignin, pectin, hemicellulose, cellulose. For the improvement of feed utilization and digestibility and animal performance, cellulase and hemicellulase are used in processing of feed of ruminants. Cellulase treatment of feed of farm animals causes more consumption of diet and 5–25% more production of milk (Murad & Azzaz, 2010).

## *10.4.7 Lactase (EC 3.2.1.108)*

Lactose (milk sugar) is a disaccharide in which galactose and glucose units are attached to each other by β-1,4 glycosidic bond. Lactase (β-galactosidase) breaks the glycosidic bond of the lactose and releases galactose and glucose. Milk is used extensively by dairy industries for the generation of many food products; therefore, lactose hydrolysis is the prime concern for these industries. Microorganisms (yeast, fungi, and bacteria) are exploited most for the generation of lactase because they produce enzyme in more quantity and in cost-effective manner.

### 10.4.7.1 Sources

Among bacterial sources *Klebsiella oxytoca* ZJUH1705, *B. longum* BCRC 15708, *B. infantis* CCRC 14633, and *E. coli* and among fungal sources *A. oryzae, A. niger, Kluyveromyces lactis,* and *K. fragilis* are exploited industrially for the production of lactase (Okpara, 2022). Because of the high cost and toxicity concerns lactases from *E. coli* are not used in the food industries. Lactases of *Kluyveromyces lactis, A. oryzae,* and *A. niger* are safe for use in food industries. pH optima for fungal, yeast, and bacteria lactases are 2.5–4.5, 6–7, and 6.5–7.5, respectively. Therefore, yeast and bacteria lactases are ideal for hydrolysis of milk and sweet whey, while fungal lactases are ideal for hydrolysis of acid whey (Singh et al., 2019).

### 10.4.7.2 Applications

Normally lactase is secreted by the intestinal cells of mammals including humans. Individuals having less amount of intestinal lactase or lactase deficient are not able to digest the lactose present in the milk. Undigested lactose reaches the colon where gut microbiota causes the fermentation of lactose, consequently, short chain fatty acids and gases are generated in the colon. This results in complications such as diarrhea, tissue dehydration, bloating, gas, and pain. Therefore, lactose-intolerant individuals are advised to consume low-lactose or lactose-deficient milk (lactase-treated milk) instead of normal milk. Therefore, in dairy industries, lactases are used in the preparation of lactose-deficient or low-lactose milk or dairy products for lactose-intolerant people. Lactase-treated milk shows improved solubility, sweetness, and digestibility in lactose-intolerant people.

Lactases are also used in making frozen desserts, yogurt, and ice cream to enhance their creaminess, sweetness, and digestibility. Lactose crystallization in dairy products can be avoided by the use of lactases. Moreover, cheese prepared from lactase-treated milk ripens fast when compared to cheese prepared from normal milk. But high cost associated with lactase restricts its use in dairy industries.

### *10.4.8 Pectinase (EC 3.2.1.15)*

Pectinase catalyzes the breakdown of pectin, complex structural polysaccharides found in middle lamella and cell walls of plant cells to produce simpler carbohydrates such as galacturonic acid. A different category of these enzymes utilizes pineapple, tomato, apple, orange, orange peel, lemon pulp, and other citrus fruits as their substrate. On the basis of their mechanism of action they can be categorized as polygalacturonases—hydrolyzes α-1,4 glycosidic bond; pectolyase—cleaves α-1,4-D-galacturonan methyl ester; and pectin esterases—removes acetyl and methoxy group from pectin (Sudeep et al., 2020).

#### 10.4.8.1 Sources

Pectinases can be obtained mostly from fungi for industrial applications. Some bacterial strains (*Bacillus subtilis* ABDR01) produce high amount of pectinases (Yadav et al., 2020). *Penicillium* spp., *Streptomyces* spp., *Moniliella* SB9 (Jaradat et al., 2008), *A. niger* MTCC (Anand et al., 2017), *Aspergillus* spp. Gm, *Fusarium* spp. C, *Aspergillus* spp. T, and *Penicillium* spp. Lco (Sudeep et al., 2020). *Aspergillus kawachii* (Esquivel & Voget, 2004), *Aspergillus fumigatus* (Okonji et al., 2019) are some fungal species that are exploited most for the generation of pectinase at industrial scale.

#### 10.4.8.2 Applications

In fruit juice industries pectinases are used for removal of pectin from cell walls of fruits, consequently, pulp pressability, juice extraction, filterability, flavor, clarification, and yield get improved. In the cell wall of unripe fruit pectin and cellulose microfibrils are attached to each other. This bound pectin is water insoluble and provides hardness or rigidity to the cell wall of fruit. Upon fruit ripening pectin structure gets altered enzymatically, consequently, pectin becomes water soluble and its attachment with the cell wall get weaken and ultimately fruit softens. When fruit is pressed pectin is released into the juice that contributes to increase in the viscosity of fruit juice and pulp particles, while some pectin molecules remain attached to cellulose microfibrils that help in retention of water (Pifferi et al., 1989). Manual pressing of pectin-rich fruits produces viscous and cloudy juice that remains attached to pulp as a jellified mass (lower juice yield). Therefore, in fruit industries the use of pectinase is highly recommended. Pectinases break the jelly like structures in the pulp and increase the juice yield by improving the pulp pressability. Furthermore, pectinases treated juice is found to be less viscous and cloudy, and juice palatability increases (Kashyap et al., 2001). Pectinase treated fruits and vegetables mash provides a high juice extraction and pulp with good pressing characteristics (Soares et al., 2021).

Citrus fruit processing industries release pectinaceous substances into the wastewater. These substances become difficult to get decomposed by microorganisms during activated-sludge treatment. Therefore, pectinolytic enzymes are used to degrade the pectins present in wastewater of food processing industries and make it appropriate for decomposition by sludge treatment (Sharma et al., 2012; Praveen & Suneetha, 2014). Use of pectinolytic organisms during the treatment of activated sludge is an eco-friendly, cheap, and time-saving procedure (Samanta, 2021). *Erwinia carotovora* FERM P-7576 (soft-rot pathogen) is known to produce endo-pectate lyase. Treatment of pectinaceous wastewater with this strain has been found to be effective in removal of pectic substances (Tanabe et al., 1986). Alkalophilic *Bacillus* sp. GIR621 is also known to release endo-pectate lyase at pH 10.0. Removal of pectic substances from wastewater by this strain has been achieved by Tanabe et al. (1987). Due to the safety concerns pectolytic enzymes extracted from bacteria are used to remove pectic substances from the wastewater. Pectinases are also employed in wine industries to enhance the efficiency of process and quality of wine produced.

Traditional methods for isolation of oils from *Canola*, coconut germ, sunflower seeds, palm, kernel, olive, etc. commonly use hexane (potential carcinogen) as organic solvents. Therefore, in oil industries pectinases and other cell wall degrading enzymes are employed to disintegrate the cell wall of oil crops to extract the oil. Pectinase treatment improves oil extraction from different sources such as olives, dates, flaxseeds, and many more (Anand et al., 2020). An enzyme preparation (Olivex) taken from *A. aculeatus* has enzymes for the degradation of cellulose (cellulase), hemicellulose (hemicellulase), and pectin (pectinase). Oil extracted from this enzyme preparation has higher yield, improved stability and rich in vitamin E and polyphenols (Kashyap et al., 2001).

Alkaline pectinases have a very important role in pectin degradation and removing mucilaginous coats from the coffee bean. Pectinolytic microbes are employed to degrade the mucilaginous coat from coffee during fermentation, thereby improving the coffee quality (Sharma et al., 2012; Praveen & Suneetha, 2014; Bhardwaj et al., 2017). Enzyme preparation having cellulolytic, hemicellulolytic, and pectinolytic activities is employed to degrade the mucilage coat and pulpy layers of the coffee beans, thereby enhancing the coffee quality. Pretreatment of coffee beans with commercial enzyme preparations reduces the fermentation time of coffee production. In tea industries pectinases of fungi are employed for quicker fermentation of tea leaves. Foam-forming properties of properties of instant tea powders can be reduced by treatment with pectinase. This treatment improves the quality, aroma, color, and market price of tea (Kashyap et al., 2001; Praveen & Suneetha, 2014; Hassan & Ali, 2016). Along with other enzymes pectinases are employed in feed industries to process the feed rich in pectin to enhance digestibility of nutrients and performance of animals.

### 10.4.9 Xylanases (EC 3.2.1.8)

Xylan is a structural heteropolysaccharide found in the cell wall of plants. It makes up the significant part of dry weight (approx. 25–35%) of woody tissues and lignified tissues in dicots and monocots, respectively. In some grasses and cereal grain tissues it makes up approx. 50% of dry weight. Xylan is made up of linear chain of xylose units which are attached to each other by β-1,4 glycosidic bond. Arabinose, methyl glucuronic acid, glucuronic acid, and *O*-acetyl groups are also found attached to main chain of xylan as branches (Ebringerová & Heinze, 2000; de Vries & Visser, 2001; Heinze et al., 2004; Mussatto et al., 2008). Xylanases degrade polysaccharide xylan into monosaccharide xylose through breakdown of β-1,4 glycosidic bond. Majorly three enzymes are needed for the complete catalysis of xylan into xylose. (1) Endoxylanases make the cut in between the chain of xylan to xylo-oligosaccharides (XOS). (2) Exoxylanases remove the xylose from the non-reducing end of xylan. (3) β-xylosidases cleave XOS and xylobiose and produce xylose. Depolymerization of xylan for industrial applications requires a few more xylanolytic enzymes such as *p*-coumaric esterase, α-1-arabinofuranosidase, ferulic acid esterase, and acetylxylan esterase and α-glucuronidase (Collins et al., 2005; Walia et al., 2017; Bhardwaj et al., 2019).

#### 10.4.9.1 Sources

Xylanases can be obtained from actinomycetes, bacteria, and fungi. Commercially bacteria are used frequently for the isolation of xylanases. Xylanases work effectively at pH in between 5 and 9 and temperature in between 35 and 60 °C. *Clostridium acetobutylicum* (Walia et al., 2017), *Streptomyces* sp., *Pediococcus acidilactici* GC25 (Adiguzel et al., 2019), *B. subtilis* ABDR01 (Yadav et al., 2020), *B. licheniformis* DM5 (Ghosh et al., 2019), and *B. pumilus* (Chakdar et al., 2016) are some bacterial strains that are exploited most for the generation of xylanases at industrial scale. Fungi produce extracellular xylanases with a higher yield. Also fungal xylanases show higher activity as compared to bacterial and yeast xylanases. *Aspergillus japonicus*, *Penicillium occitanis* Pol6 (Driss et al., 2012), *Fusarium* sp., *Pichia pastoris* (de Queiroz Brito Cunha et al., 2018), etc. are some fungal sources of xylanases.

#### 10.4.9.2 Applications

In food industries xylanases find many applications such as fruit juice clarification, enhancing the overall quality cookies and breads, synthesis of prebiotics (xylooligosaccharides and arabinoxylooligosaccharides), etc.

Along with different enzymes such as α-amylase, arabinofuranosidase, laccase, and glucanase, xylanases are used to enhance the rheological properties of the bakery products. During making of bread xylanases are used to transform the water insoluble hemicellulose (arabinoxylan) into water soluble hemicellulose (arabinoxylan). Formation of water soluble arabinoxylan in the dough causes the increase in its volume. Moreover, more uniform and finer crumbs are formed and hardness of dough reduces. Also, xylanases treated dough does not stick to machine parts while making dough (Butt et al., 2008). Degradation of xylan by xylanases causes redistribution of water in the wheat flour. It makes the kneading easier and dough softer. Also it delays the formation of crumb and causes the dough to rise during baking of bread (Polizeli et al., 2005). Xylanases work as anti-staling agent and increases the shelf-life and quality of bread (Harris & Ramalingam, 2010). In 2018, xylanases were produced from genetically modified *Pichia pastoris* having xynBS27 gene of *Streptomyces* sp. It was observed that xylanases decreased the sugar content, hardness of bread, and increased the volume of bread (de Queiroz Brito Cunha et al., 2018). Xylanases are also used to enhance texture, taste, aroma, and quality of biscuits.

Naturally extracted juices are hazy or turbid due to existence of cellulose, hemicellulose, protein, pectin, lipid, and other components. Therefore, in juice industries various enzymes that degrade cell wall such as pectinase, cellulase, hemicellulose are employed. Treatment with cell wall degrading enzymes improves the clarity, extraction, yield, and palatability of fruit juices. Xylanases are used to degrade hemicelluloses present in raw juice. consequently, extraction, clarification, quality, and yield of juices get improved (Bhardwaj et al., 2019). Xylanases produced from *Bacillus stearothermophilus* were used for the clarification of citrus fruit juices (Dhiman et al., 2011). *Pediococcus acidilactici* GC25 derived endoxylanase was employed for clarification of different fruit juices (Adiguzel et al., 2019).

In the making of papad from black gram xylanases are used. In black gram content of arabinoxylan remains high that causes hardness in the papad dough, consequently, process of papad making becomes difficult. Addition of xylanases to black gram flour causes the breakdown of arabinoxylan that leads to reduction of papad hardness and requirement of water. Also oil requirement for frying of papad decreases after xylanases treatment while other characteristics (color, taste, or texture) of papad remains unaffected (Awalgaonkar et al., 2015).

Degradation of xylan by xylanases produces XOS. XOS are considered as prebiotics and added to many food products as food supplement. Prebiotics are known to confer many beneficial effects to the host. Prebiotics also known to increase the growth of probiotics (good bacteria) in the gut of humans. Therefore, prebiotics are recommended as functional food supplements to sustain healthy lifestyle.

Xylanases are also used in the processing of ruminants and non-ruminant animal feed. Due to high fiber content (cellulose and hemicellulose) it is difficult for ruminant and non-ruminant animals to digest plant-based feed. Therefore, xylanases are used to degrade hemicellulose present in the feed to enhance feed digestibility by these animals (Bhat, 2000).

During the production of beer xylanases are employed for the degradation of cell wall of barley. Xylanases treatment degrades the xylan present in the cell wall of barley into arabinoxylans and oligosaccharides. This treatment causes beer clarification.

### 10.4.10 Glucose Oxidase (EC 1.1.3.4)

Glucose oxidase (GOX) is the one of the members of oxidoreductase family of enzymes. GOX is made up of 2 identical subunits of molecular weight 80 kDa each. It is a flavoprotein having FAD at its active site. It uses $O_2$ as an electron acceptor and catalyzes the oxidation of D-glucose into D-glucono-δ-lactone and $H_2O_2$. D-Glucono-δ-lactone is hydrolyzed to D-gluconic acid by enzyme lactonase, whereas catalase breaks down $H_2O_2$ into $O_2$ and $H_2O$.

#### 10.4.10.1 Sources

Among the microbial sources fungi are exploited commonly for the generation of glucose oxidase commercially. GOX was firstly isolated from *A. niger* in 1928 (Wong et al., 2008). Different species of *Aspergillus* such as *A. carbonarius, A. niger, A. nidulans, A. tubingensis, A. oryzae,* and *A. terreus* are known for the generation of glucose oxidase (Kornecki et al., 2020). Many *Penicillium* species like *P. amagasakiense, P. purpurogenum, P. glaucum, P. notatum*, and *P. adametzil* are known for the generation of glucose oxidase (Khatami et al., 2021). Some other glucose oxidase producing fungal sources include *Mucor circinelloides* (Kornecki et al., 2020) and *Cladosporium neopsychrotolerans* (Ge et al., 2020). Many bacterial species are also exploited for production of glucose oxidase.

#### 10.4.10.2 Applications

Glucose oxidase is employed for various purposes in food industries. GOX is used to remove glucose and oxygen from the food products, thereby improve their shelf-life. In food processing industries GOX is used to produce D-glucono-δ-lactone from glucose molecules available in the food products. After production D-glucono-δ-lactone acts as preservative; therefore, shelf-life, stability, quality, and flavor of food products get improved. GOX is also used to decrease the content of glucose from the drinks, consequently, GOX treated drinks become appropriate for diabetic patients.

For the generation of wine with lower alcohol contents GOX and catalase can be used in combination. Combined activity of these enzymes decreases the alcohol content to 2% within 30 h but aroma profile and organoleptic properties of the wine affected significantly. GOX activity reduces the glucose content, thereby availability of glucose for the production of alcohol through anaerobic fermentation decreases. GOX and catalase in combination are also used to eliminate molecular oxygen from wine to enhance its shelf-life (Röcker et al., 2016).

In baking industries, GOX are used for making of strong dough and to enhance the bread volume. A novel GOX produced by researchers was also effective in enhancing the bread volume (Ge et al., 2020). Glucose oxidase is also used in food packaging industries as it has property to remove oxygen.

## *10.4.11 Laccase (EC 1.10.3.2)*

Laccases are copper metalloenzymes and have four copper atoms per molecule. They belong to oxidoreductase class of enzymes and to blue oxidase subgroup. They cause the oxidation of phenolic compounds, ascorbic acids, and aromatic amines. They are produced by bacteria, fungi, plants, soil algae, and some insects. Laccase of microbial sources degrades the lignin of wood to produce cellulose and hemicellulose, while laccase of plants synthesizes the lignin in plants.

### 10.4.11.1 Sources

Fungi are most exploited for production of laccase at industrial scale. *Funalia trogii*, *Trametes versicolor*, *Pleurotus eryngii*, *P. flabellatus*, *P. lampas* (Struch et al., 2016), *Abortiporus biennis* (Yin et al., 2017), *Pleurotus ostreatus* (Lettera et al., 2016) are some fungal sources of laccase. Some strains of bacteria, viz. *Bacillus licheniformis* are employed for the generation of recombinant laccase for use in industries.

### 10.4.11.2 Applications

Wastewater from olive-oil industries has high concentration of phenols (1.5–8.0 g/L). Therefore, laccase taken from fungal sources is utilized to oxidize the different phenolic compounds released into wastewater from olive-oil industries (Osma et al., 2010). In wine industries laccase is employed to eliminate the $O_2$ from the wine to improve its shelf-life (Okpara, 2022). Laccases are also used for stabilization of wine through control of phenolic compounds (Osma et al., 2010). During production of beer laccase is used for the oxidation of polyphenols, consequently, formation of haze is minimized, and beer clarification and stabilization get improved. Moreover, laccase acts as preservative through eliminating the molecular oxygen during the process of beer production. Laccases are also used in fruit juice industries either alone or in combination with cellulase and pectinase to enhance the juice yield and clarification. In 2016, a group of researchers have reported the reduction of phenol content in fruit juice by 45% through the use of immobilized laccase taken from fungal sources (Lettera et al., 2016). In 2017, another group of researchers have reported the efficient clarification of litchi juice through the use of thermostable laccase taken from *Abortiporus biennis* strain J2 (Yin et al., 2017). In baking industry, laccase is utilized to crosslink biological polymers, consequently, dough strength, stability, and rheological properties get improved (Manhivi et al., 2018). In dairy industries, laccase is utilized to crosslink milk proteins in skim milk to enhance the quality of yogurt (Mokoonlall et al., 2016; Struch et al., 2016).

### *10.4.12 Naringinases (EC 3.2.1.40)*

Naringin, limonin, and neohesperidin are most bitter substances found in the citrus fruits. In grapes naringin is the major component, and most bitter compound. Naringinases, an enzyme complex, have α-L-rhamnosidase and β-D-glucosidase activities. Naringinases catalyze the breakdown of naringin to simpler compounds like prunin, rhamnose, naringenin, and glucose due to its activities. Naringin is hydrolyzed first by α-rhamnosidase to produce prunin and rhamnose. Thereafter, prunin is hydrolyzed by β-glucosidase to produce naringenin and glucose. Therefore, naringinases are used in industries as debittering enzymes during the generation of citrus fruit juices and to enhance the taste, flavor, and aroma of fruit juices.

#### 10.4.12.1 Sources

Microorganisms mainly fungi and bacteria are extensively exploited for the isolation of naringinases commercially. Fungal sources are utilized largely for the production of high yield of enzyme. *Thermomicrobium roseum*, *Bacillus amyloliquefaciens*-D1 (Pegu et al., 2021), *Burkholderia cenocepacia* (Patil et al., 2019), *B. amyloliquefaciens* strain 11568 (Zhu et al., 2017), *Thermotoga maritima* MSB8, *Thermoclostridium stercorarium* DSM 8532, *Thermotoga neapolitana* Z2706-MC24, *Caldicellulosiruptor bescii* DSM 6725 (Baudrexl et al., 2019), *Cryptococcus albidus* (Borzova et al., 2018) are some examples of bacterial strains producing naringinase. *A. usamii*, *A. niger*, *A. oryzae*, *A. flavus*, *Penicillium decumbens*, *Rhizoctonia solani*, *Cochliobolus miyabeanus*, *Lasiodiplodia theobromae*, *Rhizopus nigricans*, *Coniothyrium diplodiella* (Patil et al., 2019) are some examples of fungal species producing naringinase.

#### 10.4.12.2 Applications

Citrus fruit juice industries use the naringinase as debittering enzyme and to enhance the taste, flavor, and aroma of fruit juice. In 2017, a group of researchers isolated the naringinases from *B. amyloliquefaciens* strain 11568 and used it to decrease the bitterness of citrus fruit juice through degradation of naringin present in citrus fruits (Zhu et al., 2017). Naringinases are also used by various food processing industries to synthesize different food additives such as sweeteners to enhance the taste and flavor of food. Kinnow peel is the major waste product of citrus fruit processing industries. Naringin is the major component in the peel of kinnow. Therefore, α-L-rhamnosidase activity of naringinase can be utilized to produce the L rhamnose

(Puri et al., 2011). L-Rhamnose has applications in pharmaceutical industries and can be used as a plant protective agent. Naringinase is used in wine industries along with arabinosidase for improving the aroma of wine.

## *10.4.13 Esterase (EC 3.1)*

Esterase is one of the members of hydrolase family of enzymes. They degrade the esters into alcohols and acids in aqueous solution. Esterases degrade the short chain acylglycerols into their components like glycerol and fatty acids rather than long chain acylglycerol, which makes them different from lipases. Feruloyl esterase is the one of the members of esterases family of enzymes. This enzyme produces the ferulic acid through the degradation of ester bond between ferulic acid (hydroxycinnamic acid) and different polysaccharides present in the plant cell wall. These enzymes are also useful in the management of waste as they can degrade the lignocellulosic biomass produced from the plants.

### 10.4.13.1 Sources

Bacteria are exploited most for production of esterases at industrial scale. *Lactobacillus acidophilus*, *L. farciminis*, *L. amylovorus*, *L. fermentum* (Xu et al., 2017), *Bacillus licheniformis* (Alvarez-Macarie & Baratti, 2000) are exploited to produce feruloyl esterase.

### 10.4.13.2 Applications

Esterases are mainly used in beverage industries for the making of beer, wine, alcohol, and fruit juices. Esterases and lipases are used to convert the low-value fat or oil into high value fat or oil through transesterification reactions. To improve the flavor and aroma in cheese and its related products esterase and lipase isolated from *L. casei* CL96 are used to hydrolyze the milk fat (Choi & Lee, 2001). Ferulic acid produced by the action of feruloyl esterase has found many applications in food industries. Ferulic acid is utilized to synthesize vanillin, an aroma compound. Vanillin being a major ingredient of vanilla is used to enhance the flavor of beverages (Gallage et al., 2014). Moreover, ferulic acid is employed in food industries as an additive in functional foods. In juice industries esterases are utilized to enhance the flavor and aroma of various fruit juices through modification of oil and fat in juices (Panda & Gowrishankar, 2005). A protease resistant feruloyl esterase was isolated from the microbes of cow rumen and was utilized to produce ferulic acid through hydrolysis of wheat straw. This enzyme is of great industrial significance as it has high pH resistance, thermal resistance, and protease resistance (Cheng et al., 2012).

## *10.4.14 Glucoamylase (EC 3.2.1.3)*

Glucoamylase is a type of exoamylase that releases the D-glucose from the non-reducing end of starch. It is also known as saccharifying enzyme and is found in all living organisms. D-Glucose generated by the action of glucoamylase can be used as substrate for various fermentation processes in food and beverage industries.

### 10.4.14.1 Sources

*Rhizopus oryzae* F-923 (Fadel et al., 2020), *A. niger* (Bagheri et al., 2014), and *Aspergillus awamori* (Coutinho & Reilly, 1997; Blanco et al., 2014) are common fungal sources for production of glucoamylase at industrial scale.

### 10.4.14.2 Applications

In confectionary industries glucoamylases are employed for making of glucose and/or fructose syrup for the making of candies. In baking industries glucoamylases are used to generate simple sugars from starch. Fermentation of these sugars by yeast produces $CO_2$ that causes the dough to rise. They also acts as anti-staling agent and enhance the quality of fluor, bread crust colour and high fibre baked products. In brewing industries glucoamylases are used for the production of simple sugars that are ultimately fermented by *S. cerevisiae* to produce ethanol.

## *10.4.15 Phospholipase (EC 3.1.1.4)*

Phospholipase causes the breakdown of phospholipids to produce fatty acids and other lipophilic compounds. Phospholipases are divided into two categories, acyl hydrolases and phosphodiesterases. Phospholipase A1, A2, and B are types of acyl hydrolase, while phospholipase C and D are types of phosphodiesterase.

### 10.4.15.1 Sources

Fungi are exploited most for the isolation of phospholipase commercially. Some fungal sources of production of phospholipase are *Fusarium oxysporum* (Su et al., 2017), *B. cereus* (Elena et al., 2017), *Streptomyces chromofuscus* (Cerminati et al., 2019), and *A. oryzae* (Wang et al., 2021).

#### 10.4.15.2 Applications

Phospholipases are mostly used in oil, dairy, and bakery industries. In dairy industries phospholipases are employed for the generation of cheese. They decrease the loss of milk fat in whey, consequently, cheese texture, flavor, aroma, and yield get improved (Lilbaek et al., 2006). Alike lipases, phospholipases are employed as degumming agent during the production of refined oil to improve its yield and quality. A chimeric enzyme named Lecitase ultra was prepared by fusing phospholipase A1 gene and lipase gene for degumming of vegetable oils. Phospholipase A1 was taken from *Fusarium oxysporum* while lipase gene was taken from *Thermomyces lanuginosus* (Virgen-Ortíz et al., 2019). Degumming of crude soybean oil by using alkaline cold active phospholipase C of *Aspergillus oryzae* has been reported elsewhere (Wang et al., 2021). A mutant version of phospholipase C (F66Y) isolated from *Bacillus cereus* has been used for degumming of soybean oil. This enzyme was able to remove almost 90% of phosphatidylethanolamine (Elena et al., 2017). Phospholipases are also employed to enhance the quality and shelf-life of sauces, mayonnaise, and baked products. Phospholipases are also employed to enhance the nutritive value of soya lecithin (fat containing animal feed).

### *10.4.16 Phytase (EC 3.1.3.8)*

Phytase is a member of phosphatase class of enzymes and removes the phosphorus from phytic acid present in grains and oilseeds.

#### 10.4.16.1 Sources

Phytase is widely distributed enzyme and found in all living organisms. However, microbes are exploited most for the generation of phytase at industrial scale. *A. ficuum*, *A. fumigatus*, *A. niger*, *K. oxytoca*, *K. terrigena*, *E. coli*, *B. amyloliquefaciens*, *B. subtilis*, *Schizosaccharomyces pombe* are some microbial species that are exploited for the isolation of phytase (Pandey et al., 2001; Ciofalo et al., 2003; Selle & Ravindran, 2008).

#### 10.4.16.2 Applications

In grains and oilseeds, minerals such as calcium, zinc, iron sometimes remain bound to phytic acid, consequently, bioavailability of these minerals to monogastric animals gets reduced. Therefore, to improve the bioavailability of minerals bound to phytic acid, phytases are mixed with the feed of monogastric animals. This leads to increase in nutritive value of animal feed (Ciofalo et al., 2003; Selle & Ravindran, 2008).

## 10.5 Conclusion and Future Perspectives

Enzymes are used routinely in food industries for the generation of wine, bread, juices, ice cream, cheese, beer, refined oil, gluten-free food, etc. Sometimes maintenance of optimum conditions for the working of enzymes during industrial production of products seems difficult and tedious. Therefore, innovative methods are being employed for the design of improved/new biocatalysts with desired features such as less sensitive to change in temperature and pH, less susceptible to presence of inhibitory agents, less or no requirement of cofactor and/or coenzyme, without affecting the activity of novel enzymes. Development of different technologies such as protein engineering and recombinant DNA technology has made a huge impact on the development of improved or novel biocatalyst.

## References

Abada, E. A. (2018). Application of microbial enzymes in the dairy industry. In M. Kuddus (Ed.), *Enzymes in food biotechnology: Production, applications, and future prospects* (pp. 61–72). Academic Press. https://doi.org/10.1016/B978-0-12-813280-7.00005-0

Adiguzel, G., Faiz, O., Sisecioglu, M., Sari, B., Baltaci, O., Akbulut, S., Genc, B., & Adiguzel, A. (2019). A novel endo-β-1,4-xylanase from *Pediococcus acidilactici* GC25; purification, characterization and application in clarification of fruit juices. *International Journal of Biological Macromolecules, 129*, 571–578.

Alvarez-Macarie, E., & Baratti, J. (2000). Short chain flavour ester synthesis by a new esterase from *Bacillus licheniformis*. *Journal of Molecular Catalysis B: Enzymatic, 10*(4), 377–383.

Anand, G., Yadav, S., Gupta, R., & Yadav, D. (2020). Pectinases: From microbes to industries. In P. Chowdhary, A. Raj, D. Verma, & Y. Akhter (Eds.), *Microorganisms for sustainable environment and health* (pp. 287–313). Elsevier. https://doi.org/10.1016/B978-0-12-819001-2.00014-0

Anand, G., Yadav, S., & Yadav, D. (2017). Production, purification and biochemical characterization of an exo-polygalacturonase from *Aspergillus niger* MTCC 478 suitable for clarification of orange juice. *3 Biotech, 7*(2), 122. https://doi.org/10.1007/s13205-017-0760-3

Aravindan, R., Anbumathi, P., & Viruthagiri, T. (2007). Lipase applications in food industry. *Indian Journal of Biotechnology, 6*(2), 141–158.

Awalgaonkar, G., Sarkar, S., Bankar, S., & Singhal, R. S. (2015). Xylanase as a processing aid for papads, an Indian traditional food based on black gram. *LWT - Food Science and Technology, 62*(2), 1148–1153.

Azmi, A. S., Malek, M. I. A., & Puad, N. I. M. (2017). A review on acid and enzymatic hydrolysis of sago starch. *International Food Research Journal, 24*, 265–273.

Azzopardi, E., Lloyd, C., Teixeira, S. R., Conlan, R. S., & Whitaker, I. S. (2016). Clinical applications of amylase: Novel perspectives. *Surgery, 160*(1), 26–37.

Bagheri, A., Khodarahmi, R., & Mostafaie, A. (2014). Purification and biochemical characterisation of glucoamylase from a newly isolated *Aspergillus niger*: Relation to starch processing. *Food Chemistry, 161*, 270–278.

Baudrexl, M., Schwarz, W. H., Zverlov, V. V., & Liebl, W. (2019). Biochemical characterisation of four rhamnosidases from thermophilic bacteria of the genera *Thermotoga, Caldicellulosiruptor* and *Thermoclostridium*. *Science Reports, 9*, 15924. https://doi.org/10.1038/s41598-019-52251-0

Bekhit, A. A., Hopkins, D. L., Geesink, G., Bekhit, A., & Franks, P. (2014). Exogenous proteases for meat tenderization. *Critical Reviews in Food Science and Nutrition, 54*(8), 1012–1031.

Bhardwaj, N., Kumar, B., & Verma, P. (2019). A detailed overview of xylanases: An emerging biomolecule for current and future prospective. *Bioresources and Bioprocessing, 6*, 40. https://doi.org/10.1186/s40643-019-0276-2

Bhardwaj, V., Degrassi, G., & Bhardwaj, R. K. (2017). Microbial pectinases and their applications in industries: A review. *International Research Journal of Engineering and Technology, 4*(8), 829–836.

Bhat, M. K. (2000). Cellulases and related enzymes in biotechnology. *Biotechnology Advances, 18*, 355–383.

Blanco, C. A., Caballero, I., Barrios, R., & Rojas, A. (2014). Innovations in the brewing industry: Light beer. *International Journal of Food Sciences and Nutrition, 65*, 655–660.

Borzova, N., Gudzenko, O., & Varbanets, L. (2018). Purification and characterization of a naringinase from *Cryptococcus albidus*. *Applied Biochemistry and Biotechnology, 184*, 953–969.

Butt, M. S., Tahir-Nadeem, M., Ahmad, Z., & Sultan, M. T. (2008). Xylanases and their applications in baking industry. *Food Technology and Biotechnology, 46*(1), 22–31.

Cerminati, S., Paoletti, L., Aguirre, A., Peirú, S., Menzella, H. G., & Castelli, M. E. (2019). Industrial uses of phospholipases: Current state and future applications. *Applied Microbiology and Biotechnology, 103*, 2571–2582.

Chakdar, H., Kumar, M., Pandiyan, K., Singh, A., Nanjappan, K., Kashyap, P. L., & Srivastava, A. K. (2016). Bacterial xylanases: Biology to biotechnology. *3 Biotech, 6*, 150. https://doi.org/10.1007/s13205-016-0457-z

Cheng, F., Sheng, J., Cai, T., Jin, J., Liu, W., Lin, Y., Du, Y., Zhang, M., & Shen, L. (2012). A protease-insensitive feruloyl esterase from China Holstein cow rumen metagenomic library: Expression, characterization, and utilization in ferulic acid release from wheat straw. *Journal of Agricultural and Food Chemistry, 60*(10), 2546–2553.

Choi, Y. J., & Lee, B. H. (2001). Culture conditions for the production of esterase from *Lactobacillus casei* CL96. *Bioprocess and Biosystems Engineering, 24*, 59–63.

Ciofalo, V., Barton, N., Kretz, K., Baird, J., Cook, M., & Shanahan, D. (2003). Safety evaluation of a phytase, expressed in *Schizosaccharomyces pombe*, intended for use in animal feed. *Regulatory Toxicology and Pharmacology, 37*, 286–292.

Collins, T., Gerday, C., & Feller, G. (2005). Xylanases, xylanase families and extremophilic xylanases. *FEMS Microbiology Reviews, 29*(1), 3–23.

Cong, S., Tian, K., Zhang, X., Lu, F., Singh, S., Prior, B., & Wang, Z.-X. (2019). Synthesis of flavor esters by a novel lipase from *Aspergillus niger* in a soybean-solvent system. *3 Biotech, 9*(6), 244. https://doi.org/10.1007/s13205-019-1778-5

Coutinho, P. M., & Reilly, P. J. (1997). Glucoamylase structural, functional, and evolutionary relationships. *Proteins, 29*(3), 334–347.

Crabb, W. D., & Shetty, J. K. (1999). Commodity scale production of sugars from starches. *Current Opinion in Microbiology, 2*(3), 252–256.

de Queiroz Brito Cunha, C. C., Gama, A. R., Cintra, L. C., Bataus, L. A. M., & Ulhoa, C. J. (2018). Improvement of bread making quality by supplementation with a recombinant xylanase produced by *Pichia pastoris*. *PLoS One, 13*(2), e0192996. https://doi.org/10.1371/journal.pone.0192996

de Vries, R. P., & Visser, J. (2001). Aspergillus enzymes involved in degradation of plant cell wall polysaccharides. *Microbiology and Molecular Biology Reviews, 65*, 497–522.

Deng, L., Wang, Z., Yang, S., Song, J., Que, F., Zhang, H., & Feng, F. (2016). Improvement of functional properties of wheat gluten using acid protease from *Aspergillus usamii*. *PLoS One, 11*(7), e0160101. https://doi.org/10.1371/journal.pone.0160101

Dhiman, S. S., Garg, G., Sharma, J., & Mahajan, R. (2011). Characterization of statistically produced xylanase for enrichment of fruit juice clarification process. *New Biotechnology, 28*(6), 746–755.

Driss, D., Bhiri, F., Siela, M., Ghorbel, R., & Chaabouni, S. E. (2012). Purification and properties of a thermostable xylanase GH 11 from *Penicillium occitanis* Pol6. *Applied Biochemistry and Biotechnology, 168*(4), 851–863.

Ebringerová, A., & Heinze, T. (2000). Xylan and xylan derivatives—Biopolymers with valuable properties, 1. Naturally occurring xylans structures, isolation procedures and properties. *Macromolecular Rapid Communications, 21*, 542–556.

Elena, C., Cerminati, S., Ravasi, P., Rasia, R., Peiru, S., Menzella, H. G., & Castelli, M. E. (2017). *B. cereus* phospholipase C engineering for efficient degumming of vegetable oil. *Process Biochemistry, 54*, 67–72.

Elyasi Far, B., Ahmadi, Y., Yari Khosroshahi, A., & Dilmaghani, A. (2020). Microbial alpha-amylase production: Progress, challenges and perspectives. *Advanced Pharmaceutical Bulletin, 10*(3), 350–358.

Esquivel, J. C. C., & Voget, C. E. (2004). Purification and partial characterization of an acidic polygalacturonase from *Aspergillus kawachii*. *Journal of Biotechnology, 110*, 21–28.

Fadel, M., AbdEl-Halim, S., Sharada, H., Yehia, A., & Ammar, M. (2020). Production of glucoamylase, α-amylase and cellulase by *Aspergillus oryzae* F-923 cultivated on wheat bran under solid state fermentation. *Journal of Advances in Biology & Biotechnology, 23*, 8–22.

Gallage, N. J., Hansen, E. H., Kannangara, R., Olsen, C. K., Motawia, M. S., Jørgensen, K., Holme, I., Hebelstrup, K., Grisoni, M., & Møller, B. L. (2014). Vanillin formation from ferulic acid in Vanilla planifolia is catalysed by a single enzyme. *Nature Communications, 5*, 4037. https://doi.org/10.1038/ncomms5037

Ge, J., Jiang, X., Liu, W., Wang, Y., Huang, H., Bai, Y., Su, X., Yao, B., & Luo, H. (2020). Characterization, stability improvement, and bread baking applications of a novel cold-adapted glucose oxidase from *Cladosporium neopsychrotolerans* SL16. *Food Chemistry, 310*, 125970. https://doi.org/10.1016/j.foodchem.2019.125970

Ghosh, A., Sutradhar, S., & Baishya, D. (2019). Delineating thermophilic xylanase from *Bacillus licheniformis* DM5 towards its potential application in xylooligosaccharides production. *World Journal of Microbiology and Biotechnology, 35*, 34. https://doi.org/10.1007/s11274-019-2605-1

Guerrand, D. (2017). Lipases industrial applications: Focus on food and agroindustries. *OCL-Oilseeds & Fats Crops and Lipids, 24*, D403. https://doi.org/10.1051/ocl/2017031

Haq, I. U., Muhammad, M. J., Uzma, H., & Fazal, A. (2010). Kinetics and thermodynamic studies of alpha amylase from *Bacillus licheniformis* mutant. *Pakistan Journal of Botany, 42*(5), 3507–3516.

Harris, A. D., & Ramalingam, C. (2010). Xylanases and its application in food industry: A review. *Journal of Experimental Sciences, 1*(7), 1–11.

Hassan, B., & Ali, S. (2016). A review on biotechnological impact of pectinases in industries. *Journal of Pharmaceutical, Chemical and Biological Sciences, 1*(2), 1–6.

Heinze, T., Koschella, A., & Ebringerová, A. (2004). Chemical functionalization of xylan: A short review. In *Hemicelluloses: Science and technology* (ACS symposium series) (pp. 312–325). American Chemical Society.

Jaradat, Z., Dawagreh, A., Ababneh, Q., & Saadoun, I. (2008). Influence of culture conditions on cellulase production by *Streptomyces sp.* (strain J2). *Jordan Journal of Biological Sciences, 1*(4), 141–146.

Jegannathan, K. R., & Nielsen, P. H. (2013). Environmental assessment of enzyme uses in industrial production—A literature review. *Journal of Cleaner Production, 42*, 228–240.

Kahveci, D., & Xu, X. (2011). Repeated hydrolysis process is effective for enrichment of omega 3 polyunsaturated fatty acids in Salmon oil by *Candida rugose* lipase. *Food Chemistry, 129*(4), 1552–1558.

Kashyap, D. R., Vohra, P. K., Chopra, S., & Tewari, R. (2001). Applications of pectinases in the commercial sector: A review. *Bioresource Technology, 77*, 215–227.

Khatami, S. H., Vakili, O., Ahmadi, N., Fard, E. S., Mousavi, P., Khalvati, B., Maleksabet, A., Savardashtaki, A., Taheri-Anganeh, M., & Movahedpour, A. (2021). Glucose oxidase: Applications, sources, and recombinant production. *Biotechnology and Applied Biochemistry, 69*(3), 939–950.

Kornecki, J. F., Carballares, D., Tardioli, P. W., Rodrigues, R. C., Berenguer-Murcia, A., Alcántarae, A. R., & Fernandez-Lafuente, R. (2020). Enzyme production of D-gluconic acid and glucose oxidase: Successful tales of cascade reactions. *Catalysis Science & Technology, 10*, 5740–5771.

Lettera, V., Pezzella, C., Cicatiello, P., Piscitelli, A., Giacobelli, V. G., Galano, E., Amoresano, A., & Sannia, G. (2016). Efficient immobilization of a fungal laccase and its exploitation in fruit juice clarification. *Food Chemistry, 196*, 1272–1278.

Lilbaek, H. M., Broe, M. L., Høier, E., Fatum, T. M., Ipsen, R., & Sørensen, N. K. (2006). Improving the yield of Mozzarella cheese by phospholipase treatment of milk. *Journal of Dairy Science, 89*, 4114–4125.

Littlechild, J. A. (2015). Enzymes from extreme environments and their industrial applications. *Frontiers in Bioengineering and Biotechnology, 3*, 161. https://doi.org/10.3389/fbioe.2015.00161

Liu, L., Yang, H., & Shin, H. D. (2013). How to achieve high-level expression of microbial enzymes strategies and perspectives. *Bioengineered, 4*(4), 212–223.

López-Otín, C., & Bond, J. S. (2008). Proteases: Multifunctional enzymes in life and disease. *The Journal of Biological Chemistry, 283*(45), 30433–30437.

Lyons, T. P. (1982). Proteinase enzymes relevant to the baking industry. *Biochemical Society Transactions, 10*(4), 287–290.

Mamo, J., Getachew, P., Kuria, M. S., & Assefa, F. (2020). Application of milk-clotting protease from *Aspergillus oryzae* DRDFS13 MN726447 and *Bacillus subtilis* SMDFS 2B MN715837 for Danbo cheese production. *Journal of Food Quality, 2020*, 8869010. https://doi.org/10.1155/2020/8869010

Manhivi, V. E., Amonsou, E. O., & Kudanga, T. (2018). Laccase-mediated crosslinking of gluten-free amadumbe flour improves rheological properties. *Food Chemistry, 264*, 157–163.

Michael Eskin, N. A., & Douglas Goff, H. (2013). Milk. In N. A. Michael Eskin & F. Shahidi (Eds.), *Biochemistry of foods* (3rd ed., pp. 187–214). Academic Press. https://doi.org/10.1016/B978-0-08-091809-9.00004-2

Mokoonlall, A., Sykora, L., Pfannstiel, J., Nobel, S., Weiss, J., & Hinrichs, J. (2016). A feasibility study on the application of a laccase-mediator system in stirred yoghurt at the pilot scale. *Food Hydrocolloids, 60*, 119–127.

Morr, C. V. (1975). Chemistry of milk proteins in food processing. *Journal of Dairy Science, 58*(7), 977–984.

Murad, H. A., & Azzaz, H. H. (2010). Cellulase and dairy animal feeding. *Biotechnology, 9*(3), 238–256.

Mussatto, S. I., Fernandes, M., Milagres, A. M. F., & Roberto, I. C. (2008). Effect of hemicellulose and lignin on enzymatic hydrolysis of cellulose from brewer's spent grain. *Enzyme and Microbial Technology, 43*, 124–129.

Okada, T., & Morrissey, M. T. (2007). Production of n-3 polyunsaturated fatty acid concentrate from sardine oil by lipase-catalyzed hydrolysis. *Food Chemistry, 103*(4), 1411–1419.

Okonji, R. E., Itakorode, B. O., Ovumedia, J. O., & Adedeji, O. S. (2019). Purification and biochemical characterization of pectinase produced by *Aspergillus fumigatus* isolated from soil of decomposing plant materials. *Journal of Applied Biology and Biotechnology, 7*(3), 1–8.

Okpara, M. O. (2022). Microbial enzymes and their applications in food industry: A mini-review. *Advances in Enzyme Research, 10*, 23–47.

Osma, J. F., Toca-Herrera, J. L., & Rodríguez-Couto, S. (2010). Uses of laccases in the food industry. *Enzyme Research, 2010*, 918761. https://doi.org/10.4061/2010/918761

Panda, T., & Gowrishankar, B. S. (2005). Production and applications of esterases. *Applied Microbiology and Biotechnology, 67*, 160–169.

Pandey, A., Szakacs, G., Soccol, C. R., Rodriguez-Leon, J. A., & Soccol, V. T. (2001). Production, purification and properties of microbial phytases. *Bioresource Technology, 77*, 203–214.

Parker, K., Salas, M., & Nwosu, V. C. (2010). High fructose corn syrup: Production, uses and public health concerns. *Biotechnology and Molecular Biology Reviews, 5*(5), 71–78.

Patel, A. K., Singhania, R. R., & Pandey, A. (2017). Production, purification, and application of microbial enzymes. In G. Brahmachari (Ed.), *Biotechnology of microbial enzymes:*

*Production, biocatalysis and industrial applications* (pp. 13–41). Academic Press. https://doi.org/10.1016/B978-0-12-803725-6.00002-9

Patil, S. V., Koli, S. H., Mohite, B. V., Patil, R. P., Patil, R. R., Borase, H. P., & Patil, V. S. (2019). A novel screening method for potential naringinase-producing microorganisms. *Biotechnology and Applied Biochemistry, 66*, 323–327.

Pegu, B. K., Kardong, D., & Gogoi, D. K. (2021). Purification and characterization of α-L-rhamnosidase from *Bacillus amyloliquefaciens*-D1. *Asian Journal of Biological and Life Sciences, 10*, 454–461.

Pifferi, P. G., Buska, G., Manenti, I., Lo Presti, A., & Spagna, G. (1989). Immobilization of pectinesterase on c-alumina for the treatment of juices. *Belgian Journal of Food Chemistry and Biotechnology, 44*, 173–182.

Polizeli, M. L. T. M., Rizzatti, A. C. S., Monti, R., Terenzi, H. F., Jorge, J. A., & Amorim, D. S. (2005). Xylanases from fungi: Properties and industrial applications. *Applied Microbiology and Biotechnology, 67*, 577–591.

Prakash, O., & Jaiswal, N. (2010). Alpha-amylase: An ideal representative of thermostable enzymes. *Applied Biochemistry and Biotechnology, 160*(8), 2401–2414.

Pravda, L., Berka, K., Svobodová Vařeková, R., Sehnal, D., Banáš, P., Laskowski, R. A., Koča, J., & Otyepka, M. (2014). Anatomy of enzyme channels. *BMC Bioinformatics, 15*(1), 379. https://doi.org/10.1186/s12859-014-0379-x

Praveen, K. G., & Suneetha, V. (2014). A cocktail enzyme-pectinase from fruit industrial dump sites: A review. *Research Journal of Pharmaceutical, Biological and Chemical Sciences, 5*(2), 1252–1258.

Puri, M., Kaur, A., Schwarz, W. H., Singh, S., & Kennedy, J. F. (2011). Molecular characterization and enzymatic hydrolysis of naringin extracted from kinnow peel waste. *International Journal of Biological Macromolecules, 48*, 58–62.

Quin, M. B., & Schmidt-Dannert, C. (2011). Engineering of biocatalysts—From evolution to creation. *ACS Catalysis, 1*(9), 1017–1021.

Ramarethinam, S., Latha, K., & Rajalakshmi, N. (2002). Use of a fungal lipase for enhancement of aroma in black tea. *Food Science and Technology Research, 8*(4), 328–332.

Research and Markets. (2021). *Food enzymes market, global forecast, impact of COVID-19, industry trends, by types, growth, opportunity company analysis.* Retrieved from https://www.researchandmarkets.com/reports/5411844/food-enzymes-market-global-forecast-impact-of#rela3-4801548

Robertson, G. H., Wong, D. W. S., Lee, C. C., Wagschal, K., Smith, M. R., & Orts, W. J. (2006). Native or raw starch digestion: A key step in energy efficient biorefining of grain. *Journal of Agricultural and Food Chemistry, 54*(2), 353–365.

Robinson, P. K. (2015). Enzymes: Principles and biotechnological applications. *Essays in Biochemistry, 59*, 1–41.

Röcker, J., Schmitt, M., Pasch, L., Ebert, K., & Grossmann, M. (2016). The use of glucose oxidase and catalase for the enzymatic reduction of the potential ethanol content in wine. *Food Chemistry, 210*, 660–670.

Samanta, S. (2021). Microbial pectinases: A review on molecular and biotechnological perspectives. *Journal of Microbiology, Biotechnology and Food Sciences, 2021*, 248–266.

Selle, P. H., & Ravindran, V. (2008). Phytate-degrading enzymes in pig nutrition. *Livestock Science, 113*, 99–122.

Sharma, N., Rathore, M., & Sharma, M. (2012). Microbial pectinase: Sources, characterization and applications. *Reviews in Environmental Science and Biotechnology, 12*(1), 45–60.

Sidkey, N. M., Maha, A., Reham, B., Reham, S., & Ghadeer, B. (2011). Purification and characterization of α-amylase from a newly isolated *Aspergillus flavus* F2Mbb. *International Research Journal of Microbiology, 2*(3), 96–103.

Silva, M. T. S. L., Santo, F. E., Pereira, P. T., & Poseiro, C. P. (2006). Phenotypic characterization of food waste degrading *Bacillus* strains isolated from aerobic bioreactors. *Journal of Basic Microbiology, 46*, 34–46.

Singh, R., Singh, A., & Sachan, S. (2019). Enzymes used in the food industry: Friends or foes? In M. Kuddus (Ed.), *Enzymes in food biotechnology* (pp. 827–843). Academic Press. https://doi.org/10.1016/B978-0-12-813280-7.00048-7

Sivaramakrishnan, S., Gangadharan, D., Nampoothiri, K. M., Soccol, C. R., & Pandey, A. (2006). α-Amylases from microbial sources—An overview on recent developments. *Food Technology and Biotechnology, 44*, 173–184.

Soares, M. M. C. N., Da Silva, R., Carmona, E. C., & Gomes, E. (2021). Pectinolytic enzyme production by *Bacillus* species and their potential application on juice extraction. *World Journal of Microbiology and Biotechnology, 17*, 79–82.

Struch, M., Krahe, N., Linke, D., Mokoonlall, A., & Berger, R. G. (2016). Dose dependent effects of a milk ion tolerant laccase on yoghurt gel structure. *LWT - Food Science and Technology, 65*, 1144–1152.

Su, L., Ji, D., Tao, X., Yu, L., Wu, J., & Xia, Y. (2017). Recombinant expression, characterization, and application of a phospholipase B from *Fusarium oxysporum*. *Journal of Biotechnology, 242*, 92–100.

Sudeep, K. C., Upadhyaya, J., Joshi, D. R., Lekhak, B., Chaudhary, D. K., Pant, B. R., Bajgai, T. R., Dhital, R., Khanal, S., Koirala, N., & Raghavan, V. (2020). Production, characterization, and industrial application of pectinase enzyme isolated from fungal strains. *Fermentation, 6*(2), 59. https://doi.org/10.3390/fermentation6020059

Tanabe, H., Kobayashi, Y., & Akamatsu, I. (1986). Pretreatment of pectic wastewater from orange canning by soft-rot *Erwinia carotovora*. *Journal of Fermentation Technology, 64*, 265–268.

Tanabe, H., Yoshihara, K., Tamura, K., Kobayashi, Y., Akamatsu, I., Niyomwan, N., & Footrakul, P. (1987). Pretreatment of pectic wastewater from orange canning process by an alkalophilic Bacillus sp. *Journal of Fermentation Technology, 65*(2), 243–246.

Vaillant, F., Millan, A., Dornier, M., Decloux, M., & Reynes, M. (2001). Strategy for economical optimization of the clarification of pulpy fruit juices using crossflow microfiltration. *Journal of Food Engineering, 48*, 83–90.

Van Der Maarel, M. J. E. C., Van Der Veen, B., Uitdehaag, J. C. M., Leemhuis, H., & Dijkhuizen, L. (2002). Properties and applications of starch-converting enzymes of the α-amylase family. *Journal of Biotechnology, 94*(2), 137–155.

Virgen-Ortíz, J. J., dos Santos, J. C. S., Ortiz, C., Berenguer-Murcia, A., Barbosa, O., Rodrigues, R. C., & Fernandez-Lafuente, R. (2019). Lecitase ultra: A phospholipase with great potential in biocatalysis. *Molecular Catalysis, 473*, 110405. https://doi.org/10.1016/j.mcat.2019.110405

Walia, A., Guleria, S., Mehta, P., Chauhan, A., & Parkash, J. (2017). Microbial xylanases and their industrial application in pulp and paper biobleaching: A review. *3 Biotech, 7*(1), 11. https://doi.org/10.1007/s13205-016-0584-6

Wang, L., Hu, T., Jiang, Z., Yan, Q., & Yang, S. (2021). Efficient production of a novel alkaline cold-active phospholipase C from *Aspergillus oryzae* by molecular chaperon co-expression for crude oil degumming. *Food Chemistry, 350*, 129212. https://doi.org/10.1016/j.foodchem.2021.129212

Wong, C. M., Wong, K. H., & Chen, X. D. (2008). Glucose oxidase: Natural occurrence, function, properties and industrial applications. *Applied Microbiology and Biotechnology, 78*, 927–938.

Xu, Z., He, H., Zhang, S., Guo, T., & Kong, J. (2017). Characterization of feruloyl esterases produced by the four *Lactobacillus* species: *L. amylovorus*, *L. acidophilus*, *L. farciminis* and *L. fermentum*, isolated from ensiled corn stover. *Frontiers in Microbiology, 8*, 941. https://doi.org/10.3389/fmicb.2017.00941

Yadav, A., Mahaboob Ali, A. A., Ingawale, M., Raychaudhuri, S., Gantayet, L. M., & Pandit, A. (2020). Enhanced co-production of pectinase, cellulase and xylanase enzymes from *Bacillus subtilis* ABDR01 upon ultrasonic irradiation. *Process Biochemistry, 92*, 197–201.

Yin, L., Ye, J., Kuang, S., Guan, Y., & You, R. (2017). Induction, purification, and characterization of a thermo and pH stable laccase from *Abortiporus biennis* J2 and its application on the clarification of litchi juice. *Bioscience, Biotechnology, and Biochemistry, 81*, 1033–1040.

Zhu, Y., Jia, H., Xi, M., Xu, L., Wu, S., & Li, X. (2017). Purification and characterization of a naringinase from a newly isolated strain of *Bacillus amyloliquefaciens* 11568 suitable for the transformation of flavonoids. *Food Chemistry, 214*, 39–46.

# Chapter 11
# Sustainable Use of Microbes in Beverage Production

**Maryam Fatima, Sultan Mehmood Ghani, Samia Khanum, Abdel Rahman Mohammad Said Al-Tawaha, Arun Karnwal, Abhijit Dey, Abdel Razzaq M. Al Tawaha, and Iftikhar Ali**

**Abstract** Microorganisms are significant in the deterioration and spoiling of foods and beverages. The traits of spoiled food include unpleasing flavor, odor, and texture. Microbes are significant in preparing fermented foods and drinks at home and in industrial sectors, even though they are harmful. To ferment dairy products and create alcoholic drinks, microbes are utilized. Microbes are necessary to create dairy products, including yogurt, curd, sour cream, buttermilk, and cheese. Fermented foods, probiotics, and alcoholic beverages are gaining popularity because of their delicious and healthful properties. This chapter highlights the numerous microorganisms employed in the industrial sector of food and beverage manufacturing and demonstrates the advantages of utilizing the following bacteria in the beverage industry.

M. Fatima · S. M. Ghani
National Institute of Food Science and Technology, University of Agriculture Faisalabad, Faisalabad, Pakistan

S. Khanum
Department of Botany, University of the Punjab, Lahore, Pakistan

A. R. Mohammad Said Al-Tawaha (✉)
Department of Biological Sciences, Al Hussein Bin Talal University, Ma'an, Jordan
e-mail: Abdel-al-tawaha@ahu.edu.jo

A. Karnwal
Department of Microbiology, Lovely Professional University, Phagwara, Punjab, India

A. Dey
Department of Life Sciences, Presidency University, Kolkata, India

A. R. M. Al Tawaha
Department of Crop Science, Faculty of Agriculture, Universiti Putra Malaysia, Serdang, Selangor, Malaysia

I. Ali
School of Life Science, Chinese University of Hong Kong, Ma Liu Shui, Hong Kong

Centre for Plant Science and Biodiversity University of Swat, Charbagh, Pakistan



A. Karnwal, A. R. Mohammad Said Al-Tawaha (eds.), *Food Microbial Sustainability*, https://doi.org/10.1007/978-981-99-4784-3_11

**Keywords** Beverage · Fermentation dairy product · Fermented functional food · Yeast · Dairy microbiology

## 11.1 Introduction

Humans consume beverages for various purposes, including quenching thirst, providing hydration, and supplying nutrients to the body. Beverages are classified based on their composition, such as water, milk-based, fruit juices, carbonated soft drinks, energy drinks, and alcoholic beverages. Beverages can be consumed hot or cold and may contain various ingredients such as water, sugar, flavorings, vitamins, minerals, and preservatives. Some beverages may also have specific health benefits, such as tea, known for its antioxidant properties, and milk, a source of calcium and other nutrients. Overall, beverages play a vital role in keeping a healthy and balanced diet and are an integral aspect of human life and culture (Sui et al., 2016).

Liquids that are intended for consumption collectively are known as beverages (fermented and non-fermented). Alcoholic drinks are often divided into five categories, beers, wines, hard liquors, liqueurs, etc. Similarly, there are three categories of nonalcoholic beverages: hot, nonalcoholic, and carbonated. Some examples of nonalcoholic beverages include juice, carbonated beverages, tea, coffee, and bottled water. Recently, the use of numerous modern technologies has led to a significant development of several beverage companies. Despite the fact that customers still prefer age-old traditional beverages and drinks, new value-added juices, multifunctional fermented beverages, alcoholic and nonalcoholic beverages with microencapsulation, and nutraceutical and value-added herbal drinks manufactured using current biotechnological technologies have all formed business development in the beverage industry for industrial expansion (Malakar et al., 2020a, b).

Broad ranges of ingredients are utilized to produce a variety of beverages. Beverage manufacturing is divided into numerous classes depending on the composition of raw materials utilized and the technique employed to process the ingredients. The fermentation and distillation processes are used to produce alcoholic drinks. Microorganisms use sugar-containing raw materials to produce alcohol (ethanol) and carbon dioxide. Microbes also play an essential role in producing distinctive flavoring properties (Maryam et al., 2017; Paul et al., 2014). The fermentation process improved the nutritional content of foods along with value addition. The nutritious value of food is increased by the traditional food-preserving method of fermentation. Fermented drinks have gained popularity for their health-improving properties in different regions around the globe. In recent years, the creation of non-dairy probiotic fermented drinks from various substrates, including soymilk, whey, cereals, and the juices of vegetables and fruits, has been the primary focus of both creative problem-solving strategies and the commercialization of traditional beverages. In light of recent developments, fermented drinks are anticipated to play an essential role in functional foods (Simatende et al., 2015).

The primary motivation for producing fermented foods and beverages was to lengthen the shelf life of perishable raw agricultural commodities. However, modern bioprocessing technology aims to generate products with the intended quality characteristics regarding shelf life, texture, taste, mouthfeel, flavor, and color by using microorganisms and their enzymes to acidify, alcoholize, proteolyze, and/or convert amino acids. This is done to manufacture goods with the desired quality characteristics. Furthermore, producing foods with valuable properties in an ecologically sound way, following nature, and being reasonably priced is also an absolute need (De Roos & De Vuyst, 2018).

## 11.2 Microorganisms Involved in Beverage Industries

### *11.2.1 Yeast*

Yeasts are unicellular fungal organisms that thrive in a variety of environments. They are typically discovered in the soil, water, and plant flowers and leaves. These can also be found as parasitic or symbiotic microorganisms on the animal's body skin and occasionally in the digestive tract. These organisms either divide by sexual or asexual reproduction. Asexual division occurs by budding, as in *Saccharomyces*, by fission, or filament growth in Schizosaccharomyces (mycelium) (De Chiara et al., 2022). The Egyptians first employed yeast about 6000 BC to manufacture beer, wine, and bread. The Romans eventually copied this practice. Yeast is now employed in the beverage production for alcoholic fermentation. Their unrestricted ability to convert basic sugars into ethanol, lipids, enzymes, and foreign proteins makes industrial applications possible. In the process of making wine, yeast causes ferment of the sugars in grape juice, which results in products like carbon dioxide. Most of the time, the yeast is present on the grape skin and is adequate for fermentation. Soy sauce preparation is another fermentation method in which yeast could be employed (Zhao et al., 2015).

While making fermented milks like kefir and koumiss and cheese, yeast is primarily employed to improve flavor and texture in dairy products. In addition, they are used as supplementary starting cultures to enhance lactic acid bacteria development and increase food item fragrance. *S. cerevisiae, D. hansenii,* and *K. marxianus* are some of the most often utilized species. Yeasts for producing certain fermented goods, such as cheese, bread, and vinegar, are included in starter cultures. Yeasts are an optional anaerobe that may thrive with or without oxygen. Distillery yeast produces alcohol and spirits used in industry, including brandy, rum, and tequila. The ability of intestinal pathogens such as *E. coli, Salmonella*, and *Shigella* to survive has been demonstrated and supported by research on the probiotic qualities of yeasts. More specifically, *S. boulardii* is a thermophile non-pathogenic yeast that has been used for more than 50 years as a probiotic supplement for various gastrointestinal diseases, including diarrhea (Nain et al., 2020).

### 11.2.2 Bacteria

Prokaryotic microbes called bacteria are extensively distributed in the local habitat. Bacteria multiply rapidly. Certain bacteria can survive in extreme environments. The ideal temperature for bacterial development is 37 °C, while the lowest water activity needed for microbial multiplication is 0. Certain bacterial species are harmful or can cause sickness, while others can cause food to deteriorate (Bangar et al., 2022). Food intoxications have been linked to *Bacillus cereus, Clostridium perfringens,* and *Clostridium botulinum.*

Bacteria are essential to produce food and drink in domestic and industrial scenarios. Several fermented beverages and commodities primarily rely on lactic acid bacteria. These bacteria participate in the production of lactic acid, the breakdown of proteins and lipids, the production of alcoholic drinks, and the preparation of curd, yogurt, and fermented milk. Also, they aid in improving the taste, texture, and nutritional benefits of fermented products (sauerkraut, kimchi). The main LAB genera used to manufacture food and beverages are *Lactobacillus, Leuconostoc, Pediococcus*, and *Streptococcus* (Arqués et al., 2015). Many bacteria possess traits that can facilitate the production and processing of foodstuff. Different food microbes create various fermented substances prepared from raw animals and plants. The acidic nature and sensory characteristics of fermentation products cause the microbes to undergo fermentation.

Along with providing a significantly longer shelf life than the essential ingredients, fermentation microorganisms improve the flavor and aroma of the foods like matured cheeses, fermented sausages, sauerkraut, and pickles. The primary actors in dairy-based fermentation are lactic acid bacteria. Before starters were available, milk fermentations relied on the LAB in raw milk (Ali, 2010).

## 11.3 Fermented Functional Foods and Beverages

The human diet is enriched by fermented foods because they provide and retain various nutrients in a complex combination of flavor, aroma, and consistency. The overall availability of nutritive substances is improved by traditional fermentation. Microbes for fermented products of amino acids, vitamins, and minerals with high therapeutic benefits greatly influence consumer health. Most fermented foods are produced by natural microbes that are present as natural microbiota in raw plants, containers, utensils, and the environment. These bacteria change the biochemical components of raw materials, enhancing some fermented foods' flavor, digestibility, fragrance, nutritional value, and medicinal properties (Tamang et al., 2015). Most of the traditional preparation techniques are kept secret and passed down from generation to generation. They frequently pertain to specific tribes and castes in various provinces and are produced on a small scale at home employing back-slopping. Each ethnic group region has its own culinary diversity and cultural beliefs, including fermented foods representing the society's agroeconomic, socio-cultural, and historical characteristics. Certain fermented foods are marketed widely as delicious staple food for their medicinal, functional, and nutraceutical benefits (Kavitake et al., 2018) (Tables 11.1 and 11.2).

## 11.3.1 Fermented Milk Products

With a market value of more than €46 billion globally, the fermented milk products sector is large and accounts for all milk-based drinks and yogurts. Traditional fermented milk products are produced in many nations around the globe using various plant or animal sources, including goats, camels, bovine, sheep, coconut, and soymilk. The resulting milk can either be pasteurized or skimmed, followed by applying certain probiotic strains as the starter or the dairy substrate that can ferment naturally (Wolfe et al., 2014). However, the microbial makeup of these conventionally fermented milk products has not yet been established with the proper characterization and metabolic profiling (Marsh et al., 2014). Milk source, processing conditions, cleanliness, starter culture variety, fermentation time and temperature affect milk-based products' fermentation and microbiological composition.

**Table 11.1** Major fermented beverage from different parts of the world

| Source | Name of the fermented beverage | Name of the distilled beverage |
|---|---|---|
| Barley | Beer | Ale Scotch whisky |
| Rye | Rye beer | Rye whisky |
| Corn | Corn beer | Bourbon whiskey |
| Wheat | Wheat beer | Wheat whisky, Korn (Germany) |
| Rice | Sake sonti | Shochu (Japan), soja (Korea) |
| Juice of fruits, other than apples or pears | Wine (most commonly thought of from grapes) | Brandy, cognac (France), branntwein (Germany), pisco (Peru/Chile) |
| Juice of apples | Hard cider | Applejack, apple brand, calvados |
| Juice of pears | Perry, or pear cider | Pear brandy |
| Juice of sugar cane, or molasses | Basi, betsa-betsa (regional) | Rum, cachaça, aguardiente, guaro |
| Juice of agave | Pulque | Tequila, mezcal |
| Juice of plums | Plum wine | Slivovitz, tzuica, palinca |
| Pomace | Pomace wine | Grappa (Italy), trester (Germany), marc (France) |
| Honey | Mead | Distilled mead (mead brandy or honey brandy) |
| Potato and/or grain | Potato beer | Vodka (potato mostly used in Ukraine, otherwise grain) |
| Milk | Kumis | Araka |

Source: Marshall and Mejia (2011)

**Table 11.2** Milk, cereal, and other fermented beverages famous around the globe

| Product | Region | Microflora |
|---|---|---|
| Kefir | Eastern Europe, Africa | Bacteria: *Lactococcus, Lactobacillus, Leuconostoc, and Acetobacter*; yeast: *Naumovozyma, Kluyveromyces and Kazachstania* |
| Kumis | South America (Columbia) | Bacteria: *Lb. cremoris, L. lactis, and Enterococcus (E. faecalis, E. faecium)*; yeast: *Galactomyces geotrichum, Pichia kudriavzevii, Clavispora lusitaniae, and Candida tropicalis* |
| Koumiss/ Airag | Asia/Russia | LAB: *Lactobacillus; yeast: Kluyveromyces, Saccharomyces, and Kazachstania* |
| Water Kefir | Mexico, worldwide | Bacteria: *Lactobacillus (Lb. casei, Lb. hilgardii, Lb. brevis, Lb. plantarum), L. lactis, Leu. Mesenteroides and Zymomonas;* yeast: *Dekkera (D. anomola, D. bruxellensis), Hanseniaspora (H. valbyensis, H. vineae) Saccharomyces cerevisiae, Lachancea fermentati, and Zygosaccharomyces.* |
| Amazake | Japan | Fungi: *Aspergillus* spp. |
| Boza | Turkey, Bulgaria | Bacteria: *Leuconostoc (Leu. paramesenteroides, Leu. sanfranciscensis, Leu. mesenteroides), Lactobacillus (Lb. plantarum, Lb. acidophilus, Lb. fermentum);* yeast: *Saccharomyces (S. uvarum, S. cerevisiae), Pichia fermentans, Candida* spp. |
| Bushera | Africa (Uganda) | Bacteria: *Lactobacillus, Streptococcus, Enterococcus. Uncharacterized fungal component* |
| Koko Sour Water | Africa (Ghana) | Bacteria: *Weissella confusa, Lb. fermentum, Lb. salivarius, Pediococcus* spp. *Uncharacterized fungal component* |
| Kombucha | China, worldwide | Bacteria: *Gluconacetobacter* (G. xylinus), Acetobacter, Lactobacillus; yeast: Zygosaccharomyces, Candida, Hanseniaspora, Torulaspora, Pichia, Dekkera, saccharomyces |

Source: Marsh et al. (2014)

Back-slopping involves using a tiny part of milk that previously undergone fermentation to start a renewed fermentation, a standard method for creating artisanal fermented milk beverages (Ashaolu, 2019).

This is how bacterial cultures from the lactic acid bacteria that occur naturally in raw milk are passed down from one family member to the next. Fermented milk drinks are dominated by lactic acid bacteria (LAB), especially Leuconostoc, lactobacilli, and lactococci. When beverages are fermented at cooler temperatures, mesophilic bacteria like Lactococcus and Leuconostoc flourish, while thermophilic bacteria like Lactobacillus and Streptococcus proliferate (Fayemi et al., 2023). Over time, kefir has grown popular. In North America the market of barely V78.7 million, although shepherds in the Caucasus Mountains originally ate it. Kefir "grains" are polysaccharide matrices with symbiotic bacteria and yeast that ferment. Unlike kefir, this beverage ferments milk by back-slopping or letting it ferment spontaneously (Rahman et al., 2009). Shubat, a fermented camel milk widely consumed throughout Asia, has some therapeutic benefits. In Africa, where the craft of producing fermented goods is handed down to generations, fermented milk beverages are prevalent there. The fermented kinds of milk with different names are relatively similar and collectively referred to as naturally fermented milk, given that most kinds of milk are produced by the inherent bacteria in milk that spontaneously ferments milk. Moreover, yogurts are sometimes diluted with water to create palatable fermented milks such doogh, ayran, chaas, and lassi, which typically have a microbiological makeup comparable to that of yogurt (Shiby & Mishra, 2013).

### *11.3.2 Fermented Beverages from Non-dairy Alternatives*

Cereal-based fermented drinks, prevalent in tropical areas and particularly on the African continent, are another significant category of alcoholic beverages. Like milk-based beverages, natural microbes ferment grains like maize, barley, oats, wheat, rice, or sorghum. The kernels are frequently blended, heated, and occasionally screened. Although back-slopping often occurs once more, less is known about the microbial communities that ferment these beverages. For example, Boza is a popular beverage in Bulgaria and Turkey that is made by fermenting a range of grains, such as barley, oats, rye, millet, maize, wheat, or rice. The final beverage's consistency and fermenting ability vary depending on the grain's composition (Akpinar-Bayizit et al., 2010). The grain is cooked, filtered, and then combined with a source of carbohydrates. The combination can then ferment naturally or with the help of a backslop.

Although studies have demonstrated that Boza's microbial population is varied, the product has not yet been mass-produced. Out of all the possible fermentation combinations, it has been proposed that *Saccharomyces cerevisiae*, *Leuconostoc mesenteroides*, and *Lactobacillus confusus* make the most delicious beverage. Togwa, a nonalcoholic, sweet-and-sour cereal drink, has been studied more than most other African drinks. Select substrates, such as maize, sorghum, finger millet, and even cassava root flour, are boiled, chilled, and fermented for about 12 h to generate a porridge that is then diluted to make a beverage (Jargin, 2009). Besides milk and grain, fermented drinks come in various flavors and textures.

A good example of this is kombucha, a sweetened fermented tea originally popular in China. A synergetic community of bacteria (frequently acetic acid bacteria with trace levels of lactic acid bacteria) and yeast is encased in a cellulose matrix and float on top of the fermentate, much like vinegar mother cultures. Kombucha's beneficial effects may largely be attributed to its acidic and alkaline physiochemical properties. The tea component also means it is loaded with antioxidants and vitamins (Marsh et al., 2014). Kefir made from water is theoretically comparable to kefir made from milk in that both are fermented by a symbiotic connection between bacteria and yeast that is contained inside grains. Yet it is believed that these grains, composed of dextran and have the physical appearance of transparent crystals, originated in Mexico, where they first took the form of brittle granules and fermented from the fluid present on the buds of the opuntia cactus (Amoutzopoulos et al., 2013). Before fermentation begins, it is customary to add figs and lemon to the sweetened water to offer flavor and nutrients. Kefir made from water may have many components; nevertheless, it is often known to contain lactic acid bacteria, such as Lactobacillus and Bifidobacterium (Laureys & De Vuyst, 2014). Kefir and koumiss are only examples of fermented beverages that developed appeal beyond their original purpose of preserving food. Despite unsubstantiated assertions, surprisingly, little study has been done on traditional fermented beverages. Any health claims made for such a drink should preferably be backed up by rigorous scientific evidence from several well-controlled human intervention trials. Such data collecting is expensive and unpleasant to businesses, yet it is necessary for proof-of-concept and further research. Sadly, this proof is not often seen in drinks of this type (particularly non-dairy varieties) (Amoutzopoulos et al., 2013).

## 11.4 Microbial Interventions in Beverage Production

### *11.4.1 Fermentation*

Fermentation is decomposing of carbohydrates into alcohols and organic acids by microorganisms and enzymes (Swain et al., 2014). Through biochemical and biotechnological methods, it is also possible to genetically modify microbes used as cultures in food processing, which improves the quality attributes of both conventional and modern fermented products and their enzyme activities and flavor improvement. To start and complete the appropriate fermentation in raw resources under regulated conditions, the growth of appropriate microorganisms deliberately introduced to the base material is essential to producing fermented beverages. Food and beverage industries frequently use lactic acid (LA) fermentation to maintain and enhance foods' nutritional and sensory qualities, including milk, vegetables, and fruits (Sathe & Mandal, 2016; Di Cagno et al., 2013). Lactic acid bacteria isolated from traditionally fermented foods were the main microflora of fermented food products. In the production of beverages, microbes like bacteria, yeast, etc., change carbohydrates to alcohol, organic acids, and carbon dioxide.

#### 11.4.1.1 Lactic Acid Fermentation

Most of the lactic acid bacteria (LAB) employed is thought to belong to the genus *Lactobacillus*, and variants used in the industrial manufacture of lactic acid have subsequently become patent. Vitamin levels in food, notably vitamin B, can be enhanced by LAB (Di Cagno et al., 2013). By converting 3-carbon pyruvate to 3-carbon lactic acid ($C_3H_6O_3$) through anaerobic metabolism carried out by bacteria (Lactobacillus and others), LA fermentation (such as fermented milks and cereals) allows glycolysis to continue producing ATP under low oxygen circumstances. Lactic acid fermentation and biochemical modifications are often involved in lactic acid biotechnological manufacturing techniques. *Leuconostoc, Streptococcus,* and

*Lactobacillus* bacteria ferment LA, transforming sugar molecules into lactic acid (Ghaffar et al., 2014).

#### 11.4.1.2 Alcoholic Fermentation

Alcoholic fermentation is a microbiological process that converts glucose into ethyl alcohol and carbon dioxide using yeast, bacteria, and other microbes. Aqueous mono-saccharide solutions (raw materials) serve as the culture medium that mainly uses yeast as a culture for beverage preparation (Stanbury et al., 2013). Yeast typically performs aerobic fermentation in alcohol production but can also ferment the essential ingredients anaerobically. Alcoholic fermentation occurs in the yeast cytoplasm in an oxygen-deprived environment (Huang et al., 2015). For instance, the first step in ethanol production is the conversion of pyruvate to the intermediate molecule acetaldehyde, which results in carbon dioxide emission. This step is followed by converting acetaldehyde into ethanol under anaerobic circumstances.

#### 11.4.1.3 Solid-State Fermentation

A fermentation method known as "solid-state fermentation" (SSF) is one in which bacteria develop on solid substrates without the presence of liquid. It uses microorganisms like fungi primarily for food processing and forming enzymes. Microbes are cultured in the humid solid substrate in the SSF method (Singhania et al., 2009). While this process occurs in a solid matrix and is almost entirely devoid of water, the substrates must be moist enough to support the growth and metabolism of the bacteria involved. Several biotechnologically induced solid-state fermentations are being investigated in beverage production.

#### 11.4.1.4 Submerged Fermentation

In contrast to a solid medium, submerged fermentation involves the growth of microbes in a liquid media that is aggressively agitated and aerated. The fermentation process is relatively quick because of free-flowing substrates like molasses and broths (Saqib et al., 2010). Microbes that need a substantial moisture content to develop throughout the fermentation are ideal for this type of fermentation. Selecting the suitable substrate is crucial because organisms respond differently to various substrates, impacting production. Typical substrates for submerged fermentation include molasses, fruit and vegetable juices, wastewater/effluent, and soluble sugar (Paul & Sahu, 2014). Various factors, including temperature, pH, oxygenation, incubation duration, and inoculation rate, among other parameters present in submerged cultures and other components of the medium, influence fermented drink production.

In this fermentation method, organisms can establish batch- or continuous-style fermentation in a strongly aerated and agitated liquid (Paul et al., 2014; Tang et al., 2009). The microorganism is cultured in a fixed volume of culture media throughout the batch fermentation process, and metabolic biosynthesis is permitted for a certain time. The second batch of fermentation, known as fed-batch fermentation, is initiated after cleaning and re-sterilizing a changed and improved form of a standard closed fermenter (Hashemi et al., 2011). The contents of the culture media and the cell solute concentrations will often fluctuate due to the cell's metabolic activities (Paul & Sahu, 2014; Speight & Harmon, 2010).

### 11.4.2 Starter Cultures Technology

Starter culture is the preparation of microbes that can impact the processing using biotechnological methods. They may be used to help manage the first stage of a fermentation process and include huge numbers of specialized or changeable organisms (Marsh et al., 2014). The creation of culture dairy and food items drinks like yogurt, dahi, lassi, and buttermilk, primarily used starter cultures are composed of different lactic acid bacteria, including lactobacilli, lactococci, propionic, and pediococci bacteria. Most conventionally fermented foods are made using solid substrate fermentation techniques, where the substrate is left to ferment spontaneously or with the addition of starter microorganisms (Sathe & Mandal, 2016). Microbes often decompose the raw materials' carbohydrates, proteins, and lipids by delivering enzymes into the media. The microbial concentration used as starter culture improves the effectiveness of fermentation by serving as inoculants in spontaneous fermentation (Tyagi et al., 2017).

Back-slopping is a method for making fermented foods that use samples from the previous batch as inoculum. This method is usually used to make the right starter cultures. Defined starter cultures consisting of single or mixed strains of pure microflora are used to produce dairy and other food products, such as kefir, yogurt, dahi, cheeses, and alcoholic beverages (Hugenholtz, 2013). The multifunctional starter cultures satisfy the necessary conditions for spontaneous fermentation to increase the bacteriocin, flavor, and acid production during fermentation to prevent spoilage and the growth of harmful bacteria and provide additional health-promoting functions (Corbo et al., 2014). So, starter cultures that can do more than one thing could help keep food fresh and improve its nutritional value. In addition, the newest developments in metabolic modification methods, genetics, and bioinformatics should help improve starting cultures in the future so that the food and beverage processing industries can make more money.

### *11.4.3 Bio-preservation of Food and Beverages*

The "bio-preservation" method improves beverages' shelf life by employing microbial metabolites and microflora (Giraffa et al., 2010). Alternative food preservation technologies, including bio-preservation, are replacing traditional methods of managing microbial deterioration and safety threats in foods to prolong their shelf life and ensure safe and wholesome food to balance consumer needs with the essential safety requirements (Bigliardi & Galati, 2013).

#### 11.4.3.1 Lactic Acid Bacteria as Bio-preservatives

Fermentation using LAB such as *lactobacilli, lactococci, streptococci, leuconostoc, and pediococcus* was traditionally used to preserve food products (Papagianni, 2016). These bacteria are used extensively in the production of starter cultures, which are then used to manufacture milk, fruit, vegetable, and beverage products. Because LAB is capable of producing active metabolites, such as organic acids (lactic, acetic, formic, propionic, and butyric acids), which are more effective when the pH of the medium is lower, LAB is able to preserve food. Ethanol, fatty acids, acetone, hydrogen peroxide, diacetyl, and chemicals that inhibit fungal growth are also examples of other active metabolites (Cousin et al., 2017). Additionally, it has been shown (Awojobi et al., 2016) that the lactic and acetic acids produced by lactic acid bacteria have an effect on a number of the fungal infections that are caused by *Aspergillus flavus*, hence reducing the likelihood that these diseases will spread.

Food items are historically preserved by fermentation using LAB, such as lactobacilli, lactococci, streptococci, leuconostoc, and pediococci. Widespread starter cultures of these bacteria are employed to make milk, fruits, and vegetable products and beverage. LAB's active metabolites, such as organic acids (lactic, acetic, formic, propionic, and butyric acids), ethanol, fatty acids, acetone, hydrogen peroxide, diacetyl, and antifungal substances, enhance its preservative activity by lowering the medium's pH (Awojobi et al., 2016; Crowley et al., 2013; Fatima & Fernanda, 2016). Moreover, it has been shown that the lactic and acetic acids from LAB have inhibitory activity on several fungal pathogens of *Aspergillus flavus* during the preparation and preservation of juice. Since LAB competes for resources and produces active metabolites, like organic acids, hydrogen peroxide, and antimicrobial peptides, these substances have antagonist and inhibitory characteristics (Rani et al., 2016).

#### 11.4.3.2 Bacteriocins as Bio-preservatives

Bacteriocins are a diverse group of potent antimicrobial peptides predominantly active against gram-positive bacteria and other species. They are produced by ribosome synthesis during the first stage of growth (Zacharof & Lovitt, 2012). The production of a particular immunity protein that is transcribed in the bacteriocin operon is how LAB strains make bacteriocins defend against their toxins. Bacteriocins have become increasingly popular in the beverage sector for improving safety and increasing shelf life since they serve an essential function as innovative food preservative agents (Rocha et al., 2017). Food and beverage preservation includes different substances like bacteriocins to preserve them and improve their shelf stability. Several bacteriocins are generated by LAB, but their potential use as bio-preservatives has not yet been extensively investigated. Several food scientists have carried out significant studies on the bactericidal characteristics of several bacteriocins. The bacteriocin produced by *Lactococcus lactis subsp. lactis BZ,* which has a broad inhibitory spectrum, has the potential to be used as a bio-preservative in food items. A number of bacteriocins have been created recently, and some of them have even received patents for use in food (Bali et al., 2016).

### *11.4.4 Probiotic and Prebiotic Functional Beverages*

Probiotics are formulations that include isolated or mixed cultures of microbes when given to humans or other animals in the proper proportion have positive health effects. The production of traditional non-dairy fermented beverages is widespread, and many of them are nonalcoholic drinks made by encapsulating probiotic and prebiotic microbes (Saad et al., 2013). Yogurt and fermented milks have long been employed as probiotic carrier foods in dairy industry. Yogurt has a long history of being linked to human health and lifespan (Granato et al., 2010). Conventional yogurt has undergone several attempts to increase its positive benefits by value addition of different components such as probiotics, prebiotics, and other plant based components. Probiotic yogurt is one of these fermented foods with value addition that has had significant market success over the past 20 years (Champagne et al., 2018). Although probiotic cells are better protected from unfavorable environmental factors by the solid texture, high fat content, and pH values, the probiotic cheese industry is far from reaching its commercial potential. Dairy products with probiotics have been available on the market for a long time (Özer & Kırmacı, 2011).

Synbiotics refers to probiotics and prebiotics taken together. Probiotic foods and beverages commonly include the bacteria *Lactobacillus casei, Lb. delbrueckii var. bulgaricus, Lb. acidophilus, Lb. brevis, Lb. lactis, Lb. plantarum, and Lb. fermentum Bifidobacterium breve, Bifidobacterium lactis, Bifidobacterium longum, Bifidobacterium adolescentis*, and more species (Yerlikaya, 2014). Many possible probiotic sources are currently being used via biotechnological applications, probiotic species including Lactobacillus, Enterococcus, and Bifidobacterium are widely employed in functional foods and encapsulating supplements. Recombinant DNA-based genetic modification and sequencing of common probiotics are all contributing to the fast production of probiotic microbe strains due to breakthroughs in the uses of biotechnological methods in food processing. The most recent studies in this field support the advancement of unique, modified probiotics with improved nutritional and functional efficiency for protecting human health and creating new products. Moreover, probiotic bacteria that can survive the severe process conditions are being developed using current techniques in addition to these (Sekhon & Jairath, 2010).

It is widely acknowledged that fermented milk products are excellent probiotic carriers. Probiotic LAB can withstand bile salts and acidic environments, and it also generates bacteriocins that are effective toward food spoilage and pathogenic microbes and might have possible applications for enhancing the quality and safety of food production (Panesar et al., 2013). Several foods and beverages can spontaneously ferment and contain probiotic LAB, which has GRAS designation and is usually considered harmless. The therapeutic benefit of LAB was one of the main drivers of probiotic adoption in the dairy and beverage sectors. Probiotics have been studied as a potential bio-therapeutic against intestinal problems and lactose intolerance, changed nutrient composition of milk, antagonistic action against different microbial pathogens, and anti-mutagenic and anticarcinogenic properties. Despite these studies, the therapeutic benefit of LAB was one of the main drivers of probiotic adoption (Bali et al., 2016; Fleet & Rahman, 2017).

## 11.5 Genetically Modified Microbial Strains Utilized in Beverage Production

The procedure of changing a living organism's genome orientation or composition to serve a primary purpose is known as genetic modification. Genetic modification aims to transmit copies of certain genes with known features from one organism to a different host organism to introduce the desired traits. The term "genetically modified food" describes foods prepared by various biotechnology procedures and applications, in which the organisms are created by having particular alterations made to their DNA using genetic engineering methods. Many genetically modified foods, microorganisms, and natural resources for preparing food and beverages have already begun to appear due to this field's development (Agrawal et al., 2013).

The production of fermented beverages has benefited from the use of genetically modified yeast strains since these yeasts have been shown to improve fermentation processes, increase yeast ethanol tolerance, nitrogen absorption, and sugar utilization, and modify sensory qualities. Because the malate permease gene was integrated into some of the genetically modified yeast strains created to improve wine fermentation, those strains may also be able to carry out malolactic fermentation. This was made possible by the introduction of malolactic wild strains (Malakar et al., 2020a, b). This is important since it lessens the acidity of beverages while keeping their microbiological stability intact, and it cuts the amount of time needed for the fermentation process down by a significant amount. Malate is converted to lactic acid with the help of technology; however, the resulting acid does not have the same flavor profile as lactic acid bacteria (Varela et al., 2012). Using genetically modified yeast strains during wine fermentation reduces hydrogen sulfides ($H_2S$), which hurt wine taste, aroma, and other elements of alcoholic drink production. The fermentation characteristics of microorganisms like yeast are not drastically changed by selective breeding, but their metabolic activities are somewhat modified (Chambers & Pretorius, 2010).

## 11.6 Health Benefits of Microbes in Beverage Production

When food and beverages are fermented, different strains of bacteria, yeast, and fungi are utilized, producing different kinds of cultured goods with improved flavor, taste, and aroma.

- Fermented milk products are common in dairy products. Many forms of cheese, such as soft unripened and ripened as well as numerous hard types, are generated during the coagulation of milk. The majority of the microorganisms utilized ferment lactic acid.
- Alcohol-containing drinks include wine, beer, cider, and vinegar. Several yeast strains are employed during the fermentation of grains and cereals to produce alcoholic beverages. Molds may grow on decaying grapes to produce wine. Several food sectors, i.e., fermentation industries, utilize these microbes in different ways. For instance, brewers and vinegar producers cultivate their own strains and inoculum.
- In the food business, some microbes are employed to make processing aids, such as lactase made from *Aspergillus niger, Aspergillus oryzae*, and *Kluyveromyces lactis* strains. For those who are lactose intolerant, it is mostly employed in the manufacture of reduced or lactose-free products. The benefit of lactase-treated milk is that the sweetness of the milk is increased, eliminating the need to add sugar when making flavored milk. In order to improve sweetness, creaminess, and digestibility, lactase can also be employed in the ice cream, yogurt, and frozen dessert sectors.
- The shelf life of food and beverage items may also be extended with the help of microorganisms, which ultimately results in the preservation of the food goods. Bacteriocins, organic acids, hydrogen peroxides, carbon dioxide, and diacetyls are only a few examples of the antimicrobial chemicals formed by microorganisms, namely lactic acid-producing bacteria, that increase the shelf life of food and drinks. Other antimicrobial compounds include lactic acid, which is produced by lactic acid bacteria.
- Probiotics can endure the harsh conditions of the digestive system because they are acid-tolerant and bile-resistant. These microorganisms should be safe to use, productive, and able to lengthen the food's shelf life. Probiotics are live microorganisms that come from a variety of diverse families. Most often used are lactic acid bacteria like Lactobacillus and Enterococcus, as well as Bifidobacterium strains. Cheese, yogurt, and ice cream are just few of the dairy products that often

include probiotics into their production processes. Probiotics, such as the Lactobacillus bacteria found in fermented dairy products, have been around for quite some time (Bintsis, 2018; Nain et al., 2020).

## 11.7 Conclusion

Microbes may be controlled in the food and beverage sector to create modified products, such as fermented foods and drinks, or, if not, they can be a significant factor in food degradation following processing. To lower the danger of spoilage and contamination, raw food products should be thoroughly cleaned before processing. Prevent contamination from external sources while monitoring and providing ideal conditions for the food-producing bacteria throughout processing. Several ways to get rid of bacteria that cause spoiling include applying high temperatures, changing the pH, or creating aseptic conditions. There are multiple uses for beneficial microorganisms, and various strains of these bacteria should be managed independently. Nowadays, fermentation and other manufacturing methods used in the beverage industry are being upgraded by applying current biotechnological advancements. The numerous bacteria utilized in fermentation are improved by contemporary technology, which results in substances that kill other bacteria that cause food poisoning and product spoilage, enhancing the product's nutritional and taste profiles. Improving safety standards and shelf life through bio-preservation also significantly lowers the likelihood of product spoiling. The main bio-preservatives utilized in the beverage industry are LAB and bacteriocins, which may be extracted or generated using biotechnological methods. There are several ways to develop improved strains of microbes, including genetic manipulation.

## References

Agrawal, G. K., Timperio, A. M., Zolla, L., Bansal, V., Shukla, R., & Rakwal, R. (2013). Biomarker discovery and applications for foods and beverages: Proteomics to nanoproteomics. *Journal of Proteomics, 93*, 74–92. https://doi.org/10.1016/j.jprot.2013.04.014

Akpinar-Bayizit, A., Yilmaz-Ersan, L., & Ozcan, T. (2010). Determination of boza's organic acid composition as it is affected by raw material and fermentation. *International Journal of Food Properties, 13*(3), 648–656.

Ali, A. A. (2010). Beneficial role of lactic acid bacteria in food preservation and human health: A review. *Research Journal of Microbiology, 5*(12), 1213–1221.

Amoutzopoulos, B., Löker, G. B., Samur, G., Çevikkalp, S. A., Yaman, M., Köse, T., & Pelvan, E. (2013). Effects of a traditional fermented grape-based drink 'hardaliye' on antioxidant status of healthy adults: A randomized controlled clinical trial. *Journal of the Science of Food and Agriculture, 93*(14), 3604–3610.

Anandharaj, M., Sivasankari, B., & Rani, R. P. (2020). Corrigendum to "effects of probiotics, prebiotics, and synbiotics on hypercholesterolemia: A review". *Chinese Journal of Biology, 2020*, 1–8.

Arqués, J. L., Rodríguez, E., Langa, S., Landete, J. M., & Medina, M. (2015). Antimicrobial activity of lactic acid bacteria in dairy products and gut: Effect on pathogens. *BioMed Research International, 2015*, 584183.

Ashaolu, T. J. (2019). A review on selection of fermentative microorganisms for functional foods and beverages: The production and future perspectives. *International Journal of Food Science & Technology, 54*(8), 2511–2519.

Awojobi, K., Adeyemo, S., & Sanusi, O. (2016). Biosynthesis of antimicrobial compounds by lactic acid bacteria and its use as biopreservative in pineapple juice. *Frontiers in Science, 6*(1), 17–24.

Bali, V., Panesar, P. S., Bera, M. B., & Kennedy, J. F. (2016). Bacteriocins: Recent trends and potential applications. *Critical Reviews in Food Science and Nutrition, 56*(5), 817–834.

Bangar, S. P., Suri, S., Trif, M., & Ozogul, F. (2022). Organic acids production from lactic acid bacteria: A preservation approach. *Food Bioscience, 46*, 101615.

Bigliardi, B., & Galati, F. (2013). Innovation trends in the food industry: The case of functional foods. *Trends in Food Science & Technology, 31*(2), 118–129.

Bintsis, T. (2018). Lactic acid bacteria: Their applications in foods. *Journal of Bacteriology & Mycology, 6*(2), 89–94.

Chambers, P. J., & Pretorius, I. S. (2010). Fermenting knowledge: The history of winemaking, science and yeast research. *EMBO Reports, 11*(12), 914–920.

Champagne, C. P., da Cruz, A. G., & Daga, M. (2018). Strategies to improve the functionality of probiotics in supplements and foods. *Current Opinion in Food Science, 22*, 160–166.

Corbo, M. R., Bevilacqua, A., Petruzzi, L., Casanova, F. P., & Sinigaglia, M. (2014). Functional beverages: The emerging side of functional foods: Commercial trends, research, and health implications. *Comprehensive Reviews in Food Science and Food Safety, 13*(6), 1192–1206.

Cousin, F. J., Le Guellec, R., Schlusselhuber, M., Dalmasso, M., Laplace, J.-M., & Cretenet, M. (2017). Microorganisms in fermented apple beverages: Current knowledge and future directions. *Microorganisms, 5*(3), 39.

Crowley, S., Mahony, J., & van Sinderen, D. (2013). Current perspectives on antifungal lactic acid bacteria as natural bio-preservatives. *Trends in Food Science & Technology, 33*(2), 93–109.

De Chiara, M., Barré, B. P., Persson, K., Irizar, A., Vischioni, C., Khaiwal, S., et al. (2022). Domestication reprogrammed the budding yeast life cycle. *Nature Ecology & Evolution, 6*(4), 448–460.

De Roos, J., & De Vuyst, L. (2018). Acetic acid bacteria in fermented foods and beverages. *Current Opinion in Biotechnology, 49*, 115–119.

Di Cagno, R., Coda, R., De Angelis, M., & Gobbetti, M. (2013). Exploitation of vegetables and fruits through lactic acid fermentation. *Food Microbiology, 33*(1), 1–10.

Fatima, D., & Fernanda, R. L. (2016). Characterization of bacteriocin produced by Lactobacillus brevis isolated from traditional fermented tomatoes juice. *Scientific Agriculture, 15*(3), 401–408.

Fayemi, O. E., Akanni, G. B., Sobowale, S. S., Oelofse, A., & Buys, E. M. (2023). Potential for increasing folate contents of traditional African fermented sorghum gruel (Motoho) using presumptive probiotic lactic acid bacteria. *Journal of Food Composition and Analysis, 115*, 104850. https://doi.org/10.1016/j.jfca.2022.104850

Fleet, M., & Rahman, P. K. (2017). Probiotics and their health benefits. In *Microbial functional foods and nutraceuticals* (pp. 267–279). Wiley.

Ghaffar, T., Irshad, M., Anwar, Z., Aqil, T., Zulifqar, Z., Tariq, A., et al. (2014). Recent trends in lactic acid biotechnology: A brief review on production to purification. *Journal of Radiation Research and Applied Sciences, 7*(2), 222–229.

Giraffa, G., Chanishvili, N., & Widyastuti, Y. (2010). Importance of lactobacilli in food and feed biotechnology. *Research in Microbiology, 161*(6), 480–487.

Granato, D., Branco, G. F., Cruz, A. G., Faria, J. A. F., & Shah, N. P. (2010). Probiotic dairy products as functional foods. *Comprehensive Reviews in Food Science and Food Safety, 9*(5), 455–470.

Hashemi, M., Mousavi, S., Razavi, S., & Shojaosadati, S. (2011). Mathematical modeling of biomass and α-amylase production kinetics by Bacillus sp. in solid-state fermentation based on solid dry weight variation. *Biochemical Engineering Journal, 53*(2), 159–164.

Huang, H., Qureshi, N., Chen, M.-H., Liu, W., & Singh, V. (2015). Ethanol production from food waste at high solids content with vacuum recovery technology. *Journal of Agricultural and Food Chemistry, 63*(10), 2760–2766.

Hugenholtz, J. (2013). Traditional biotechnology for new foods and beverages. *Current Opinion in Biotechnology, 24*(2), 155–159.

Jargin, S. V. (2009). Kvass: A possible contributor to chronic alcoholism in the former Soviet Union—Alcohol content should be indicated on labels and in advertising. *Alcohol & Alcoholism, 44*(5), 529.

Kavitake, D., Kandasamy, S., Devi, P. B., & Shetty, P. H. (2018). Recent developments on encapsulation of lactic acid bacteria as potential starter culture in fermented foods—A review. *Food Bioscience, 21*, 34–44.

Laureys, D., & De Vuyst, L. (2014). Microbial species diversity, community dynamics, and metabolite kinetics of water kefir fermentation. *Applied and Environmental Microbiology, 80*(8), 2564–2572.

Malakar, S., Paul, S. K., & Jolvis Pou, K. R. (2020a). 1—Biotechnological interventions in beverage production. In A. M. Grumezescu & A. M. Holban (Eds.), *Biotechnological progress and beverage consumption* (pp. 1–37). Academic Press.

Malakar, S., Paul, S. K., & Pou, K. J. (2020b). Biotechnological interventions in beverage production. In *Biotechnological progress and beverage consumption* (pp. 1–37). Elsevier.

Marsh, A. J., Hill, C., Ross, R. P., & Cotter, P. D. (2014). Fermented beverages with health-promoting potential: Past and future perspectives. *Trends in Food Science & Technology, 38*(2), 113–124.

Marshall, E., & Mejia, D. (2011). *Traditional fermented food and beverages for improved livelihoods* (FAO diversification booklet 21). FAO, FIAT PANIS.

Maryam, B. M., Datsugwai, M. S. S., & Shehu, I. (2017). The role of biotechnology in food production and processing. *Industrial Engineering, 1*(1), 24–35.

Nain, N., Kumari, K. G., Haridasan, H., & Sharma, S. G. (2020). Microbes in food and beverage industry. In *Microbial diversity, interventions and scope* (pp. 249–258). Springer.

Özer, B., & Kırmacı, H. A. (2011). Technological and health aspects of probiotic cheese. In *Cheese: Types, nutrition and consumption*. Marcel and Dekker.

Panesar, P. S., Kumari, S., & Panesar, R. (2013). Biotechnological approaches for the production of prebiotics and their potential applications. *Critical Reviews in Biotechnology, 33*(4), 345–364.

Papagianni, M. (2016). Food fermentation and production of biopreservatives. In *Handbook of animal-based fermented food and beverage technology* (pp. 126–141). CRC Press.

Paul, S. K., Dutta, H., Mahanta, C. L., & Kumar, G. P. (2014). Process standardization, characterization and storage study of a sweet potato (Ipomoea batatas L.) wine. *International Food Research Journal, 21*(3), 1113.

Paul, S. K., & Sahu, J. K. (2014). Process optimization and quality analysis of carambola (Averrhoa carambola L.) wine. *International Journal of Food Engineering, 10*(3), 457–465.

Rahman, N., Xiaohong, C., Meiqin, F., & Mingsheng, D. (2009). Characterization of the dominant microflora in naturally fermented camel milk shubat. *World Journal of Microbiology and Biotechnology, 25*(11), 1941–1946. https://doi.org/10.1007/s11274-009-0092-5

Rani, P., Saini, N. K., Gagneja, A., & Kaur, M. (2016). Biopreservation of apple and pomegranate juice using bacteriocin of Lactobacillus acidophilus NCDC 343. *Emergent Life Sciences Research, 2*, 43–49.

Rocha, M. A. M., Coimbra, M. A., & Nunes, C. (2017). Applications of chitosan and their derivatives in beverages: A critical review. *Current Opinion in Food Science, 15*, 61–69.

Saad, N., Delattre, C., Urdaci, M., Schmitter, J.-M., & Bressollier, P. (2013). An overview of the last advances in probiotic and prebiotic field. *LWT-Food Science and Technology, 50*(1), 1–16.

Saqib, A. A., Hassan, M., Khan, N. F., & Baig, S. (2010). Thermostability of crude endoglucanase from Aspergillus fumigatus grown under solid state fermentation (SSF) and submerged fermentation (SmF). *Process Biochemistry, 45*(5), 641–646.

Sathe, G. B., & Mandal, S. (2016). Fermented products of India and its implication: A review. *Asian Journal of Dairy and Food Research, 35*(1), 1–9.

Sekhon, B. S., & Jairath, S. (2010). Prebiotics, probiotics and synbiotics: An overview. *Journal of Pharmaceutical Education & Research, 1*(2).

Shiby, V., & Mishra, H. (2013). Fermented milks and milk products as functional foods—A review. *Critical Reviews in Food Science and Nutrition, 53*(5), 482–496.

Simatende, P., Gadaga, T. H., Nkambule, S. J., & Siwela, M. (2015). Methods of preparation of Swazi traditional fermented foods. *Journal of Ethnic Foods, 2*(3), 119–125.

Singhania, R. R., Patel, A. K., Soccol, C. R., & Pandey, A. (2009). Recent advances in solid-state fermentation. *Biochemical Engineering Journal, 44*(1), 13–18.

Speight, S., & Harmon, D. (2010). Batch culture evaluation of carbohydrase inhibitors to moderate rumen fermentation. *Animal Feed Science and Technology, 155*(2–4), 156–162.

Stanbury, P. F., Whitaker, A., & Hall, S. J. (2013). *Principles of fermentation technology*. Elsevier.

Sui, Z., Zheng, M., Zhang, M., & Rangan, A. (2016). Water and beverage consumption: Analysis of the Australian 2011–2012 National Nutrition and Physical Activity Survey. *Nutrients, 8*(11), 678.

Swain, M. R., Anandharaj, M., Ray, R. C., & Rani, R. P. (2014). Fermented fruits and vegetables of Asia: A potential source of probiotics. *Biotechnology Research International, 2014*, 250424.

Tamang, J. P., Tamang, N., Thapa, S., Dewan, S., Tamang, B., Yonzan, H., Rai, A. K., Chettri, R., Chakrabarty, J., & Kharel, N. (2012). Microorganisms and nutritional value of ethnic fermented foods and alcoholic beverages of North East India. *Indian Journal of Traditional Knowledge, 11*(1), 7–25.

Tamang, J. P., Thapa, N., Tamang, B., Rai, A., & Chettri, R. (2015). Microorganisms in fermented foods and beverages. In *Health benefits of fermented foods and beverages* (pp. 1–110). Routledge.

Tang, Y.-J., Zhang, W., & Zhong, J.-J. (2009). Performance analyses of a pH-shift and DOT-shift integrated fed-batch fermentation process for the production of ganoderic acid and Ganoderma polysaccharides by medicinal mushroom Ganoderma lucidum. *Bioresource Technology, 100*(5), 1852–1859.

Tyagi, R., Pandey, A., & Pilli, S. (2017). *Current developments in biotechnology and bioengineering*. Elsevier.

Varela, C., Kutyna, D., Solomon, M., Black, C., Borneman, A., Henschke, P., Pretorius, I. S., & Chambers, P. (2012). Evaluation of gene modification strategies for the development of low-alcohol-wine yeasts. *Applied and Environmental Microbiology, 78*(17), 6068–6077.

Wolfe, B. E., Button, J. E., Santarelli, M., & Dutton, R. J. (2014). Cheese rind communities provide tractable systems for in situ and in vitro studies of microbial diversity. *Cell, 158*(2), 422–433.

Yerlikaya, O. (2014). Starter cultures used in probiotic dairy product preparation and popular probiotic dairy drinks. *Food Science and Technology, 34*, 221–229.

Zacharof, M.-P., & Lovitt, R. (2012). Bacteriocins produced by lactic acid bacteria a review article. *APCBEE Procedia, 2*, 50–56.

Zhao, X., Procopio, S., & Becker, T. (2015). Flavor impacts of glycerol in the processing of yeast fermented beverages: A review. *Journal of Food Science and Technology, 52*, 7588–7598.

# Chapter 12
# Economic Importance of Microorganisms in Food Processing

**Cholakkal Shahada, Sonia Morya, Riya Farha, and Deepika Sandhu**

**Abstract** Microorganisms' relation in forming new varied products dates back long time. From the formulation of wine to the development of probiotics, microorganisms are involved with food in various forms. The utilisation of microorganisms and incorporation into the food industry had a raise recently in the production process of many products, for example, fruits and vegetable products, preservation of meat and its products, food colour, and so on. Different chemical compounds like nisin with antimicrobial properties are extracted from microorganisms. Besides, enzymes like beta-galactosidase produced by microorganisms have many advanced applications in different industries. Bio-preservation is another application of bacteria in the food industry. Utilising controlled or natural microorganisms or antimicrobials for the preservation of food and to increase its shelf life is known as bio-preservation. By carefully utilising the antimicrobial properties of naturally occurring food microbes and/or their safe-use metabolites, bio-preservation prolongs the shelf life of food. Bio-preservation also enables the production of various other food products also. Some algae and fungi are recently used for the production of colours. Introducing microorganism in the food industries enhances the development of novel and new food-producing techniques and thereby enables coping with the shortage of food. Various application is so far developed and studies are going on based on the use of microorganisms in the food industry/factories. Microorganisms play a vital role in producing various products that can be consumed by a wide variety of populations.

**Keywords** Microorganisms · Food industry · Lactic acid bacteria · Fermentation · Probiotics

---

C. Shahada · S. Morya (✉) · R. Farha · D. Sandhu
Department of Food Technology and Nutrition, School of Agriculture, Lovely Professional University, Phagwara, Punjab, India



A. Karnwal, A. R. Mohammad Said Al-Tawaha (eds.), *Food Microbial Sustainability*, https://doi.org/10.1007/978-981-99-4784-3_12

## 12.1 Introduction

Microbes are extremely minute living entities that occur all around us but are invisible to the unaided eye. They are aquatic, terrestrial, and avian organisms. Millions of these bacteria, which are also referred to as microorganisms, reside in the human body. While certain microorganisms make us ill, others are vital to our well-being (IQWiG, 2019). The most prevalent kinds include fungi, viruses, and bacteria. Protozoa are a class of microorganisms as well. These are tiny living creatures that cause toxoplasmosis and malaria, among other illnesses (Bilgili et al., 2018). Microorganisms play a very important and significant role in the production and preservation of food products. They are used in fermentation products such as bread, cheese, yoghurt, beer, and wine. Microorganisms also contribute in the development of new flavour and providing new texture in fermented foods (Cocolin et al., 2018).

Humans require nutrients that contain energy for movement, growth, repair, and maintenance of all bodily processes. Additionally, eating fosters a sense of completion, provides gastronomic pleasure, and provides a venue for social entertainment. Food is crucial to life's survival and continued existence Lin, F., Bao, Y.-W., & Wu, F.-G. (2019). Carbon Dots for Sensing and Killing Microorganisms. C, 5 (2), 33. MDPI AG. Retrieved from https://doi.org/10.3390/c5020033 (Morya et al., 2022a, b). Microorganisms swallow food in the form of macromolecules such as carbohydrates, lipids, nitrogenous chemicals, vitamins, and minerals. Organisms at different evolutionary levels eat food in diverse ways (Sharma et al., 2020). Higher eukaryotes, such as humans, on the other hand, consume a more sophisticated variety of foods such as fruits, vegetables, pulses, cereals, dairy, and meet products. The misuse and improper management of this essential resource, food, at any point in its life cycle raise substantial social, economic, and environmental concerns (Gallo et al., 2020).

However, not all microbes are advantageous for the creation of food. Salmonella and E. coli are examples of pathogenic microorganisms that must be regulated to maintain the safety of food products as they can cause foodborne illnesses (Lund et al., 2018). The food industry uses a variety of techniques, such as cleanliness, antimicrobial treatments, and temperature control, to prevent or retard the growth of hazardous microbes (Hurtado-Fernández & Montero-Calderón, 2018). Food spoilage is also caused by microbes, hence extending the shelf life of food products by decreasing and preventing the growth of germs that cause food spoilage is a crucial component of food preservation (Cervera et al., 2018). The bacteria included in food products must get past a variety of obstacles, both technological and physiological, in order to reach their intended target, the gut. Food processing may harm beneficial microorganisms in a variety of ways, which ultimately results in decreasing their vitality (Danneskiold-Samsøe et al., 2019). Additionally, after being consumed and prior to entering the stomach, microbes are subjected to other food components, a low pH, and digestive secretions, all of which can create unfavourable circumstances that reduce their viability (Cassani et al., 2020).

Research in this field is ongoing and continues to focus on identifying and controlling pathogenic microorganisms, developing new preservation methods, and improving our understanding of the complex interactions between microorganisms and food products (Garcia-Gallego et al., 2018). To restrict the growth of microbes in food, new technologies including pulsed electric field (PEF) and high-pressure processing (HPP) have been developed in addition to conventional techniques (Zhang et al., 2018). HPP involves applying high pressure to food, which inactivates bacteria without compromising the meal's quality. PDF uses an electric current to inactivate germs while also flowing through food. The potential for these innovative technologies to enhance food safety and lengthen the shelf life of food products makes them a prospective replacement for conventional preservation techniques (Jaichakan et al., 2021).

The food sector has also taken a keen interest in microbiome research as it offers a deeper understanding of the intricate relationships that exist between microbes and food goods. According to research on the microbiome, some microbes can contribute to the preservation and safety of food, while others can have the opposite effect (Vanderhoof et al., 2018). Understanding the microbiome of food products can enhance food preservation techniques and aid in the detection and management of pathogenic microbes. Given that various techniques of food production and preservation may have a negative influence on the environment, it is also crucial to take this into account (Sánchez et al., 2018). The natural effect of food products developed can be minimised by using microorganisms to digest organic waste and transform it into useful items. Each bacteria has its own definite use in the food industry. Some bacterial genera with its application are mentioned in Fig. 12.1.

**Fig. 12.1** Application of different bacteria genera in food processing

In conclusion, microbes are crucial to the production and preservation of food products in the food industry. A better understanding of the intricate interactions between microorganisms and food products, as well as new and improved preservation techniques, will result from more research on the role of microorganisms in the food industry. This will ultimately result in safer and more sustainable food production.

## 12.2 History of Microorganisms in Food Processing

For thousands of years, the food business has used microorganisms to preserve and ferment food goods. Early societies employed fermentation to make foods like cheese, bread, and beer. Before Louis Pasteur discovered that microbes were responsible for fermentation in the nineteenth century, the use of microorganisms in food production was not widely recognised (Odeyemi et al., 2020). As a result, new methods of food preservation were created, such as pasteurisation, which significantly extended the shelf life of numerous food items. For a long time, starting cultures' lengthy history of safe use served as the primary safety criterion for their use in food production (Zarzecka et al., 2020). Metchnikoff saw the positive impact of fermented milk on the health of Bulgarian peasants a century ago, and he hypothesised that the daily consumption of soured or fermented milk was responsible for the villagers' improved gastrointestinal health. This observation made a century later sparked the ground-breaking idea of using gut microbiota to treat gastrointestinal disorders linked to dysbiosis (Tamoghna et al., 2019).

In 1665, an Italian surgeon named Francesco Redi proved that maggots on rotting meat were actually the larval phases of flies and did not appear out of the blue. The notion of spontaneous generation was abandoned by this action (Kapur, 2019). Louis Pasteur made the first scientific demonstration of the microbiological cause of lactic fermentation in food in 1857. Joseph Lister isolated a pure strain of the lactic bacterium which was shown to be in charge of acidification of milk, a short time later, in 1878 (Azizi-Lalabadi et al., 2023). Nowadays, it is standard practice to isolate LAB from the outside world. This encouraged the manufacture of a range of commercial starting cultures for various industrial uses beginning in the twentieth century. Currently, they play an important and significant role in the manufacture of fermented foods, contributing to both the volume and the value of commercial starting cultures (Peng et al., 2020).

According to genetic studies from the Neolithic settlement of Jiahu in China, the first evidence of the use of microbes for major cereal crops fermentation to create an alcoholic beverage date to 7000 BC. Similar artefacts from 5400 to 5000 BC were discovered in the northern Mesopotamian Zagros Mountains (Zhang et al., 2020). Tartaric acid found in an antique jar at the Prehistoric city of Tepe in Mesopotamia, which is also estimated to 5400–5000 BC, and grape juice leftovers discovered in Dikili Tash in Greece, which are both dated to 5000 BC, provide the earliest evidence of wine manufacture (Vitorino & Bessa, 2017). The food science and industry has a long history of using food ingredients that are derived from microorganisms. Vitamins, vital amino acids, polyunsaturated fatty acids, flavour compounds, thickening and gelling agents, flavour enhancers, and acidulants are a few examples of such substances (Dufossé, 2018).

Early in the twentieth century, Russian scientist and Nobel Prize laureate Metchnikoff invented the concept of probiotics. The researcher thought that through improving gut microbial stability and lowering gastrointestinal disorders, these probiotic bacteria assist the host. Probiotics are currently understood to be live microorganisms that, when administered in therapeutic doses, boost the host's health (Kumar & Morya, 2022). The use of probiotics and other advantageous microbes in food manufacturing has grown in favour in recent years. Probiotics are living bacteria that have health advantages when taken in sufficient quantities. They are frequently included in fermented foods like yoghurt (Martín and Langella, 2019). The lengthy history of yeast use in food fermentation serves as proof of its biotechnological potential. Numerous yeast species have been identified from fermented foods, characterised, and used as starters or co-starters in the functional food industry (Rai et al., 2019). New acetic acid bacteria (AAB) species and genera have been proposed throughout the past few decades. Because of exceptional ability of microorganism to oxidise ethanol, sugar alcohols and sugar as well as in the production of crystalline and pure cellulose, a biopolymer with significant industrial relevance, bacteria play a significant role in the biotechnological field and food sectors (Gomes et al., 2018). AAB is also used to make the beverages vinegar and kombucha which contain antioxidant qualities and are good for human health (Deshmukh & Manyar, 2020).

To maintain the safety of food products, government bodies control the use of microbes in the food sector. The optimisation of microorganisms in production of food is governed by the United States Food and Drug Administration (FDA) and the United States Department of Agriculture (USDA) (Sutay Kocabaş & Grumet, 2019).

## 12.3 Economic Importance of Microorganisms in the Meat Processing Industry

Because of their high nutritional value, moisture content and neutral pH foods of animal source are very perishable. To keep these goods safe and of high quality, they need to be well preserved. Failure to do so results in disease outbreaks and human sickness. These foodborne diseases are a major, expensive public health hazard on a global scale (Jansen et al., 2019). Therefore, excellent manufacturing procedures, good hygiene practices, and other steps are typically used in the food industries to ensure the quality and security of meals. But the secret to good food safety standards is proper food preservation (Singh, 2018). All meats contain microorganisms including bacteria, yeasts, and moulds, which are crucial for the safety, flavour, and preservation of meat products. However, some microbes can also result in food decomposition or foodborne disease (Munekata et al., 2021). The meat business uses a variety of techniques, including sanitation, antimicrobial treatments, and refrigeration, to limit the growth of hazardous microbes. Nevertheless, outbreaks of foodborne illness connected to meat products continue to happen and these techniques are not always successful (Da Costa et al., 2019).

There has been a breakthrough in bio-preservation as a result of customers' increased concern for food safety, quality of product and avoiding chemical food preservatives. As a result shelf life and quality of food products are now improved through the use of beneficial microorganism, such as bacteria's and its secondary metabolites (Lahiri et al., 2022). Beneficial bacteria and their metabolites are currently the most widely used substitute for natural preservatives since they can increase the shelf life of food. Humans have traditionally used meat and various meat products as a substantial part of their diet because it supplies a variety of essential nutrients that support health and growth (Barcenilla et al., 2022). Meat and its products are among the processed foods that pose a significant challenge to the food industry because *Listeria monocytogenes* and other potentially harmful bacteria can contaminate fresh meat and meat products, which cannot be handled solely through physical means like pH lowering, freezing, or salting (Matle et al., 2020). Therefore, these issues are caused by typical food-degrading microbes. Using LAB (lactic acid bacteria) and making wise use of their antibacterial capabilities, such as producing bacteriocins and the major metabolite lactic acid can lower pH and prevent the growth of number of food spoilage organisms is one of the most popular ways to deal with such issues (Da Costa et al., 2019).

### *12.3.1 Probiotics*

The idea that altering the composition of intestinal microbiota with certain specific beneficial bacteria has potential to improve health and lengthen life was initially put out by Russian Nobel laureate Élie Metchnikoff. Probiotics are also defined as "live microorganisms that, when supplied in suitable proportions, impart a health benefit on the host" by the FAO/WHO (Zendeboodi et al., 2020). In addition to *Lactobacillus* and *Bifidobacterium,* other significant and prevalent probiotic microorganisms include *Lactococcus, Enterococcus, Streptococcus Saccharomyces,* and *Propionibacterium* yeasts (Chugh & Kamal-Eldin, 2020). Due to restrictions on dairy beverage consumption such as lactose intolerance and milk protein allergies consumers and beverage makers have turned to non-dairy beverages such as grain, vegetable, and fruit juices to supplement diets with probiotics (Nazhand et al., 2020). According to the global sales of probiotic food products, which increased to $24.2 billion from $21.6 billion during the period between 2010 and 2011, the rising human consumption of probiotics can also be attributed to increasing more consumer interest in food products that contain nutrients good for health. In 2020, it is possible to project that sales will reach $96 trillion (Baschali et al., 2017).

Probiotic meals are made up of a single or combined culture of microorganisms that enhance the gut microbial balance of the consumer and so enhance their health. Probiotic products must include live microorganisms at concentrations greater than 108 to 109 cfu/mL in order to have a positive impact on the human body (Syiemlieh & Morya, 2022). Probiotics must be able to tolerate the challenging conditions found in the body's digestive system. Probiotics experts have praised lactic acid bacteria (LABs) that have been deemed generally recognised as safe (GRAS). Probiotic bacteria have been shown to have several advantageous benefits (Markowiak & Ślizewska, 2017). The parameters used during manufacturing (pH, fermentation temperature, additives, and salt) may restrict the development of probiotic strains. In order to ensure that probiotics survive both throughout manufacture and during the product's shelf life, it is crucial. Processing, matrix-specific and microbiological parameters are exactly a few of many variables that influence the survival of the probiotic strain (Sirini et al., 2021).

### *12.3.2 Bacteriocin*

The bacteriocin-producing strains are present in both bacteria, i.e. Gram-positive and negative, and their names are often taken from the genus or species that they belong to. Bacteriocins are categorised according to factors including their mode of production (ribosomal and non-ribosomal genetics), size of the plasmid, type of sugar and protein present, molecular weight of the substance, chemistry reaction it undergoes, and killing mechanism (such as nuclease, inhibition of murein production, and pore formation) (Rahmdel et al., 2019). Based on their physiochemical characteristics and all these varied considerations, bacteriocins are divided into three main types (Table 12.1) (Silva et al., 2018). There are three different classes of LAB bacteriocins, each with a strong scientific foundation: antibiotics (class I), non-antibiotics (class II), and heat-sensitive (class III). Class IV was once thought to be complex bacteriocins with glycol- and/or lipid moieties but is no longer taken into account when classifying substances (Kumariya et al., 2019).

The American government has one of the strictest laws against *Listeria monocytogenes*, forbidding even the tiniest amount of the bacteria to exist in any kind of food. This food pathogen, which is frequently found in abattoirs, the meat processing industry, cooked meat, and even refrigerated meat has been identified as one of the causes of food poisoning in livestock (Shamloo et al., 2019). There needs to be more aggressive study into using natural and safer techniques, with bacteriocin being a good metabolite to take into consideration in order to reduce and control the occurrences of *L. monocytogenes* in meat and poultry products (Timothy et al., 2021). The effectiveness of bacteriocin, also known as sakacin, produced by *Lactobacillus sakei* CTC494 inhibited the development of *Listeria innocua*. The lactocin 705 secreted by the *Lactobacillus casei* CRL 705 regulates *L. monocytogenes*' growth (Rollini et al., 2020). In other investigations, the potential advantages of bacteriocin as a preservative in raw meat, cooked pig, and packed poultry items

**Table 12.1** Classification of bacteriocins

| Class | Sub-class | Example | Producing strain |
|---|---|---|---|
| I. | 1 | Nisin | *Lactococcus lactis* |
| | 2 | Labyrinthopeptin A1 | *Actinomadura namibiensis* |
| | 3 | Thuricin CD | *Bacillus thuringiensis* |
| II. | 1 | Pediocin PA-1 | *Pediococcus pentosaceus* |
| | | Sakacins A and P | *Lactobacillus sakei* |
| | | Leucocin A | *Leuconostoc gelidum* |
| | 2 | Lactococcin G | *Lactococcus lactis* subsp. *cremoris* |
| | | Plantaricin EF | *Lactobacillus plantarum* |
| | | Plantaricin JK | *Lactobacillus plantarum* |
| | 3 | Gassericin A | *Lactobacillus gasseri* |
| | | Enterocin AS-48 | *Enterococcus faecalis* |
| | | Garvicin ML | *Lactococcus garvieae* |
| | 4 | Bactofencin A | *Lactobacillus salivarius* |
| | | LsbB | *Lactococcus lactis* subsp. *lactis* |
| III. | 1 | Helveticin M | *Lactobacillus crispatus* |
| | 2 | Helveticin J | *Lactobacillus helveticus* |
| | 3 | Enterolysin A | *Enterococcus faecalis* |

have been demonstrated (Rhamdel et al., 2019). To gain a deeper understanding of their action, the active ingredients they contain, and the fermentation conditions for their optimum formation, all of the detected bacteriocins require in-depth investigation.

### *12.3.3 LAB (Lactic Acid Bacteria)*

Foods that need the fermentation of lactic acid such as dairy products like cheese and yoghurt, fermented vegetables like olives, sauerkraut, and pickles, fermented meats like sourdough bread and salami, all benefit from the presence of lactobacilli. Lactobacilli have been used in food products from long time. It has long history of its use in the food business and their roles in an industrial context have been thoroughly investigated (Bangar et al., 2022). The genera *Lactobacillus, Leuconostoc, Streptococcus,* and *Pediococcus* are significant contributors to this group. Due to their heterotrophic nature and a lack of many metabolic capacities, these organisms typically have complex dietary needs. As a result, the majority of species have various needs for vitamins and amino acids (A. Kumar, 2016). Phosphates and hydrocolloids are employed in the meat business to raise the quality standards of meat products. The use of exopolysaccharide (EPS)-forming LAB (lactic acid bacteria) which may manufacture EPS in situ during processing has emerged as an intriguing alternative, according to recent research results (Abarquero et al., 2022).

In the preparation of (spreadable) cured hams and raw fermented sausages, LAB (together with other starter cultures like Staphylococcus) may be utilised. Starter cultures are used in the preparation of raw fermented sausage to alter sensory qualities (such as texture and taste) and regulate the microbiological safety of the finished goods (Loeffler et al., 2020). In respect to goat meat, the antimicrobial and antilisterial properties of pediocin from *P. pentosaceus* and extract of *Murraya koenigii* berry were examined in an emulsion of goat meat. The majority of research data reveal that when used in in vitro experiments, both lactic acid bacteria and pure bacteriocins (or bacteriocin-like compounds) exhibit a consistent antibacterial action (Barcenilla et al., 2022).

The most widely utilised LAB genera are *Lactobacillus* and *Pediococcus*, which decrease the levels of native bacteria in raw meat products. It is performed by producing acetic acid and lactic acid, direct nutritional competition, and bacteriocin (Laslo et al., 2019). According to some data, lactobacilli strains may be more protected by fermented meat as a matrix than by their lyophilized counterparts while they travel through the gastrointestinal tract (Korcz, 2021).

Table 12.1 Beneficial microorganism performed in meat and meat products.

## 12.4 Economic Importance of Microorganisms in Dairy Processing Industry

Dairy and dairy products generally contain abundant nutrients that include vitamins, minerals, high-quality proteins, and fats that are high in energy. As a result, milk offers the perfect habitat for a wide range of foodborne microbes and zoonotic pathogens to develop. Theoretically, it is anticipated that the microbial quality of milk from a healthy cow at the time of milking is suitable for human consumption (Agyei et al., 2019). Milk, however, can quickly become damaged by disease causing bacteria and food poisoning derived through a variety of origins, such as manure, loam, feed, wind, water, machinery, animal skins and once it has been discharged from the udder. Producing various production from milk is an effective way of utilisation (Owusu-Kwarteng et al., 2020).

Since the beginning of human civilisation, people have eaten fermented foods alongside other processed food items. The primary metabolites and microorganisms used during food fermentation can be used to categorise the processes. It frequently intersects by the subjects of biotechnology and bioengineering depending on methods and utilisations (Admassie, 2019). Fermented dairy products, also known as functional foods or nutraceuticals, are becoming more and more popular in today's society due to their functional qualities and health-promoting qualities. Early on, lactobacilli-fermented milk was advised for its many health advantages and ability to lengthen life. Eli Metchnikoff claimed in his theory that dangerous microorganisms that would produce toxins are suppressed by fermented milk (Morya et al., 2022a, b). Probiotic bacteria can be found in fermented foods, and their metabolism is primarily responsible for their functional importance. They are believed to aid people in maintaining their health and boosting resistance to a variety of ailments. When these bacteria were presented as natural cultures that aid in digestion and health, consumers accepted probiotic cultures and products more readily (Colombo et al., 2018).

The following categories apply to dairy food products:

1. Traditional milk products: This group includes dairy items like yoghurt, cheese, butter, ice cream, and so forth.
2. Milk products with added value: This category includes low milk sugar or lactose-free items, formulae for infants with milk allergies, fortified milk with vitamins, etc.
3. Functional dairy food items: These dairy products, which can come from both dairy and non-dairy sources, have been improved with elements that offer extra health benefits beyond their fundamental nutritional composition (Hati et al., 2019).

### *12.4.1 Yogurt*

Many of these beverage's health advantages are thought to be caused by the numerous probiotic bacteria, moulds, and yeasts particularly Lactobacillus sp. And Bifidobacterium. Probiotics are living microorganisms that improve the host's health when consumed in appropriate quantity (Afzaal et al., 2019). The International Scientific Association of Probiotics and Prebiotics defines prebiotics as fermented compounds that particularly modify the make-up or activity of the gut microbiota. The host's health is positively impacted by this since it favours the helpful bacteria that make up the microbiota (Tesfaye et al., 2019). Increased enzyme activity, enhanced intestinal function, as production of bacteriocin-like substances or induction of antibacterial, defensive impact on pathogenic bacteria, pH change, competition for nutrients and physical barriers, modulation of the host immune system, cholesterol uptake, inhibition of intestinal carcinogenesis, and competitive depletion are just a few of the positive health effects of probiotics (Hati et al., 2019).

Fruits, which are the main dietary source of phenolic compounds, can be utilised to improve the phenolic content of yoghurt by adding fruit juices, powders, and extracts. Yogurt is not only a substantial source of phenolic compounds. The introduction of probiotic bacteria into dairy products as a result of the fruit and probiotic bacteria's beneficial synergistic connection has ushered in a new era of functional food innovations (Salehi, 2021). Probiotic fermented goat milk containing L. rhamnosus HN001 performed noticeably better when grape juice was added because it has a positive effect on gut microbiota regulation and grape pomace extract has a greater protective effect on the viability and various antioxidant properties of grape polyphenols (Sarkar, 2019). Strawberries are the fruit that are most frequently used to flavour yoghurt in Europe, and dairy markets in Romania sell fruit-flavoured yoghurts that also contain sour cherries, apricots, strawberries, and wild berries. Cherries, apricots, blueberries, papaya, cactus pears, and apples are the fruits most frequently used in yoghurt dishes (Moeiny et al., 2017). Probiotics and strawberries both helped strawberry yoghurt with bifidobacteria have much higher antioxidant activity than plain yoghurt. The sea buckthorn (Hippophae rhamnoides) fruit syrup-based fruit yoghurt, which is rich in antioxidants, had a higher concentration of fat, protein, carbs, and antioxidants (vitamins C and E, carotenoids, phenols, and anthocyanins) when compared to commercial yoghurt (Najgebauer-Lejko et al., 2021). Also, it has been shown that natural extracts rich in antioxidants, such as apple polyphenols, grape, and grape callus extracts, can be added to yoghurt. Two thermophilic LABs, Lactobacillus delbrueckii subs. Bulgaricus and Streptococcus thermophiles, are essential starting microorganisms used in the production of yoghurt and various types of cheese (Fardet & Rock, 2018). Since it enables both bacteria to thrive in milk, this indirect beneficial contact is known as proto-cooperation. This symbiotic relationship frequently promotes the growth of bacteria, the formation of lactic acid, and the synthesis of aromatic compounds. Lactic acid production lowers pH, which prevents the growth of hazardous or rotting bacteria (Admassie, 2019).

### *12.4.2 Lassi*

By churning curd, also known as dahi, and water until granules of butter are formed and separated, lassi is historically manufactured as a by-product of butter. It can also be created by mixing sugar, spices, or, in rare instances, fruits with yoghurt or dahi (Mallappa et al., 2021). Although dahi is the basis for lassi, it also contributes to the nutritional content of the drink, and lassi's microbiological profile and biofunctional features are quite similar to those of dahi (Pradhan et al. 2019).

In some parts of India, lassi is sometimes referred to as "buttermilk," but the end product which is technically referred to as "buttermilks" is enriched with phospholipids and is made by churning sweet or matured sour cream or curd. In lassi, phospholipids have the ability to bind water molecules, quenching thirst. Good probiotics can be found in buttermilk (Patil et al., 2017). Whole milk cannot be consumed by those who have lactose sensitivity due to the possibility of bloating, diarrhoea, and gas. Adults who are lactose intolerant have low amounts of lactase in their small intestines or none at all. However, buttermilk can be used as a replacement for whole milk by those who are lactose intolerant (Darshane & Vidyapeeth, 2021). Lactose can be metabolised by the bacteria found in buttermilk. Lactose is converted into lactic acid, which is easily digested, by the metabolic activity of bacteria. As physiologically active ingredients in Lassi, developed lactic acid, microbial metabolites, and natural milk components protect the body from different GI illnesses including diarrhoea, dysentery, bowel movement, etc. (Pradhan et al. 2019).

### 12.4.3 Whey Beverage

Fermentation is most affordable technology for food preservation, nutritional value development, and sensory attribute enhancement. Cheese whey or milk is an excellent starting material for the lactic acid bacteria and yeast used to make fermented beverages. It has cheese whey, WPC, or WPI added to it (Pires et al., 2021). Similar to other dairy-fermented drinks, LAB can improve the nutritional (producing bioactive peptides and protein breakdown), sensory (producing lactic acid and different fragrance compounds), and shelf life (reducing pH prevents the growth of spoilage bacteria) aspects of whey-based drinks (Zotta et al., 2020). The allergenic protein found in milk and whey-based products, lactoglobulin, can also be broken down by some strains. Yet, the market for whey beverages faces a challenge with the usage of probiotic strains. The species primarily used for producing functional whey beverages are *Lactobacillus acidophilus, Lactobacillus rhamnosus, Lactobacillus casei, and Lactobacillus reuteri* (Turkmen et al., 2019). Different probiotic lactobacilli, such as L. casei NCDC-12, L. acidophilus NCDC-15, L. casei RTS, L. acidophilus NCDC-15, and L. rhamnosus ATCC 7469, were able to produce whey drinks with satisfactory sensory attributes and in some of the cases with antimicrobial effects on different foodborne pathogens (such as *E. coli, Salmonella typhi, Klebsiella pneumonia, and Staphylococcus aureus*) (Silva e Alves et al. 2018).

The probiotic bifidobacteria and LAB combos B. lactis Bl-07/L. acidophilus La-14, B. animalis subsp. Lactis Bb-12/L. rhamnosus GG, B. animalis subsp. Lactis Bb-12/L. acidophilus La-5/S. thermophilus, and B. bifidum have all been used in the past (Abdul Alim et al., 2018). Kefir grains, which include *Lactococcus, Lactobacillus, Leuconostoc, Streptococcus* spp., *Saccharomyces* spp., *Kluyveromyces,* and *Candida* are fermented using a mixture of LAB and yeasts to create kefir-like whey beverages (Łopusiewicz et al., 2019). Due to these products' unique aroma profiles, sensory attributes, and antioxidant capability, they could serve as the foundation for whey beverage production (Zotta et al., 2020). Although it is rare to find unfermented goods with LAB cultures and/or probiotics, fermented whey beverages might have greater advantages.

### 12.4.4 Koumiss

Functional dairy beverages with a probiotic, prebiotic, or symbiotic basis were the first and continue to dominate the industry. As "live microorganisms that impart health advantages on the body when fed in inadequate numbers," probiotics were initially defined by the FAO/WHO in 2002 (Samedi & Linton Charles, 2019). Probiotic drinkable yoghurt is made using commercial strains of *Bifidobacterium* spp. and *Lactobacillus* spp., including *B. animalis subsp. lactis* strain Bb-12, B. bifidum strain BB536, and others. Commercial Lactobacillus spp. strains include *L. acidophilus* strain La-5, *L. casei* strain Shirota, *L. rhamnosus* strain, *L. casei* strain Dan (Turkmen et al., 2019). Even conventional dairy drinks like kefir and Koumiss and others (Table 12.2) are employed as dairy-based functional beverages in many regions of the world to treat conditions like hepatitis, stomach and colon diseases, and tuberculosis (Wurihan et al., 2019). Koumiss is one of the fermented food that is rich in vitamin and minerals including vitamin C, as well as minerals like calcium and phosphorus. Significant levels of the vitamins A, B12, B2, E, and pantothenic acid are also present. High lactose concentrations promote bacterial fermentation because the original cultures convert milk's lactose to lactic acid (Bayat, 2020). Koumiss also contains essential fatty acids including linolenic acid and linoleic. Koumiss has a favourable effect on the kidneys, endocrine glands, intestines, liver, neurological, circulatory, and immunological systems as well as maintaining blood pressure (Afzaal et al., 2021).

## 12.5 Economic Importance of Microorganism in Baking Industry

Sourdough fermentation is the most efficient and natural method for ensuring the best functional and sensory properties in bread. It is important that selecting a LAB with unusual characteristics can aid in preventing the growth of mould and can lead to the production of acrylamide in bread (Bartkiene et al., 2017). Aspergillus niger glucoamylase in combination with an appropriate and specific LAB strain may be one of the factors controlling dough acidification and lowering the acrylamide level

**Table 12.2** Beneficial microorganism in dairy and non-dairy products

| Product name | Product characteristics | Microbial diversity | References |
|---|---|---|---|
| Yogurt | A product made by bacterial fermentation of milk | *Lactobacillus bulgaricus, Streptococcus thermophilus* | Sarkar (2019) |
| Lassi | Milk that has been made sour by churning diluted curd or dahi and adding sugar, spices, etc. | *Lactobacillus delbrueckii, Lb. johnsonii, Lb. helveticus, Lb. fermentum, streptococcus thermophiles, macrococcus caseolyticus, and enterococcus faecalis* | Mallappa et al. (2021) |
| Whey beverage | A liquid, formed as a by-product during the process of cheese production | *Bifidobacterium lactis, lactobacillus acidophilus* | Syiemlieh and Morya (2022) |
| Cereal-based beverage | A sorghum based whey beverage was produced | *Lactobacillus rhamnosus, lactobacillus casei, lactobacillus acidophilus* | Morya et al. (2017) |
| Koumiss | Produced from mare's milk, fermented milk | *Lactobacilli, non-lactose-fermenting yeasts* | Afzaal et al. (2021) |

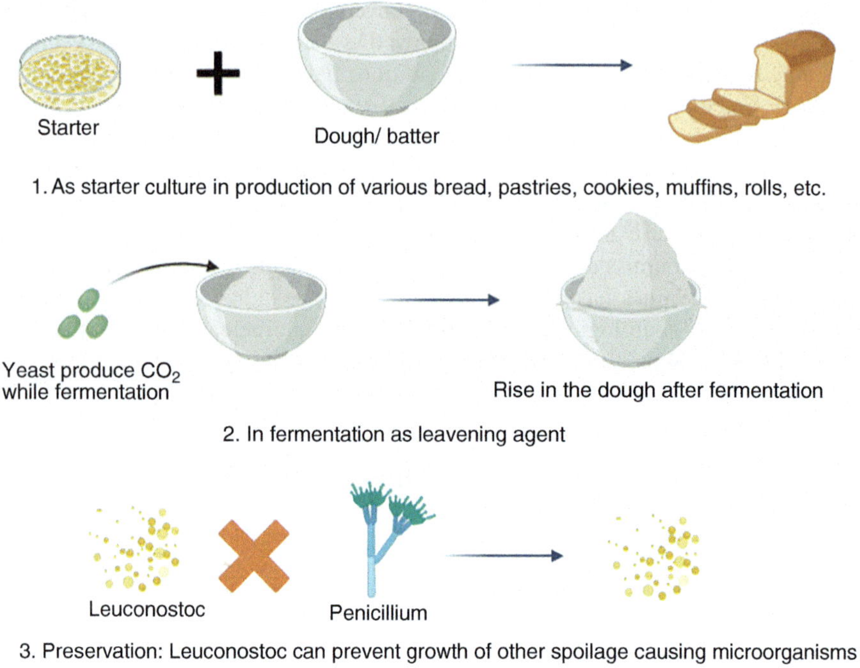


**Fig. 12.2** Application of microorganisms in baking industry

in bread (Hu et al., 2022). The various applications of the microorganisms in baking industry are shown in Fig. 12.2.

## 12.5.1 As Starter Culture

In order to create sourdough starters with the potential to produce breads with phytase activity and significant EPS generation, a pool of isolates was chosen (including *L. citreum* PB220 and L. brevis LD66). These enhanced suppleness and sensory qualities, as well as improved mineral availability and digestion (Milanovic et al., 2020). Additionally, some cultures with the potential to produce functional bread rich in fibre and high in β-glucan were discovered. *Leuconostoc* and *L. citreum* in particular have a lot of potential as multifunctional starting cultures for sourdough fermentations (Müller et al., 2021). Through its oxidising impact and strengthening of gluten structures, lactase produced by fungi is used in the bakery sector to improve the texture, volume, and flavour of baking (Drozłowska, 2019). The oxidation process it uses has a positive impact on the ingredients, enhancing the integrity of the gluten in the dough. As a result, baking volume increases and strength stability improves. Additionally, it lessens viscosity, which enhances the dough's workability. This enables the use of low-quality, poor-flour baking (De Vuyst et al., 2021).

### 12.5.2 In Fermentation

Without the help of any external oxidisers, organic materials are utilised to produce energy during the metabolic process of fermentation. For this, aero-tolerant Gram-positive cocci or bacilli belonging to the lactic acid bacteria class are used. They are exclusively fermentative, organotrophic, and their main fermentation by-product is lactic acid (Zhou et al., 2021). Since the beginning of time, fermentation has been used and employed to preserve a variety of foods. However, given that food fermentation increases the quantities of bioactive chemicals and encourages the hydrolysis of anti-nutrients to reduce their levels, it is now more important for nutrition and health (Marti-Quijal et al., 2020). Microbial biotechnology known as fermentation produces value-added products like enzymes, acids, alcohols, polymers, and more from naturally renewable substrates. End-products of fermentation like ethanol and lactic acid operate as proton sinks, converting NADH back into NAD+ so that the cell can continue generating energy through glycolysis by phosphorylation (Ali Al-Maqtari et al., 2019). To maintain an equilibrium in their energy production, bacteria produce a variety of by-products. A study by Falasconi et al., 2020 led to the development of a blend of wheat and sorghum flour with added bakery yeast and LAB, producing a dough with characteristics that are the same as those of all doughs made with wheat flour. The presence of lactic acid bacteria has a significant impact on the finished dough (LAB). The ability to produce exopolysaccharides (EPS), quinoa, acidify chickpea and buckwheat flour extracts, and ferment them were among the selection criteria used to identify 19 LAB strains from the genera Leuconostoc, Lactobacillus, Streptococcus, and Pediococcus (Vasile & Bahrim, 2020). When proteolytic preparations are added to flour in the bakery, the quality of the final goods is significantly improved. The addition of amylase (enzyme produced by fungi) to cakes causes the starch to break down into dextrin and speeds up yeast fermentation, both of which improve the structure and volume of baking (Drozłowska, 2019).

### 12.5.3 Preservation

When the anti-bacillus activity of the LAB isolated from sourdough was examined using the spot-on-the-lawn method, the majority of the LAB show a strong inhibitory capability against *Bacillus cereus DSM31, B. licheniformis DSM13, B. subtilis LMG7135, and B. subtilis* S15.20. LAB isolated from sourdoughs exhibit promising features for usage as natural preservatives in wheat-based bakery products as a result of improved functional activities (Fraberger et al., 2020). The LAB culture developed from different strains for the production of non-glutenous sourdough even exhibits the antibacterial activity against bacteria and moulds that cause food spoiling (Vasile & Bahrim, 2020). A test of 99 LAB strains for antifungal activities revealed that nine Leuconostoc strains with activity against Aspergillus sp.,

Penicillium sp., and Cladosporium sp. were very successful at suppressing bakery-relevant moulds (Müller et al., 2021). Amylase obtained from bacteria plays a significant part in improving the product's quantity, flavour, and scent in the baking business. When manufacturing bread, amylase and lipase enzymes are added to decrease crystallisation and increase shelf life (Singh et al., 2018).

## 12.6 Recent Trends

Agriculture, harvesting, manufacturing, and distribution are all part of the food sector, which has historically been a crucial source of income. The food industry offers vast scope and opportunities for innovative technology development and research (Morya et al., 2022a, b). Recent developments in the food industry have highlighted the economic significance of microorganisms in the production of fermented foods, functional foods, and innovative food additives. One of the most relevant subjects at the moment is the use of probiotics in food items. Probiotics contain live microorganisms that enhance health and is to be consumed adequately (Johansen et al., 2019). They are utilised in many different foods, including yoghurt, cheese, and fermented beverages. Fermented dairy products with probiotics have been found in studies to improve human health and the microbial flora in the stomach (Kailasapathy, 2018). Indigenous fermented foods (IFFs) are ingrained in cultural norms and practices in Africa and have a long history there. They significantly help the continent achieve its sustainable development objectives, which are centred on ensuring food security, reducing poverty, and promoting gender equality. Foods' organoleptic qualities, variety, nutritional value, digestibility, and safety are all enhanced by fermentation, which increases the value of food (Anyogu et al., 2021). Biofilm produced by microorganisms is used in edible and biodegradative packaging. A biofilm is a layer of complex polysaccharide matrix created by parasitic microorganisms that attach to solid surfaces and cause serious chronic bacterial infections as well as drug resistance (Pham et al., 2019). According to the type of environment used for food manufacture and the species that are colonising it, biofilms are intricate microbial ecosystems made up of one or more species submerged in an extracellular environment with a variety of compositions. Bacteria and fungus are two types of microbes that can be found in these biofilms (Gebreyohannes et al., 2019). Multiple species of bacteria in a biofilm can aid in the biofilm's ability to adhere to a surface, which has significant ecological benefits. This may also happen in some species without the presence of specific fimbriae. Mixed biofilms exhibit greater resilience to biocides and other disinfectants such as quaternary ammonium compounds (Galié et al., 2018).

Another development in the food sector is functional foods. These are foods that offer additional health advantages over and above simple dietary needs. By creating bioactive substances including enzymes, vitamins, and minerals, microorganisms contribute to the creation of functional foods (Reque & Brandelli, 2021). According to a review article published in the Journal of Food Science and Technology in

2018, customers are becoming more accustomed to functional meals because of the health advantages they offer (Vinod, 2018). Carbon dots, or C-dots, are a new class of fluorescently small nanomaterials with particles less than 10 nm (Lin et al., 2019). To prevent the growth of pathogenic and spoilage bacteria in foods and to lessen chemical deterioration, C-dots are used in food packaging materials as anti-oxidant, antibacterial, photoluminescent, and UV-light inhibitor agents. This is done in order to create intelligent and active food packaging materials (Moradi et al., 2021). In the modern food industry, plant secondary metabolites are frequently used as colouring, flavouring, or texturizing agents, for example. Antioxidant-rich metabolites in particular are used as preservatives or anti-browning agents (Kumar et al., 2021). The primary source of these metabolites today is extraction from plant material, but advancements in metabolic engineering have also made it possible for microorganisms to produce these important molecules, which is a more economical and environmentally friendly option (Kallscheuer et al., 2019).

Moreover, it has been shown that various microbial cultures of bacteria, fungi, and algae (including yeast) can create protein, making them attractive meat alternatives or protein-rich fortified meals for consumption by both humans and animals with high fiber and mineral contents, and with a low reliance on arable land and water, as well as a small carbon footprint (Sharif et al., 2021). Advanced optimisation-aided design and microbial protein that is also known as single cell protein (SCP) (Fig. 12.3), which is produced under controllable fermentation or

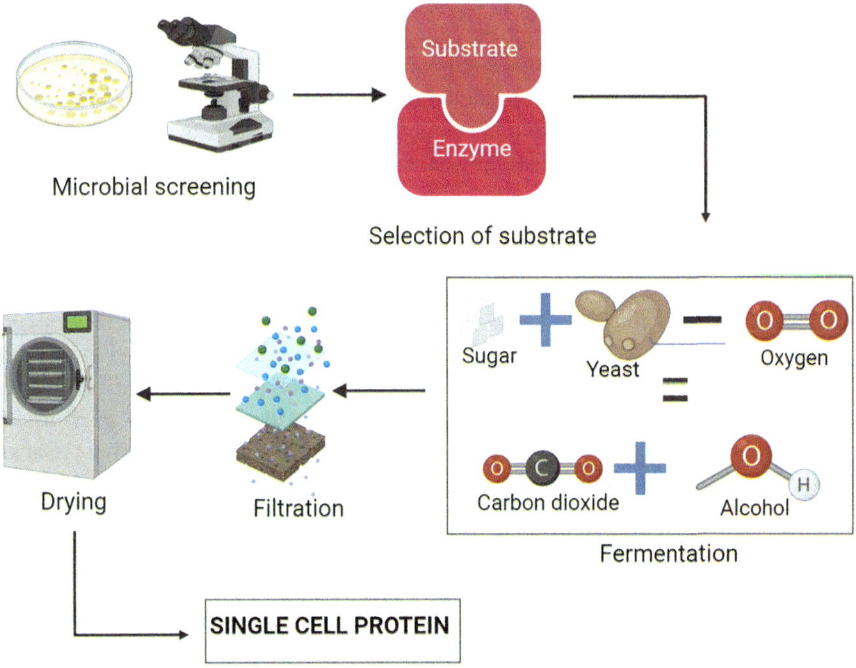


**Fig. 12.3** Production of single cell protein

photosynthesis, provide for rapid technical improvement for upcoming protein supply without compromising environmental sustainability and is a promising development in food industry (Banks et al., 2022). One of the naturally autotrophic microorganisms, such as the cyanobacterium Arthrospira platensis or the alga Chlorella vulgaris, which are already used as food additives or nutritional supplements, could be used as a substitute for S. cerevisiae. Yet above and beyond them, yeast is the most suggested microbe for expanding food space. With a yeast collection that has been bioengineered for one-carbon metabolism, enough nutrition, and a variety of textures, flavours, perfumes, and colours, it may be able to create a flexible platform for food production. Yeast is a common element in innovative foods, and research is being done to create more yeast-based products (Llorente et al., 2022).

One of the novel fermentation techniques to emerge in recent years is high-throughput fermentation. High-throughput fermentation allows for the speedy screening of a large number of microorganisms, making it easier to identify strains with desirable characteristics for application in food production (El-Dalatony et al., 2019). According to the study in 2018 in the Journal of Biotechnology, high-throughput fermentation is an effective method for identifying novel microorganisms with qualities that are suitable for use in food fermentation (Almeida et al., 2018). Also, there has been a rise in the utilisation of unusual bacteria in food fermentation. These microorganisms, which are not typically employed in food fermentation include filamentous fungus, bacteria, and yeasts (Ali Al-Maqtari et al., 2019). It has been discovered that these microbes possess special qualities that can be utilised to create new and improved food products. According to a 2018 study in the Journal of Applied Microbiology, unconventional microorganisms are quite useful for producing fermented food products (Bartoszewicz & Wieczorek, 2018).

One recent topic of research in the field of food and food products is the extraction of food colours from microbial sources. There are many ways to give a certain species of microorganism a distinguishable colour (Mussagy et al., 2019). Before extraction, microbial colours are tested for efficacy and safety, and it is found that they are both non-toxic and healthy. A dependable supply for colour extraction and culinary usage is provided by fungal cells (Mazhar et al., 2022).

Contrarily, new terminology has appeared in the literature to describe bioactive substances that do not fit in the traditional definitions (symbiotics, prebiotics, and probiotics), as well as to describe the important and beneficial effects of probiotics in a specific system or disease, as in the case of par probiotics and postbiotics (Scarpellini et al., 2021). Recent studies have shown that non-viable microbes or bacterial metabolic by-products can also exert biological activity at the host (Barros et al., 2020). Edible microbial biomass generated from yeasts, filamentous fungi, bacteria, or microalgae is produced as a possible alternative to conventional sources of feed and food. Microorganisms are also a good source of both protein and vitamins and in certain cases they can also include beneficial lipids (Linder, 2019a, b). In locations where it would not interfere with agricultural output, microorganisms can grow on simple organic substrates, allowing for the industrial-scale cultivation of consumable microbial biomass. Very few microbial food products are now available for human consumption (Linder, 2019a, b). In conclusion, the utilisation of probiotics, functional foods, and novel fermentation techniques such as high-throughput fermentation and non-conventional microorganisms has increased significantly in the food business. These changes are being driven by customer desire for food products that are more nutritious and useful.

## 12.7 Conclusion

In food processing, microorganisms play an essential role and their utilisation can also improve the flavour, texture, and nutritional content of food. They are employed in the creation of fermented foods like pickles, cheese, yoghurt, and bread. Using microorganisms like yeast and bacteria, fermentation is a process that turns sugars and other carbohydrates into alcohol or organic acids. This process creates useful molecules like vitamins and amino acids, which not only enhance the flavour of food but also increase its nutritional worth. For instance, yoghurt, which is high in calcium and probiotics, is created when lactic acid bacteria digest milk. Similar to this, the fermentation of soybeans by the bacterium *Rhizopus oryzae* results in the production of tempeh, a traditional cuisine of Indonesia that is rich in protein and B vitamins. Additionally, microorganisms are employed in food preservation. They generate organic acids, alcohol, and antibacterial substances that prevent the development of potentially dangerous pathogens. For instance, the fermentation of fish by lactic acid bacteria yields well-known, durable fish products like pickled fish and fish sauce, which are high in omega-3 fatty acids. Particularly in comparison to this, the fermentation of meat by lactic acid bacteria yields traditional fermented meat products, such as salami and sausages, with a long shelf life and rich flavour.

Not all microbes, it is crucial to remember, are helpful in food processing. Some microorganisms, including bacteria and fungi, can ruin food and even make people sick. In order to stop the growth of dangerous germs, it is crucial to apply the right food preservation methods, such as refrigeration, pasteurisation, and irradiation. Food is heated to a high temperature for a brief amount of time during the pasteurisation process in order to kill hazardous germs. To create stable and safe milk products, this method is frequently utilised in the dairy sector. For instance, milk is heated to 63 °C for 30 min or 72 °C for 15 s to kill microorganisms like Salmonella and Mycobacterium tuberculosis.

Ionising radiation is used to irradiate food in order to destroy dangerous germs. To create products that are stable and safe, this method is commonly employed in the meat and poultry industry. Ionizing radiation, for instance, is used to get rid of Salmonella and E. coli in cattle and poultry.

The employment of microbes can improve the texture, flavour, and nutritional content and value of food. Microorganisms are essential to food preparation. In addition to being used to preserve meat and fish, they are utilised to make fermented goods including cheese, yoghurt, bread, and pickles. By disassembling complicated chemicals and generating useful substances, these microbes can also assist in enhancing the nutritional content and safety of food. Not all microbes, it is crucial to remember, are helpful in food processing. Some microorganisms, including bacteria and fungi, can ruin food and even make people sick. Using the proper food preservation techniques, such as pasteurisation, irradiation, and refrigeration, is essential to guarantee the safety of the food products and inhibit the growth of hazardous microorganisms, utilising good production methods, as well.Conflict of InterestThe authors declare no conflict of interest.

## References


Abarquero, D., Renes, E., Fresno, J. M., & Tornadijo, M. E. (2022). Study of exopolysaccharides from lactic acid bacteria and their industrial applications: a review. *International Journal of Food Science and Technology, 57*(1), 16–26. https://doi.org/10.1111/ijfs.15227

Abdul Alim, T. S., Zayan, A. F., Campelo, P. H., & Bakry, A. M. (2018). Development of new functional fermented product: Mulberry-whey beverage. *Journal of Nutrition, Food Research and Technology, 1*(3), 64–69. https://doi.org/10.30881/jnfrt.00013

Admassie, M. (2019). A review on food fermentation and the biotechnology of lactic acid bacteria. *World Journal of Food Science and Technology, 2*, 18–24. https://doi.org/10.11648/j.wjfst.20180201.13

Afzaal, M., Khan, A. U., Saeed, F., Ahmed, A., Ahmad, M. H., Maan, A. A., Tufail, T., Anjum, F. M., & Hussain, S. (2019). Functional exploration of free and encapsulated probiotic bacteria in yogurt and simulated gastrointestinal conditions. *Food Science and Nutrition, 7*(12), 3931–3940. https://doi.org/10.1002/fsn3.1254

Afzaal, M., Saeed, F., Anjum, F., Waris, N., Husaain, M., Ikram, A., Ateeq, H., Muhammad Anjum, F., & Suleria, H. (2021). Nutritional and ethnomedicinal scenario of koumiss: a concurrent review. *Food Science and Nutrition, 9*(11), 6421–6428. https://doi.org/10.1002/fsn3.2595

Agyei, D., Owusu-Kwarteng, J., Akabanda, F., & Akomea-Frempong, S. (2019). Indigenous African fermented dairy products: processing technology, microbiology and health benefits. *Critical Reviews in Food Science and Nutrition, 1–16*, 991. https://doi.org/10.1080/10408398.2018.1555133

Ali Al-Maqtari, Q., AL-Ansi, W., & Ali Mahdi, A. (2019). Microbial enzymes produced by fermentation and their applications in the food industry-a review. *International Journal of Agriculture Innovations and Research, 8*(1), 2319–1473.

Almeida, A. S., Teixeira, M. T., & Teixeira, P. (2018). High-throughput fermentation: a valuable tool for discovery of new microorganisms. *Journal of Biotechnology, 280*, 21–28. https://doi.org/10.1016/j.jbiotec.2018.01.014

Anyogu, A., Olukorede, A., Anumudu, C., Onyeaka, H., Areo, E., Adewale, O., Odimba, J. N., & Nwaiwu, O. (2021). Microorganisms and food safety risks associated with indigenous fermented foods from Africa. *Food Control, 129*, 108227. https://doi.org/10.1016/j.foodcont.2021.108227

Azizi-Lalabadi, M., Moghaddam, N. R., & Jafari, S. M. (2023). 9—pasteurization in the food industry. In *Thermal Processing of Food Products by Steam and Hot Water*. Woodhead Publishing. https://doi.org/10.1016/B978-0-12-818616-9.00009-2; ISBN 9780128186169.

Bangar, S. P., Suri, S., Trif, M., & Ozogul, F. (2022). Organic acids production from lactic acid bacteria: a preservation approach. *Food Bioscience, 46*, 101615.

Banks, M., Johnson, R., Giver, L., Bryant, G., & Guo, M. (2022). Industrial production of microbial protein products. *Current Opinion in Biotechnology, 75*, 102707. https://doi.org/10.1016/j.copbio.2022.102707

Barcenilla, C., Ducic, M., López, M., Prieto, M., & Álvarez-Ordóñez, A. (2022). Application of lactic acid bacteria for the biopreservation of meat products: a systematic review. *Meat Science, 183*, 108661. https://doi.org/10.1016/j.meatsci.2021.108661

Barros, C. P., Guimarães, J. T., Esmerino, E. A., Duarte, M. C. K. H., Silva, M. C., Silva, R., Ferreira, B. M., Sant'Ana, A. S., Freitas, M. Q., & Cruz, A. G. (2020). Paraprobiotics and postbiotics: concepts and potential applications in dairy products, current opinion in food. *Science, 32*, 1. https://doi.org/10.1016/j.cofs.2019.12.003; ISSN 2214-7993.

Bartkiene, E., Bartkevics, V., Pugajeva, I., Krungleviciute, V., Mayrhofer, S., & Domig, K. (2017). The contribution of P. acidilactici, L. plantarum, and L. curvatus starters and L-(+)-lactic acid to the acrylamide content and quality parameters of mixed rye - wheat bread. *Lwt, 80*, 43–50. https://doi.org/10.1016/j.lwt.2017.02.005

Bartoszewicz, M., & Wieczorek, P. (2018). Non-conventional microorganisms in food fermentation. *Journal of Applied Microbiology, 124*(4), 879. https://doi.org/10.1111/jam.13706

Baschali, A., Tsakalidou, E., Kyriacou, A., Karavasiloglou, N., & Matalas, A. L. (2017). Traditional low-alcoholic and non-alcoholic fermented beverages consumed in European countries: a neglected food group. *Nutrition Research Reviews, 30*(1), 1–24. https://doi.org/10.1017/S0954422416000202

Bayat, G. (2020). The oldest fermented Turkish beverage in traditional Turkish cuisine: Koumiss (Kımız). *Journal of tourism and gastronomy studies, 8*(2), 816–824. https://doi.org/10.21325/jotags.2020.581

Bilgili, E., Ciftci, O., & Kocabaş, N. (2018). Microorganisms used in food fermentation. *Journal of Food Science and Technology, 55*(3), 799–811.

Cassani, L., Gomez-Zavaglia, A., & Simal-Gandara, J. (2020). Technological strategies ensuring the safe arrival of beneficial microorganisms to the gut: from food processing and storage to their passage through the gastrointestinal tract. *Food Research International, 129*, 108852. https://doi.org/10.1016/j.foodres.2019.108852; ISSN 0963-9969.

Cervera, M., Plaza, L., & García-Gimeno, R. M. (2018). Microorganisms in food spoilage: an overview. *Food Research International, 107*, 17–33.

Chugh, B., & Kamal-Eldin, A. (2020). Bioactive compounds produced by probiotics in food products. *Current opinion in food science, 32*, 76–82. https://doi.org/10.1016/j.cofs.2020.02.003; ISSN 2214–7993.

Cocolin, L., Del Bubba, M., & Gobbetti, M. (2018). Understanding the microbiological aspects of fermented foods. *Journal of Applied Microbiology, 124*(6), 1471–1482.

Colombo, M., Todorov, S. D., Eller, M., & Nero, L. A. (2018). The potential use of probiotic and beneficial bacteria in the Brazilian dairy industry. *Journal of Dairy Research, 85*(4), 487–496. https://doi.org/10.1017/S0022029918000845

Da Costa, R. J., Voloski, F. L. S., Mondadori, R. G., Duval, E. H., & Fiorentini, Â. M. (2019). Preservation of meat products with Bacteriocins produced by lactic acid bacteria isolated from meat. *Journal of Food Quality, 2019*, 1. https://doi.org/10.1155/2019/4726510

Darshane, P., & Vidyapeeth, T. M. (2021). Impact Factor: 4.656 Available Online at www.journalcmpr.com. *5*(10). https://doi.org/10.24327/23956429.ijcmpr201910748.

De Vuyst, L., Comasio, A., & Van Kerrebroeck, S. (2021). Sourdough production: fermentation strategies, microbial ecology, and use of non-flour ingredients. *Critical Reviews in Food Science and Nutrition, 1*, 1–33. https://doi.org/10.1080/10408398.2021.1976100

Deshmukh, G., & Manyar, H. (2020). Production pathways of acetic acid and its versatile applications in the food industry. In *Biomass*. IntechOpen.

Drozłowska, E. (2019). The use of enzymatic fungal activity in the food industry—review. *World Scientific News, 116*, 222–229.

Dufossé, L. (2018). Red colourants from filamentous fungi: are they ready for the food industry? *Journal of Food Composition and Analysis, 69*, 156–161. https://doi.org/10.1016/j.jfca.2017.11.002

El-Dalatony, M. M., Salama, E. S., Kurade, M. B., Kim, K. Y., Govindwar, S. P., Kim, J. R., Kwon, E. E., Min, B., Jang, M., Oh, S. E., Chang, S. W., & Jeon, B. H. (2019). Whole conversion of microalgal biomass into biofuels through successive high-throughput fermentation. *Chemical Engineering Journal, 360*, 797–805. https://doi.org/10.1016/j.cej.2018.12.042

Falasconi, I., Fontana, A., Patrone, V., Rebecchi, A., Garrido, G. D., Principato, L., Callegari, M. L., Spigno, G., & Morelli, L. (2020). Genome-assisted characterization of lactobacillus fermentum, Weissella cibaria, and Weissella confusa strains isolated from sorghum as starters for sourdough fermentation. *Microorganisms, 8*(9), 1–16. https://doi.org/10.3390/microorganisms8091388

Fardet, A., & Rock, E. (2018). In vitro and in vivo antioxidant potential of milks, yoghurts, fermented milks and cheeses: a narrative review of evidence. *Nutrition Research Reviews, 31*(1), 52–70. https://doi.org/10.1017/S0954422417000191

Fraberger, V., Ammer, C., & Domig, K. J. (2020). Functional properties and sustainability improvement of sourdough bread by lactic acid bacteria. *Microorganisms, 8*(12), 1–17. https://doi.org/10.3390/microorganisms8121895

Galié, S., García-gutiérrez, C., Miguélez, E. M., Villar, C. J., Lombó, F., & Bonaventura, G. D. (2018). Biofilms in the food industry: health aspects and control. *Methods, 9*, 1–18. https://doi.org/10.3389/fmicb.2018.00898

Gallo, M., Ferrara, L., Calogero, A., Montesano, D., & Naviglio, D. (2020). Relationships between food and diseases: what to know to ensure food safety. *Food Research International, 137*, 109414. https://doi.org/10.1016/j.foodres.2020.109414

Garcia-Gallego, R., Murphy, J., & Montero-Calderon, M. (2018). Microorganisms and food safety: an overview. *Food Research International, 107*, 34–53.

Gomes, R. J., de Borges, M. F., de Rosa, M. F., Castro-Gómez, R. J. H., & Spinosa, W. A. (2018). Acetic acid bacteria in the food industry: systematics, characteristics and applications. *Food Technology and Biotechnology, 56*(2), 139–151. https://doi.org/10.17113/ftb.56.02.18.5593

Hati, S., Das, S., & Mandal, S. (2019). Technological advancement of functional fermented dairy beverages. In *Engineering tools in the beverage industry* (Vol. 3). The science of beverages. Elsevier Inc. https://doi.org/10.1016/B978-0-12-815258-4.00004-4

Hu, Y., Zhang, J., Wang, S., Liu, Y., Li, L., & Gao, M. (2022). Lactic acid bacteria synergistic fermentation affects the flavor and texture of bread. *Journal of Food Science, 87*(4), 1823–1836.

Hurtado-Fernández, E., & Montero-Calderón, M. (2018). Microorganisms in food: an overview of their interactions with food components and preservation methods. *Food Research International, 107*, 1–16.

Jaichakan, P., Nakphaichit, M., Rungchang, S., Weerawatanakorn, M., Phongthai, S., & Klangpetch, W. (2021). Two-stage processing for xylooligosaccharide recovery from rice by-products and evaluation of products: promotion of lactic acid-producing bacterial growth and food application in a high-pressure process. *Food Research International, 147*, 110529. https://doi.org/10.1016/j.foodres.2021.110529

Jansen, W., Mueller, A., Grabowski, N. T., Kehrenberg, C., Muylkens, B., & Al Dahouk, S. (2019). Foodborne diseases do not respect borders: Zoonotic pathogens and antimicrobial resistant bacteria in food products of animal origin illegally imported into the European Union. *The Veterinary Journal, 244*, 75–82.

Kailasapathy, K. (2018). Probiotics in fermented foods. *Journal of Dairy Science, 101*(3), 2054–2061. https://doi.org/10.3168/jds.2017-13486

Kallscheuer, N., Classen, T., Drepper, T., & Marienhagen, J. (2019). Production of plant metabolites with applications in the food industry using engineered microorganisms. *Current opinion in biotechnology, 56*, 7–17. https://doi.org/10.1016/j.copbio.2018.07.008; ISSN 0958–1669.

Kapur, Radhika. (2019). History and scope of food microbiology.

Korcz, E. (2021). Exopolysaccharides from lactic acid bacteria: techno-functional application in the food industry. *Trends in Food Science & Technology, 110*, 375–384. https://doi.org/10.1016/j.tifs.2021.02.014

Kumar, A. (2016). Food & industrial. *Microbiology, 2*(2), 10–11. https://doi.org/10.4172/2572-4134.1000e101

Kumar, M. K., & Morya, S. (2022). Probiotics: role of bioactives in foods and dietary supplements. *The pharma innovation, 11*(7), 2592–2597. https://doi.org/10.22271/tpi.2022.v11.i7ag.14368

Kumar, V., Ahluwalia, V., Saran, S., Kumar, J., Patel, A. K., & Singhania, R. R. (2021). Recent developments on solid-state fermentation for production of microbial secondary metabolites: challenges and solutions. *Bioresource Technology, 323*, 124566.

Lahiri, D., Nag, M., Sarkar, T., Ray, R. R., Shariati, M. A., Rebezov, M., Bangar, S. P., & Lorenzo, J. M. (2022). Microbes for meat preservation and fortification.

Laslo, É., György, É., & Czikó, A. (2019). Meat starter cultures: isolation and characterization of lactic acid bacteria from traditional sausages. *Acta Universitatis Sapientiae Alimentaria, 12*(1), 54–69. https://doi.org/10.2478/ausal-2019-0004

Linder, T. (2019a). Making the case for edible microorganisms as an integral part of a more sustainable and resilient food production system. *Food Security, 11*, 265–278. https://doi.org/10.1007/s12571-019-00912-3

Linder, T. (2019b). Edible microorganisms—an overlooked technology option to counteract agricultural expansion. *Frontiers in Sustainable Food Systems, 3*, 8–11. https://doi.org/10.3389/fsufs.2019.00032

Llorente, B., Williams, T. C., Goold, H. D., et al. (2022). Harnessing bioengineered microbes as a versatile platform for space nutrition. *Nature Communications, 13*, 6177. https://doi.org/10.1038/s41467-022-33974-7

Loeffler, M., Hilbig, J., Velasco, L., & Weiss, J. (2020). Usage of in situ exopolysaccharide-forming lactic acid bacteria in food production: meat products—a new field of application? *Comprehensive Reviews in Food Science and Food Safety, 19*(6), 2932–2954. https://doi.org/10.1111/1541-4337.12615

Łopusiewicz, Ł., Drozłowska, E., Siedlecka, P., Mężyńska, M., Bartkowiak, A., Sienkiewicz, M., Zielińska-Bliźniewska, H., et al. (2019). Development, characterization, and bioactivity of non-dairy kefir-like fermented beverage based on flaxseed oil cake. *Foods, 8*(11), 544. https://doi.org/10.3390/foods8110544

Lund, B. M., Mølbak, K., & Aarestrup, F. M. (2018). Food safety challenges and interventions in the food production chain. *Microbiology and Molecular Biology Reviews, 82*(3), e00034–e00017.

Mallappa, R. H., Balasubramaniam, C., Nataraj, B. H., Ramesh, C., Kadyan, S., Pradhan, D., Muniyappa, S. K., & Grover, S. (2021). Microbial diversity and functionality of traditional fermented milk products of India: current scenario and future perspectives. *International Dairy Journal, 114*, 104941. https://doi.org/10.1016/j.idairyj.2020.104941

Markowiak, P., & Śliżewska, K. (2017). Effects of probiotics, prebiotics, and synbiotics on human health. *Nutrients, 9*(9). https://doi.org/10.3390/nu9091021

Marti-Quijal, F. J., Remize, F., Meca, G., Ferrer, E., Ruiz, M.-J., & Barba, F. J. (2020). Fermentation in fish and by-products processing: an overview of current research and future prospects. *Current opinion in food science, 31*, 9–16. https://doi.org/10.1016/j.cofs.2019.08.001; ISSN 2214–7993.

Matle, I., Mbatha, K. R., & Madoroba, E. (2020). A review of listeria monocytogenes from meat and meat products: epidemiology, virulence factors, antimicrobial resistance and diagnosis. *Onderstepoort Journal of Veterinary Research, 87*(1), 1–20.

Mazhar, S., Yasmeen, R., Chaudhry, A., & Summia, K. (2022). Role of microorganisms in modern food industry. *International Journal of Food Science & Technology, 4*, 65–77.

Milanovic, V., Osimani, A., Garofalo, C., Belleggia, L., Maoloni, A., Cardinali, F., Mozzon, M., Foligni, R., Aquilanti, L., & Clementi, F. (2020). Selection of cereal-sourced lactic acid bacteria as candidate starters for the baking industry. *PLoS One, 15*, 1–21. https://doi.org/10.1371/journal.pone.0236190

Moeiny, P., Sayahpour, S., Raofie, F., Aminikhah, A., & Kharazifard, M. J. (2017). Assessment of the effect of fruit (apple) and plain yoghurt consumption on plaque pH. *Original Research, 6*(3), 117–124.

Moradi, M., Molaei, R., Kousheh, S. A., Guimarães, J. T., & McClements, D. J. (2021). Carbon dots synthesized from microorganisms and food by-products: active and smart food packaging applications. *Critical Reviews in Food Science and Nutrition, 63*, 1943. https://doi.org/10.1080/10408398.2021.2015283

Morya, S., Chandra, R., & Seelam, B. S. (2017). Microbial characteristics of whey and sorghum based low fat probiotic beverage. *International Journal of Chemical Studies, 5*(4), 403–406.

Morya, S., Singh, N., & Awuchi, C. G. (2022a). Environmental management technologies. In *Health hazards of food allergens and related safety measures* (pp. 99–114). https://doi.org/10.1201/9781003239956-7

Morya, S., Awuchi, C. G., Neumann, A., Napoles, J., & Kumar, D. (2022b). Chapter 7—advancement in acidophilus milk production technology. In *Advances in dairy microbial products*. Woodhead Publishing. https://doi.org/10.1016/B978-0-323-85793-2.00016-3; ISBN 9780323857932.

Müller, D. C., Mischler, S., Schönlechner, R., & Miescher Schwenninger, S. (2021). Multiple techno-functional characteristics of leuconostoc and their potential in sourdough fermentations. *Microorganisms, 9*(8), 1–15. https://doi.org/10.3390/microorganisms9081633

Munekata, P. E., Pateiro, M., Bellucci, E. R. B., Domínguez, R., da Silva Barretto, A. C., & Lorenzo, J. M. (2021). Strategies to increase the shelf life of meat and meat products with phenolic compounds. In *Advances in food and nutrition research* (Vol. 98, pp. 171–205). Academic Press.

Mussagy, C. U., Winterburn, J., Santos-Ebinuma, V. C., & Pereira, J. F. B. (2019). Production and extraction of carotenoids produced by microorganisms. *Applied Microbiology and Biotechnology, 103*(3), 1095–1114. https://doi.org/10.1007/s00253-018-9557-5

Najgebauer-Lejko, D., Liszka, K., Tabaszewska, M., & Domagała, J. (2021). Probiotic yoghurts with sea buckthorn, elderberry, and sloe fruit purees. *Molecules, 26*(8), 2345. https://doi.org/10.3390/molecules26082345

Nazhand, A., Souto, E. B., Lucarini, M., Souto, S. B., Durazzo, A., & Santini, A. (2020). Ready to use therapeutical beverages: focus on functional beverages containing probiotics, prebiotics and synbiotics. *Beverages, 6*, 26. https://doi.org/10.3390/beverages6020026

Odeyemi, O. A., Alegbeleye, O. O., Strateva, M., & Stratev, D. (2020). Understanding spoilage microbial community and spoilage mechanisms in foods of animal origin. *Comprehensive Reviews in Food Science and Food Safety, 19*(2), 311–331. https://doi.org/10.1111/1541-4337.12526

Owusu-Kwarteng, J., Akabanda, F., Agyei, D., & Jespersen, L. (2020). Microbial safety of Milk production and fermented dairy products in Africa. *Microorganisms, 8*(5), 752. https://doi.org/10.3390/microorganisms8050752

Peng, K., Koubaa, M., Bals, O., & Vorobiev, E. (2020). Recent insights in the impact of emerging technologies on lactic acid bacteria: a review. *Food Research International, 137*, 109544. https://doi.org/10.1016/j.foodres.2020.109544

Pham, J. V., Yilma, M. A., Feliz, A., Majid, M. T., Maffetone, N., Walker, J. R., Kim, E., Cho, H. J., Reynolds, J. M., Song, M. C., Park, S. R., & Yoon, Y. J. (2019). A review of the microbial production of bioactive natural products and biologics. *Frontiers in Microbiology, 10*, 1–27. https://doi.org/10.3389/fmicb.2019.01404

Pires, A. F., Marnotes, N. G., Rubio, O. D., Garcia, A. C., & Pereira, C. D. (2021). Dairy by-products: a review on the valorization of whey and second cheese whey. *Foods, 10*(5), 1067. https://doi.org/10.3390/foods10051067

Pradhan, C. K., Rahaman, S., Abdul Alim Sheikh, M., Kole, A., & Maity, T. (2019). EEG signal analysis using different clustering techniques. In *Emerging technologies in data mining and information security* (pp. 99–105). Springer.

Rai, A. K., Pandey, A., & Sahoo, D. (2019). Biotechnological potential of yeasts in functional food industry. *Trends in Food Science and Technology, 83*, 129–137. https://doi.org/10.1016/j.tifs.2018.11.016

Reque, P. M., & Brandelli, A. (2021). Encapsulation of probiotics and nutraceuticals: applications in functional food industry. *Trends in Food Science and Technology, 114*, 1–10. https://doi.org/10.1016/j.tifs.2021.05.022

Rollini, M., Musatti, A., Cavicchioli, D., Bussini, D., Farris, S., Rovera, C., & Barbiroli, A. (2020). From cheese whey permeate to Sakacin-A/bacterial cellulose nanocrystal conjugates for antimicrobial food packaging applications: a circular economy case study. *Scientific Reports, 10*(1), 1–14.

Salehi, F. (2021). Quality, physicochemical, and textural properties of dairy products containing fruits and vegetables: a review. *Food Science and Nutrition, 9*(8), 4666–4686. https://doi.org/10.1002/fsn3.2430

Samedi, L., & Linton Charles, A. (2019). Evaluation of technological and probiotic abilities of local lactic acid bacteria. *Journal of Applied & Environmental Microbiology, 7*(1), 9–19. https://doi.org/10.12691/jaem-7-1-3

Sánchez, A., García-Gimeno, R. M., & Plaza, L. (2018). Microorganisms and the environment in food production and preservation. *Food Research International, 107*, 54–66.

Sarkar, S. (2019). Potentiality of probiotic fruit yoghurt as a functional food—a review. *J Nutr Food Sci, 2*, 100009. https://doi.org/10.1108/NFS-05-2018-0139

Scarpellini, E., Rinninella, E., Basilico, M., Colomier, E., Rasetti, C., Larussa, T., Santori, P., et al. (2021). From pre- and probiotics to post-biotics: a narrative review. *International journal of environmental research and public health, 19*(1), 37. https://doi.org/10.3390/ijerph19010037

Shamloo, E., Hosseini, H., Moghadam, Z. A., Larsen, M. H., Haslberger, A., & Alebouyeh, M. (2019). Importance of listeria monocytogenes in food safety: a review of its prevalence, detection, and antibiotic resistance. *Iranian Journal of Veterinary Research, 20*(4), 241.

Sharif, M., Zafar, M. H., Aqib, A. I., Saeed, M., Farag, M. R., & Alagawany, M. (2021). Single cell protein: sources, mechanism of production, nutritional value and its uses in aquaculture nutrition. *Aquaculture, 531*, 735885. https://doi.org/10.1016/j.aquaculture.2020.735885

Sharma, P., Gaur, V. K., Kim, S.-H., & Pandey, A. (2020). Microbial strategies for bio-transforming food waste into resources, 122580. *Bioresource Technology, 299*. https://doi.org/10.1016/j.biortech.2019.122580; ISSN 0960-8524.

Singh, V. P. (2018). Recent approaches in food bio-preservation—a review. *Open Veterinary Journal, 8*(1), 104–111. https://doi.org/10.4314/ovj.v8i1.16

Singh, R., Singh, A., & Sachan, S. (2018). Enzymes used in the food industry: friends or foes? In *Enzymes in food biotechnology: production, applications, and future prospects*. Elsevier Inc. https://doi.org/10.1016/B978-0-12-813280-7.00048-7

Sirini, N., Frizzo, L. S., Aleu, G., Soto, L. P., & Rosmini, M. R. (2021). Use of probiotic microorganisms in the formulation of healthy meat products. *Current Opinion in Food Science, 38*, 141–146. https://doi.org/10.1016/j.cofs.2020.11.007

Sutay Kocabaş, D., & Grumet, R. (2019). Evolving regulatory policies regarding food enzymes produced by recombinant microorganisms. *GM Crops and Food, 10*(4), 191–207. https://doi.org/10.1080/21645698.2019.1649531

Syiemlieh, I., & Morya, S. (2022). Dairy and non-dairy based probiotics: a review. *Pharma Innov J, 11*, 2956–2964. https://doi.org/10.22271/tpi.2022.v11.i6Sak.13568

Tamoghna, G., Arun, B., Anupama, S., & Kumar, N. N. (2019). Mechanistic insights into probiotic properties of lactic acid bacteria associated with ethnic fermented dairy products. *Frontiers in Microbiology, 10*. https://doi.org/10.3389/fmicb.2019.00502

Tesfaye, W., Suarez-Lepe, J. A., Loira, I., Palomero, F., & Morata, A. (2019). Dairy and nondairy-based beverages as a vehicle for probiotics, prebiotics, and symbiotics: alternatives to health versus disease binomial approach through food. In *Milk-based beverages* (Vol. 9). The Science of Beverages. Elsevier. https://doi.org/10.1016/B978-0-12-815504-2.00014-1

Turkmen, N., Akal, C., & Özer, B. (2019). Probiotic dairy-based beverages: a review. *Journal of Functional Foods, 53*, 62–75.

Vanderhoof, J. A., Young, V. B., & Rolfe, R. D. (2018). The role of the microbiome in food safety and preservation. *Microbiology and Molecular Biology Reviews, 82*(3), e00035–e00017.

Vasile, M. A., & Bahrim, G. (2020). *Selection of wild lactic acid bacteria strains as promoters of postbiotics in gluten-free sourdoughs* (pp. 1–29).

Vinod, K. R. (2018). Functional foods: a review. *Journal of Food Science and Technology, 55*(5), 1689–1694. https://doi.org/10.1007/s13197-018-3077-8

Vitorino, L. C., & Bessa, L. A. (2017). Technological microbiology: development and applications. *Frontiers in Microbiology, 8*, 827. https://doi.org/10.3389/fmicb.2017.00827

Zarzecka, U., Zadernowska, A., & Chajęcka-Wierzchowska, W. (2020). Starter cultures as a reservoir of antibiotic resistant microorganisms. *Lwt, 127*, 109424. https://doi.org/10.1016/j.lwt.2020.109424

Zendeboodi, F., Khorshidian, N., Mortazavian, A. M., & da Cruz, A. G. (2020). Probiotic: conceptualization from a new approach. *Current Opinion in Food Science, 32*, 103–123.

Zhang, Y., Chen, X., & Wang, H. (2018). High pressure processing and pulsed electric field technology in food preservation: a review. *Journal of Food Engineering, 229*, 1–11.

Zhang, Q., Fu, C., Zhao, C., Yang, S., Zheng, Y., Xia, M., Yan, Y., Lang, F., & Wang, M. (2020). Monitoring microbial succession and metabolic activity during manual and mechanical solid-state fermentation of Chinese cereal vinegar. *Lwt, 133*, 109868. https://doi.org/10.1016/j.lwt.2020.109868

Zhou, N., Semumu, T., & Gamero, A. (2021). Non-conventional yeasts as alternatives in modern baking for improved performance and aroma enhancement. *Fermentation, 7*(3), 102. https://doi.org/10.3390/fermentation7030102

Zotta, T., Solieri, L., Iacumin, L., Picozzi, C., & Gullo, M. (2020). Valorization of cheese whey using microbial fermentations. *Applied Microbiology and Biotechnology, 104*(7), 2749–2764. https://doi.org/10.1007/s00253-020-10408-2

# Chapter 13
# Microbial Bioinformatics Approach in Food Science

**Iqra Jan, Priya Sharma, and Anu Bansal**

**Abstract** It has been studied that food is an important thing that plays a vital role in regulating different processes in the body. "Omics era has been growing rapidly, this growing progression with the computational mean bioinformatics interprets and curates" biological data, hence a widespread integration has been seen in various disciplines of life science which also include food science. Bioinformatics helps in increasing the food's nutritional value, food taste, and also helps in increasing the quality of food, as it can efficiently access all the data of metabolomics, genomics, and proteomics, which further is available to various groups of food industries and food companies. Therefore, in this chapter, we discussed the efficient role of microbial bioinformatics in food science.

**Keywords** Bioinformatics · Food processing · Omics era · Proteomics · Genomics

## 13.1 Introduction

The demand for food safety, healthy food, and food full of nutrition but with minute preservatives has increased worldwide (Priya & Satheeshkumar, 2020). Thus, it is a great challenge for food authorities, scientists, and food industries to ensure the quality and safety of food. However, food commodities must have to pass through various different levels, harvesting, distribution, processing, transportation, and storage, as during these stages the food is vulnerable to a number of microbes,

I. Jan
Byjus, Srinagar, Jammu and Kashmir, India

P. Sharma
DAV University, Jalandhar, Punjab, India

A. Bansal (✉)
School of Bioengineering and Biosciences, Lovely Professional University, Phagwara, Punjab, India
e-mail: anu.21971@lpu.co.in


A. Karnwal, A. R. Mohammad Said Al-Tawaha (eds.), *Food Microbial Sustainability*, https://doi.org/10.1007/978-981-99-4784-3_13

oxidative deterioration, and insects which later degrade the product quality, hence negatively affects the health of the consumer. Before reaching the consumer, the food products have gone through a form of processes such as fermentation to packaging in which microorganisms played a vital role. Food can be transformed by using microorganisms to the desired end product like the fermentation of food that as bread, olives, and rice, or the fermentation of alcoholics such as wine and beer. They also help in the fermentation of many dairy products like yogurt, milk, and cheese. Meat can also be fermented by using microorganisms. Microorganism not only helps in the transformation of food to the desired end product but can also spoil or contaminate the food (Smid & Kleerebezem, 2014). The effects of microorganisms whether it is desired or undesired can be predicted and assessed by bioinformatics (Garrigues et al., 2013). Bioinformatics has the potential to be crucial in many procedures of food processing and in improving food quality. We could overcome challenging conditions like food scarcity, nutritional quality, and food deterioration by using current science. Its use in food sciences is influenced by the numerous success stories of the application of bioinformatics technologies in diverse biological domains. In order to effectively use the obtained biological information in biotechnology, bioinformatics aids in the administration of biological data. The field is growing alarmingly quickly and becoming an increasingly important research tool. It has been recognized as a significant scientific discipline and is an interdisciplinary field. It causes a radical shift in a number of fields, including biotechnology, comparative genomics, molecular biology, molecular medicine, and molecular evolution. Food sciences have the potential to make a significant contribution. However, this sector has received less attention from bioinformatics approaches. In this chapter, we discussed the microbial bioinformatics approaches that can be used in the food and nutrition sciences to advance and expand research to ensure food safety and food quality through various processes.

## 13.2 Bioinformatics and Information Technology

In order to aid with biological concerns, the recently developed field of bioinformatics blends biology, information science, and mathematics. In 1968, the term "bioinformatics" was first used, and it was officially defined in 1978. The term "computational biology" has also been used to describe bioinformatics (Lander et al., 2001). The use of information technology in bioinformatics is to organize, store, and analyze the enormous amount of biological data that is available in the form of protein sequences and structures, which serve as the building blocks of living things, and nucleic acids, which carry information. Understanding various biological processes is the aim of biological informatics. Bioinformatics uses computational methods prior to biotechnology methods which later distinguish it from other sciences. Bioinformatics is basically used to find the gene assembly, sequence alignment, gene finding, drug discovery, structure, and prediction of genes. Now computational programs have made it easy for scientists to analyze data. Software and tools of bioinformatics vary from simple command line tools to complex databases and programs. SOAP and REST-based interfaces are used for a variety of purposes and applications of bioinformatics.

The need to create biological sequence databases initially generates interest in bioinformatics. Shortly after the insulin protein sequence was made public in 1956, the first database was created. Incidentally, it may be noted that insulin is the first protein to be sequenced. The sequence of insulin consisted of just 51 residues (analogous to alphabets in a sentence) which characterize the sequence. Around the mid-1960s, the first nucleic acid sequence of yeast tRNA with 77 bases (individual units of nucleic acids) was found. During this period, the three-dimensional structure of proteins was studied and the well-known Protein Data Bank was developed, as in 1972 the first database structure of the protein with only 10 entries. Now, it has reached a large number of databases and that is almost 10,000 entries. While the initial protein sequence database was maintained at the individual laboratories, work on a formal, consolidated database known as the SWISS-PROT protein sequence database, which currently contains about 70,000 protein sequences from more than 5000 model organisms, a small portion of all known organisms, was started in 1986. This massive variety of divergent data resources is now available for study and research by both academic institutions and industries. These are made available as public domain information in the immense interest of the research community through the Internet (www.ncbi.nlm.nih.gov) and CDROMs (www.rcsb.org). These databases are constantly updated with additional entries (Tidsall, 2013).

Computers and software tools are extensively used for creating these databases and to anticipate models of proteins structure and the protein's function, to determine the coding (valid) regions of nucleic acid sequences, find suitable drug compounds from a large pool, and optimize the drug development process by predicting possible targets. As the analysis involves a lot of computational effort, it is impossible to do it manually. Some of the software tools which are handy in the analysis include BLAST—commonly used for comparing sequences; Annotator—an interactive genome (nucleic acid sequence) analysis tool; GeneFinder—a tool to identify coding regions and splice sites (Baxevanis & Ouellette, n.d.). From the pharmaceutical industry's point of view, bioinformatics is the key to rational drug discovery. Using high-power computing workstations and software like Insight, it decreases the number of screen trials of drug compounds and discovering prospective treatment targets for a particular disease. Pharmacogenomics is a new branch of pharmacology as a result of the extensive use of bioinformatics in genome sequencing, where potential targets for drug development are hypothesized from the genome sequences. Molecular modeling which requires a lot of calculations has become faster due to the advances in computer processors and their architecture.

Under Common Graphical User Interfaces (GUIs) various leading bioinformatics firms developed various bioinformatics software which permits research scientists to integrate their diverse tools and data. This creates more opportunities for research and discovery, through savings in time and data coordination. It also permits scientists to share information and provides a powerful solution to archive data. Thus, bioinformatics has become synonymous with biological research.

Bioinformatics tools for efficient research will have significant implications in medical sciences and the betterment of human lives (Chun et al., 2014).

## 13.3 Bioinformatics and Food Science

Bioinformatics has a significant impact on food safety. In order to unravel the mechanistic investigation of biological processes that connect the cell, chromosome, DNA, RNA, protein, and metabolite, it develops novel methods and software that could be used as tools for processing, analyzing, and interpreting biological data obtained from various biochemical and omics approaches (genomics, transcriptomics, proteomics, and metabolomics) (Valdés et al., 2017). Therefore, the combined use of computational methods and experimental science may lead to the development of novel hypotheses regarding how living systems function. Recent studies have shown a beneficial relationship between health/disease and functional food ingredients; as a result, the use of bioinformatics technology in the fields of food science and nutrition, particularly in foodomics, has attracted significant interest globally in recent years. Moreover, foodborne pathogens have been detected and identified using bioinformatics techniques and next-generation sequencing technologies. The bioinformatics method is now frequently used to forecast the preservative and functional properties of plant-based bioactive chemicals. A variety of in silico techniques were employed to investigate the functional properties of bioactive compounds, including protein databases for the study of the primary sequence and structural data of various proteins, potential bioactivity prediction, allergenicity/toxicity prediction/analyzing, and Protein Structure Prediction Server for the study of anti-hypertensive peptide, precursor proteins, and antibacterial peptides (Prakash et al., 2020).

## 13.4 Databases in Food Sciences

Additionally, a variety of databases related to agriculture, food science literature, food allergies and food safety are accessible in the public domain and are based on the computational methodology described below (Table 13.1).

### *13.4.1 Agricola*

It offers access to millions of citations and includes bibliographic entries from the National Agricultural Library, Department of Agriculture (U.S.). The citations include journal articles, book chapters, theses, and many other documents that are all related to agriculture.

**Table 13.1** Various computational databases and tools in food science

| Database | Managed by | Established | Key features | Web link | References |
|---|---|---|---|---|---|
| | Literature based | | | | |
| CAB Abstracts (Ovid) | CABI—the Association of International Research and Development Centers for agriculture | 1973 | Provides international life science based literature | https://www.cabi.org/publishing-products/cab-abstracts/ | CAB Abstracts—CABI.Org, (n.d.) |
| Food Science and Technology Abstracts (FSTA) | IFIS—International Food Information Services | 1969 | R&D database provides information related to food and health science | https://www.ifis.org/fsta | Food science technology abstracts (FSTA)\|Penn State University libraries, (n.d.) |
| Food Studies Online | New York University | 1960 | Provides visual, text, and video content for food studies | https://guides.nyu.edu/foodstudies/databases | Gustafson, (n.d.) |
| Agricola | National Agricultural Library (NAL)-USDA | 1970 | Cataloging and indexing records from the NAL | https://agricola.nal.usda.gov/ | AGRICOLA\|EBSCO, (n.d.) |
| AGRIS, international information system for the agricultural sciences and technology | Food and agriculture Organization of the United Nations (FAO) | 1974 | Information resource for food and agriculture scientific literature | https://www.fao.org/agris/ | AGRIS the international system for agricultural science and technology—Data provider guide, (n.d.) |
| AgriRXiv | CABI | 2017 | Source of unpublished related to agriculture sciences | https://www.cabidigitallibrary.org/journal/agrirxiv | AgriRxiv\|UCSB library, (n.d.) |

(continued)

**Table 13.1** (continued)

| Database | Managed by | Established | Key features | Web link | References |
|---|---|---|---|---|---|
| American Society of Agricultural and Biological Engineers (ASABE) technical library | American Society of Agricultural and Biological Engineers | 2001 | Technical documents are available related to agriculture and food advancement | https://elibrary.asabe.org/ | Gustafson, (n.d.) |
| Food and Nutrient Database for Dietary Studies (FNDDS) | Agricultural Research Service U.S. DEPARTMENT OF AGRICULTURE | 2017 | Nutrient intake estimation for food and beverages | https://www.ars.usda.gov/northeast-area/beltsville-md-bhnrc/beltsville-human-nutrition-research-center/food-surveys-research-group/docs/fndds/ | Food and Nutrient Database for Dietary Studies (FNDDS)lag Data Commons, (n.d.) |
| The food database (FooDB) | Canadian Institutes of Health Research, Canada Foundation for Innovation, and the Metabolomics Innovation Centre (TMIC) | – | Information related to micro- and macronutrients | https://foodb.ca/ | FooDB, (n.d.) |
| EuroFIR-BASIS | European Food Informative Resources | 1990 | Repository for plant-based bioactive compounds with potential health benefits | https://www.eurofir.org/our-tools/ebasis/ | Gry et al., (2007) |
| FoodWiki database | – | 2010 | Repository for food and nutritional data | https://www.takeaway.com/foodwiki/ | Woolf et al., (2010) |
| Foodmoics database | The University of Florida, Institute of Food and Agricultural Sciences | 2016 | Repository for food molecular profile | https://borum.ifas.ufl.edu/pkt-for-health-providers/c4c-foodomics-database/ | Allen et al., (n.d.) |

| | | | | | |
|---|---|---|---|---|---|
| The Food Allergy Research and Resource Program (FARRP) | University of Nebraska, USA | 1995 | Protein allergen database | http://allergenonline.com/ | Food Allergens—International Regulatory Chart\|FARRP\|Nebraska, (n.d.) |
| Structural Database of Allergenic Proteins | The University of Texas Medical Branch, USA | 2001 | Sequence, structures, and IgE epitopes of allergenic proteins | https://fermi.utmb.edu/ | Ivanciuc et al., (2003) |
| Allermatch | WFSR—Wageningen University and Research and Bioscience and Wageningen University and Research, Netherlands | 2004 | Protein potential allergenicity information | http://www.allermatch.org/ | Fiers et al., (2004) |
| International Food Composition Table/ Database Directory | FAO, USA | 1988 | Food Composition Database | https://www.fao.org/infoods/infoods/tables-and-databases/en/ | Scrimshaw, (1997) |

### 13.4.2 Alt HealthWatch

The diverse viewpoints of integrated, complementary, and holistic approaches to healthcare and wellness are the main focus of this database of alternative health research. With thorough coverage of the whole range of topics covered by complementary and alternative medicine, Alt HealthWatch offers the most recent information and tools for the expanding practice of holistic medicine and therapies.

### 13.4.3 CabDirect

This is the premier database for literature on agricultural and applied life sciences, with excellent international coverage, and includes the CAB Abstracts and CAB Health.

### 13.4.4 CINAHL Complete

For those in the nursing and associated health fields, this is the only reliable research resource. Users can quickly and easily access the best nursing and allied health articles, evidence-based care sheets, and brief courses with CINAHL Complete.

### 13.4.5 Health Source: Nursing/Academic Edition

Researchers, allied health workers, nurses, and medical educators get access to full-text academic publications concentrating on a variety of medical fields through Health Source: Nursing/Academic Edition.

### 13.4.6 Hospitality and Tourism Index

The bibliographic database Hospitality and Tourism Index includes news from the business as well as scholarly research on all facets of hospitality and tourism. With coverage going back to 1965, this extensive collection has more than 749,000 records from around 650 different titles.

### 13.4.7 Human Kinetics Library Core Collection

The Human Kinetics Library was created with the goal of empowering all individuals with reliable information about physical activity and sport in order to develop their knowledge, improve their performance, and improve their health and fitness. The platform offers a growing library of cross-searchable ebooks and videos for coaching, sports, fitness, and exercise from the top educational publisher in the world.

### 13.4.8 IBISWorld

IBISWorld is a comprehensive database with hundreds of publications from industry market research. Agriculture, mining, utilities, construction, manufacturing, wholesale and retail trade, transportation, information, finance, real estate, technical and scientific, waste management, healthcare, recreation, and food services are among the sectors that are represented.

### 13.4.9 JoVE

By producing and disseminating movies of scientific experiments from the world's finest laboratories, JoVE: Journal of Visualized Experiments develops the best solutions for promoting research and science education. JoVE boosts STEM research productivity and student learning, saving their institutions time and money. It does this by enabling scientists, educators, and students to see the minute details of cutting-edge experiments rather than reading about them in text articles. The following JoVE video journal items are available to LSU subscribers: SE2: Basic Methods in Cellular and Molecular Biology, JoVE Chemistry Series, JoVE Environment, JoVE Bioengineering, JoVE Engineering, JoVE Immunology and Infection, JoVE Medicine, JoVE Neuroscience, and JoVE Biology.

### 13.4.10 Nursing and Allied Health Source

Nursing students, researchers, and teaching staff do not have to spend as much time seeking for practice reference materials thanks to this single devoted site. Leading nursing journals are made available alongside videos, dissertations, ebooks, and study courses, making it simpler to combine intricate theoretical knowledge with applying it to real-world situations.

### *13.4.11 Popular Medicine in America, 1800–1900*

This one-of-a-kind collection uses a wide range of content that was intended for laypeople rather than healthcare experts to demonstrate how "popular" medicine evolved in America during the nineteenth century. Investigate a variety of printed sources, such as old books, pamphlets, business cards, and eye-catching advertising detritus.

### *13.4.12 Sage Research Methods*

Beginning and experienced researchers can get assistance from SAGE Research Methods at every stage of a research project, including developing a research question, selecting a methodology, collecting and analyzing data, and writing up and publishing the findings. The book, reference, and journal content in SAGE Research Methods—from SAGE's renowned methods publishing program—and advanced features aid researchers of all levels in conducting their research by providing information on the full range of qualitative, quantitative, and mixed methods for the social and behavioral sciences as well as many methods frequently used in the hard sciences.

### *13.4.13 FooDB*

FooDB (http://foodb.ca/) is an online available comprehensive resource on food constituents, chemistry, and biology. It provides information on both macronutrients and micronutrients including many of the constituents that give color, flavor, texture, taste, and aroma to the food. Each chemical entry contains more than 100 separate data fields covering detailed biochemical, compositional, and physiological information. The data are obtained from the literature and it includes data on the compound's description, nomenclature, chemical class, information on its structure, its physicochemical data, its food source(s), its color, taste, aroma, physiological effect, presumptive health effects and concentrations in various foods (Schmidtke et al., 2013).

### *13.4.14 EuroFIR-BASIS*

It uniquely combines food composition and biological activity for bioactive compounds in plant-based foods. It is available only at https://www.eurofir.org/. The database covers multiple compound classes and 330 major food plants and their edible parts with data sourced from quality-assessed, peer-reviewed literature. The database will be a valuable resource for risk authorities, food regulatory and advisory bodies, epidemiologists and researchers interested in diet and health relationships, and product developers within the food industry (Overbeek et al., 2014).

### *13.4.15 FoodWiki Database*

It serves as a consensus-based central repository for food and nutritional data. It makes use of the vast amount of data that can be simply and effectively controlled by bioinformatics methodologies and protocols. With the goal of enhancing the nutritional value and quality of food sources, such resources will improve and develop the fields of food and nutrition science. For the development of bioinformatics in this field, this consolidated, extensive database on food and nutrition is essential (Guttman et al., 2014).

### *13.4.16 Foodomics Database*

It is a database for food molecular profiles. It released 28 of the USDA National Nutrient Databases for Standard Reference (SR28) which contains data for up to 150 food molecules. Nutrition Facts Labels (NFLs) on branded food reflect current nutrient information but usually contain less than 15 nutrients. The integration of current NFLs and SR28 is needed to obtain food comic profiles of a person's (or animal's) dietary intake and therefore precision medicine for each person can be implemented. In this way, the suitable food molecules are given in the right amount, at the right time, and to the right person having part of precision medicine (Allen et al., n.d.).

**The Food Allergy Research and Resource Program** (FARRP) (http://allergenonline.com/) Protein Allergen Database was established in year 1995 by the University of Nebraska, USA. Develop and offer the food sector reliable knowledge, professional advice, instruments, and services on allergic foods. Develop and deliver reliable information, professional opinions, tools, and services to the food and allied industries about novel foods and food ingredients, including genetically modified goods.

In order to improve the safety of food products for customers with food allergies and sensitivities, FARRP works with and collaborates with research institutes, governmental agencies, consumer groups, and scientific societies around the world using a holistic approach.

**Structural Database of Allergenic Proteins**: SDAP (https://fermi.utmb.edu/) was established in 2001 and managed by Department of Biochemistry and Molecular Biology, The University of Texas Medical Branch, USA. It provides rapid, cross-referenced access to the sequences, structures, and IgE epitopes of allergenic proteins. Its core is made up of a number of CGI scripts that accept user queries, query the database, run various calculations involving protein allergenic determinants, and set up the output in HTML format. The database component of SDAP includes quick links to the main web servers for proteins and literature as well as details on the allergen's name, source, sequence, and structure as well as information on IgE epitopes and research references. To find sections of known allergens comparable to user-supplied peptides or IgE epitopes chosen from the SDAP database, the computational component of SDAP employs a novel technique based on conserved features of amino acid side chains (Ivanciuc et al., 2003).

**Allermatch** (http://www.allermatch.org/) was established in 2004 by the WFSR—Wageningen University and Research and Bioscience in collaboration with Wageningen University and Research, Netherlands. It is an online available tool for the quick and accurate assessment of a protein's or peptide's potential allergenicity in accordance with the most recent recommendations of the FAO/WHO Expert Consultation as stated in the Codex alimentarium. It offers a current FAO/WHO-compliant webtool that is simple to use, effective, and valuable for analyzing the possible allergenicity of proteins added to genetically modified food before it is released in the market (Fiers et al., 2004).

## 13.5 Microbial Informatics

Traditional methods of microbe identification and categorization are frequently based on numerical taxonomy. In the years that followed, numerical taxonomic approaches were widely used for classification and identification, and they were made stronger by the widespread use of computers in many different scientific domains. Later, computer-assisted identification systems were created for the identification of bacterial groups, primarily using raw data from commercial kits or phenotypic data gathered using conventional methods. Microbial informatics is basically designed to study microbes that are computationally oriented, biologically motivated, and practical. During the first and second World Wars, in 1918 around 50–100 million deaths has been reported because of Influenza (Spanish flu). Nowadays, life would not exist without microbes. The setting in which microbial informatics is currently occurring adds another layer such as climate change and shifting land-use patterns brought on by a growing global population will have an increasing impact on all species, beginning with microorganisms (Talukdar et al., 2009). Computers have joined PCR machines, agar plates, and restriction enzymes as instruments for studying the species and systems, given the intricacy of processes currently being investigated (Tiwari & Sekhar, 2007). In India, DBT-CMI is an integrated "one-stop portal" for niche-specific microbial informatics with information on the biotechnological potential of microbes. The portal is the hub for scientists, industry, and policymakers to develop sustainable microbial technologies required for the nation (Langille et al., 2013).

In predicting and evaluating the desirable and unintended impacts of microorganisms on food, bioinformatics is becoming increasingly important. The impacts of microorganisms on food can be predicted using sequence-based prediction of microbial functionality and access to databases that combine data from genomes, systems biology, phenotypes, and "biomarkers" for functioning in specific taxa. Biomass yield may be enhanced via microbial bioinformatics, which will increase food output. For this, several bacteria, including some food-relevant microorganisms, have had their metabolic models created and the genetic sequence of the organism sequenced. The growth of the organism under the metabolic constraints provided by the substrate availability in the medium can be in silico simulated using comprehensive genome-scale metabolic models and algorithms like flux balance analysis. For instance, Identax, a novel software tool (www.identax.org), is given for the computer-assisted identification of microbes using just the findings of traditional biochemical assays. Identax is a multiplatform, user-friendly program and enhances existing microbiological identification software (Flores et al., 2009).

Microbes are a growing substitute for artificial fertilizers and pesticides in agriculture for improving crop yield. In order to better understand the ecological factors, molecular connections, and microbiome function, microbiome analysis is used to help the interpretation of sequencing data relevant to plants and soil. Understanding rhizosphere microbial dynamics and their implications for plant health and yield can be assessed for sustainable agriculture. Similarly, microbiome analysis for seeds would unlock doors for microbiome-assisted breeding strategies toward nutrition-enhanced foods. The microbiome studies specifically targeted evaluation for pathogen screening studies, dynamics of artificial syn-communities to improve the yield of plants, inter-cropping studies, pesticide and fertilizer effects on the natural cropland ecosystems (Di Bella et al., 2013).

## 13.6 Bioinformatics in Food Quality and Safety

Food is a necessity for survival; thus, it plays a significant role in daily life. However, before a product is consumed by a consumer, it often goes through a number of processing steps. It underwent fermenting before being packaged. Many of these processes, which later turn the food into the desired final product heavily, rely on microorganisms (Fang et al., 2010). However, the characteristics of fermented products are highly affected by the type of microorganism used in the fermentation process, such as yeast and lactic acid where the primary fermentation result of lactic acid bacteria is lactic acid, the primary product of fermentation for yeasts is ethanol. The food sector is particularly active in strain performance optimization, both in terms of controlling fermentation through the use of specific starter cultures to start the fermentation process and in terms of diversifying product attributes such as flavor and texture.

Therefore, the desired and unwanted effects that microorganisms may have on food can be anticipated using bioinformatics. However, if bioinformatics and the laboratory can be used together for a variety of specific scientific projects, it can be quite effective. There is a growing appreciation for bioinformatics in the area of food quality and safety (Kumar & Chordia, 2017) as genome sequencing projects are now focusing on foodborne pathogens and innovative ways which will help in determining the source of foodborne illnesses. The identification of spoilage and harmful microorganisms as well as the forecasting of thermal preservation stress resistance may one day be made possible with the help of molecular markers. Bioinformatics experts have developed a tool for detecting and identifying bacterial food pathogens. The FDA (Food and Drug Administration) has created this method for the molecular characterization of bacterial foodborne diseases utilizing microarrays (Desiere et al., 2001).

## 13.7 Metabolic Pathway Construction

Fermented foods have been produced via microbial metabolism. Food fermentation can be made better by comprehending microbial metabolism. The ability of desired bacteria to transform substrates (often carbohydrates) into organic compounds specifically produced to improve the flavor, structure, texture, stability, and safety of the food product is a benefit. The number of tools available to comprehend and regulate microbial metabolism is increased by the application of contemporary genomic and bioinformatic techniques. Stochiometric models, genomic sequences of bacteria, and knowledge on metabolic pathways that is available in literature and databases may be combined to characterize cellular processes and link genotype and phenotype (Dutilh et al., 2013).

Metabolic reconstruction models will be more crucial as knowledge grows in understanding the dynamic response of cells to environmental cues.

## 13.8 Bioinformatics and Food Processing

Turning agricultural product in food or changing the form of food is known as food processing. From cooking at home to sophisticated processes of factories used to produce convenience foods, processing food encompasses a wide range of food preparation techniques. Some food processing techniques are crucial in lowering food waste and enhancing food preservation, which lowers agriculture's overall environmental impact and boosts food security. There are three types of food processing: Primary, secondary, and tertiary food processing. Most foods require primary food processing before they are edible, while secondary food processing transforms raw materials into well-known delicacies like bread. Tertiary food processing has come under fire for being unhealthy with regard to the dietary requirements of people and farm animals, promoting overnutrition and obesity, including excessive amounts of sugar and salt, and insufficient amounts of fiber.

Food has a significant involvement in controlling the body's numerous processes. With the "omics" era continuing to advance, bioinformatics curates and it uses computer methods to understand biological data and therefore an extensive integration has been seen across every discipline of sciences including food sciences. Bioinformatics can be used for efficient access to all metabolomics, proteomics, and genomics data that has been found till the date, also to make use of this discovered data to every individual group industry, or company so that it can increase the nutritional value, taste, and quality of food that is produced by using this technology (Pridmore et al., 2000).

Physical, biological, and engineering science are combined in food science to study the nature of foods, the reasons why they deteriorate, the concepts behind food processing, and how to improve food quality. In many of these processes, bioinformatics plays a significant function. The use of modern science would enable us to combat difficult situations such as food scarcity, nutritional quality, and food deterioration. The various success stories of applying bioinformatics approaches in different biological fields lead to its use in food sciences. Bioinformatics helps in the management of biological data so that gathered biological information can be applied extensively in biotechnology. The field is growing quickly and becoming an increasingly important research tool (Gry et al., 2007).

The interdisciplinary field of bioinformatics has gained recognition as a significant area of science. It causes a paradigm change in a number of fields, including biotechnology, comparative genomics, molecular biology, molecular medicine, and molecular evolution. Many bioinformatics approaches have recognized that it has the potential for a significant contribution to food sciences, but it has been lesser appreciated. Here, in this article, we highlighted the bioinformatics approaches that can be used in nutrition sciences and food sciences to advance and expand research (Holton et al., 2013). In order to meet the needs of food production, food processing, improving the quality and nutritional content of food sources, and many other things, bioinformatics is essential for predicting and evaluating the expected and unintended impacts of microbes on food. In addition, bioinformatics approaches can also be used in producing good quality crops including high yield and disease resistance. There is also a variety of databases that contain data on food, its constituents, nutritive value, chemistry, and biology.

The bioinformatics application to food processing that has been used most recently is optimizing the quantitative compositional characteristics of conventional unit operations. A significant amount of energy is used in food processing to achieve safety or storage stability while maintaining a wide margin for error. This margin of error is necessary due to our limited understanding of the composition and structural complexity of biological materials, the inherent diversity of living organisms used as food processing input streams, and the reactivity of these materials to processing settings. Bioinformatics has provided us with a wealth of knowledge about biological creatures, including bacteria, viruses, plants, and animals. This knowledge will enable us to design food processes with smaller margins for all cost-critical inputs, especially energy. However, rather than just processing food to make it safer, the big future of food processing lies in fusing the biological understanding of organisms with the knowledge of biomaterial required to turn them into edible items. The biomaterials of live animals are restructured through the severe energy input of traditional food processing into simpler stable, or broadly consistent macrostructure forms of foods. Most of the time, to remove some of the constituent molecule's possibly dangerous features, the end product of food lacks the natural biological qualities of the living systems (protease inhibitors, etc.). However, a thorough description of the inherent complexity of biological macromolecules within living cells and the structural characteristics of these molecules that provide a significant portion of their functions is now possible, thanks to the development of the knowledge base of contemporary bioinformatics (Allen et al., n.d.).

However, now that we have this information, we have a once-in-a-lifetime chance to use it to make an equally exact evaluation of each molecule's biomaterial qualities in a complicated mixture. Soon, employing the intrinsic structural properties of natural food commodities, it will be possible for new foods to self-assemble with the maximum biological and nutritional value. Using the biological structure–function correlations discovered by bioinformatics of living systems, the structure–function connections of the next generation of meals will be able to be mapped, and the outcomes will be delicious. Foods are all ostensibly altered tissues. The fundamental biomaterial characteristics of food are therefore derived from the inherent biomaterial features of the molecules that comprise living creatures. However, the distinguishing characteristics of specific molecules are rarely utilized in most conventional food preparation ("Editorial Board and Contents," 2021). In order to restructure the material into more stable and/or accessible food systems, all biomolecules of a specific class, such as proteins, are instead exposed to high physical, thermal, and mechanical energy to make these qualities consistent. Such processing eliminates the minute variations within the majority of the classes of the main biomolecules and serves as the foundation for the intricate structure–function interactions of living things. The statistically average features of the major types of biomaterials, such as proteins, carbohydrates, and lipids, are substituted for biological complexity during processing.

Food quality is not improved by treating products to remove their biological structure's intricacy; the reverse is true. There were several real-world instances when the unique biological characteristics of the initial living things were crucial to the processing technique and the completed meals' organoleptic appeal. Such processes are bovine milk renneting, which causes milk caseins to aggregate and result in the cheese-making gelation events naturally. The final product uses the unique self-assembling properties of milk casein micelles, which are colloidally stabilized in milk by kappa caseins but destabilized when enzymatically severed from their solubilizing glycol-macro-peptide (Gilbert & Hughes, 2011).

An example is leavened bread, the structures, textures, and nutrients based on composite processing and biological rearrangement. Within a mechanically altered protein gel structure, the biological processes of yeast fermentation, in this instance, achieve simultaneous enzymatic removal of phytic acid during dough incubation and the biochemical generation of carbon dioxide gas as leavening. In this process, most of the biological structures in wheat seeds are broken down mechanically. In each of these situations, bread and cheese added significant value in terms of organoleptic quality as well as enhanced safety and nutritional value by utilizing the biological characteristics of the living creatures. Furthermore, these two food staples do not suffer from the inherent biological variation that hinders the standardization of more simple food processing goals. Rather, it is a wonderful benefit resulting in hundreds of different cheeses and bread varieties with distinctive flavors and textures. When the biological processes of catalysis, self-assembly, and restructuring are retained as the foundation of food preparation, products like cheese and bread serve as possible examples. The primary method for discovering bio-driven food processing was empirical trial and error. However, to reimagine food processing by relying on bimolecular activities rather than just composite biomaterial properties, the biological understanding that is being obtained through functional genomics, proteomics, and metabolomics is precisely what is required. The entire yeast protein–protein interaction map has been identified, or potential interactions between all 6000 yeast proteins (Ito et al., 2001). The structure and function characteristics of living organisms, which are rapidly expanding thanks to bioinformatics, will significantly impact the design of new meals and food processes. Process design engineers and plant bioengineers can work together once these tools are available to develop crops that are not only enhanced with a single valuable component but also redesigned with a renewed purpose to increase the myriad benefits that food has to offer in terms of improving quality of life (Notebaart et al., 2006).

## 13.9 Bioinformatics and Improving Food Texture and Flavor

Researchers had identified the genetic and molecular details of the taste receptors including (a) sour taste mediated by receptor called a degenerin-1 (identical to ion channel) (Ugawa et al., 1998), (b) bitter taste is responsive due to presence of a family of approximately 50 G protein-coupled receptors (GPCRs) (Chandrashekar et al., 2000), (c) umami taste is due to mGluR4 (brain glutamate splice variant receptor) (Matsunami et al., 2000), (d) sweet taste is responsive due to G-protein coupled receptor known as Tas1r3 (Max et al., 2001), and (e) salt taste transduction is due to ENac (epithelial ion channel) (Nagel et al., 2001). The next generation of food taste modifiers can be found using these taste receptors. The accessible known structures of these receptors combined with recent advancements in computational methods and software have enabled molecular modeling and simulations. It will be feasible to create molecules with stronger flavors as food additives, thanks to these simulations. These aid in comprehending the underlying principles of taste complementarity, antagonism, and persistence. The identification of sensors determining sour taste in animals and determining homology between sweet taste receptors and brain glutamate receptors have both been accomplished using bioinformatics sequence similarity methods (Talevi et al., 2012).

The texture and flavor characteristics of the food product are also influenced by fermentation. These traits are microorganism-specific and can be altered by fermentation, for example, introducing flavors in cheese by introducing adjunct strains during fermentations (Deetae et al., 2007; Whetstine et al., 2006). Moreover, the texture of yogurt is also improved by adding exopolysaccharide-producing organisms (Robitaille et al., 2009). Changing the fermentation conditions or the wine-starting cultures can also modify the flavor characteristics of wine (Carrau et al., 2015). Bioinformatics and data analytics may be utilized to optimize experimental designs, whereas improvements can be made by testing a range of experimental settings (Seo et al., 2013). The gene composition of these microbes can be used to infer how well they operate under specific fermentation conditions. For instance, the genome-scale metabolic model of *L. lactis* MG1363 had shown the nitrogen and carbon metabolism leads to formation of different flavor compounds (Flahaut et al., 2013).

## 13.10 Omics and Food Microbiology

In recent times analytical techniques have used an omics approach with high throughput technologies to help in quickly obtaining numerous measurements. Multilevel omic methods have recently been employed to assess the complexity of food-related biological systems. The entire genome of the organism can be examined using various bioinformatics methods. Whole-genome sequencing is a method for learning about different kinds of foodborne pathogens (Herrero et al., 2016). Other sub-tools for identifying foodborne infections include the pulse-field gel electrophoresis (PFGE) procedure and multiple-locus variable number tandem repeat analysis (MLVA) (Boxrud et al., 2010). Despite being novel, the approaches have drawbacks compared to PFGE and MLVA. The whole-genome sequencing process can be applied to a variety of samples without any discrimination. It has been noted that methods like MLVA and PFGE frequently fail to distinguish between different subtypes of a common type of pathogen (Allard et al., 2012).

## 13.11 Conclusion

In conclusion, the expanding information on fermented foods and safety motivates its consolidation into databases, which, combined with the appropriate experimental strategy, knowledge, and follow-up tests, should improve fermentation performance and safety prediction.

## References


AGRICOLA|EBSCO. (n.d.). Retrieved January 17, 2023, from https://www.ebsco.com/products/research-databases/agricola.

agriRxiv|UCSB Library (n.d.). Retrieved January 17, 2023, from https://www.library.ucsb.edu/agrirxiv.

AGRIS The international system for agricultural science and technology—data provider guide. (n.d.). Retrieved January 17, 2023, from https://www.fao.org/agris/publications/agris-international-system-agricultural-science-and-technology---data-provider-guide.

Allard, M. W., Luo, Y., Strain, E., Li, C., Keys, C. E., Son, I., Stones, R., Musser, S. M., & Brown, E. W. (2012). High resolution clustering of salmonella enterica serovar Montevideo strains using a next-generation sequencing approach. *BMC Genomics, 13*(1), 32. https://doi.org/10.1186/1471-2164-13-32

Allen, H. J., Lennon, D. J., Lukosaityte, J., & Borum, P. R. (n.d.). Foodomics database: a new tool for precision medicine and the -omic toolbox. *The FASEB Journal, 30*, 682.13-682.13. https://doi.org/10.1096/FASEBJ.30.1_SUPPLEMENT.682.13

Baxevanis, A. D., & Ouellette, B. F. F. (n.d.). *Bioinformatics: a practical guide to the analysis of genes and proteins* (p. 470). Retrieved January 10, 2023, from https://www.wiley.com/en-us/Bioinformatics%3A+A+Practical+Guide+to+the+Analysis+of+Genes+and+Proteins%2C+2nd+Edition-p-9780471461012.

Boxrud, D., Monson, T., Stiles, T., & Besser, J. (2010). The role, challenges, and support of pulsenet laboratories in detecting foodborne disease outbreaks. *Public Health Reports, 125*(Suppl 2), 57–62. https://doi.org/10.1177/00333549101250S207

CAB Abstracts—CABI.org. (n.d.). Retrieved January 17, 2023, from https://www.cabi.org/publishing-products/cab-abstracts/.

Carrau, F., Gaggero, C., & Aguilar, P. S. (2015). Yeast diversity and native vigor for flavor phenotypes. *Trends in Biotechnology, 33*(3), 148–154. https://doi.org/10.1016/J.TIBTECH.2014.12.009

Chandrashekar, J., Mueller, K. L., Hoon, M. A., Adler, E., Feng, L., Guo, W., Zuker, C. S., & Ryba, N. J. P. (2000). T2Rs function as bitter taste receptors. *Cell, 100*(6), 703–711. https://doi.org/10.1016/S0092-8674(00)80706-0

Chun, H. J. E., Khattra, J., Krzywinski, M., Aparicio, S. A., & Marra, M. A. (2014). Second-generation sequencing for cancer genome analysis. In *Cancer genomics: from bench to personalized medicine* (pp. 13–30). Academic Press. https://doi.org/10.1016/B978-0-12-396967-5.00002-5

Deetae, P., Bonnarme, P., Spinnler, H. E., & Helinck, S. (2007). Production of volatile aroma compounds by bacterial strains isolated from different surface-ripened French cheeses. *Applied Microbiology and Biotechnology, 76*(5), 1161–1171. https://doi.org/10.1007/S00253-007-1095-5

Desiere, F., German, B., Watzke, H., Pfeifer, A., & Saguy, S. (2001). Bioinformatics and data knowledge: the new frontiers for nutrition and foods. *Trends in Food Science & Technology, 12*(7), 215–229. https://doi.org/10.1016/S0924-2244(01)00089-9

Di Bella, J. M., Bao, Y., Gloor, G. B., Burton, J. P., & Reid, G. (2013). High throughput sequencing methods and analysis for microbiome research. *Journal of Microbiological Methods, 95*(3), 401–414. https://doi.org/10.1016/J.MIMET.2013.08.011

Dutilh, B. E., Backus, L., Edwards, R. A., Wels, M., Bayjanov, J. R., & Van Hijum, S. A. F. T. (2013). Explaining microbial phenotypes on a genomic scale: GWAS for microbes. *Briefings in Functional Genomics, 12*(4), 366–0380. https://doi.org/10.1093/BFGP/ELT008

Editorial Board and Contents. (2021). Trends in food science &. *Technology, 116*, i–viii. https://doi.org/10.1016/s0924-2244(21)00550-1

Fang, H., Xu, J., Ding, D., Jackson, S. A., Patel, I. R., Frye, J. G., Zou, W., Nayak, R., Foley, S., Chen, J., Su, Z., Ye, Y., Turner, S., Harris, S., Zhou, G., Cerniglia, C., & Tong, W. (2010). An FDA bioinformatics tool for microbial genomics research on molecular characterization of bacterial foodborne pathogens using microarrays. *BMC Bioinformatics, 11 Suppl 6*(Suppl 6), 1. https://doi.org/10.1186/1471-2105-11-S6-S4

Fiers, M. W. E. J., Kleter, G. A., Nijland, H., Peijnenburg, A. A. C. M., Nap, J. P., & van Ham, R. C. H. J. (2004). Allermatch™, a webtool for the prediction of potential allergenicity according to current FAO/WHO codex alimentarius guidelines. *BMC Bioinformatics, 5*, 133. https://doi.org/10.1186/1471-2105-5-133

Flahaut, N. A. L., Wiersma, A., Van De Bunt, B., Martens, D. E., Schaap, P. J., Sijtsma, L., Dos Santos, V. A. M., & De Vos, W. M. (2013). Genome-scale metabolic model for Lactococcus lactis MG1363 and its application to the analysis of flavor formation. *Applied Microbiology and Biotechnology, 97*(19), 8729–8739. https://doi.org/10.1007/S00253-013-5140-2/FIGURES/3

Flores, O., Belanche, L. A., & Blanch, A. R. (2009). New multiplatform computer program for numerical identification of microorganisms. *Journal of Clinical Microbiology, 47*(12), 4133. https://doi.org/10.1128/JCM.01250-09

Food Allergens—International Regulatory Chart. FARRP. Nebraska. (n.d.). Retrieved January 19, 2023, from https://farrp.unl.edu/IRChart.

Food and Nutrient Database for Dietary Studies (FNDDS). Ag Data Commons. (n.d.). Retrieved January 17, 2023, from https://data.nal.usda.gov/dataset/food-and-nutrient-database-dietary-studies-fndds.

Food Science Technology Abstracts (FSTA). Penn State University Libraries. (n.d.). Retrieved January 17, 2023, from https://libraries.psu.edu/databases/psu01751.

FooDB. (n.d.). Retrieved January 19, 2023, from https://foodb.ca/about.

Garrigues, C., Johansen, E., & Crittenden, R. (2013). Pangenomics—an avenue to improved industrial starter cultures and probiotics. *Current Opinion in Biotechnology, 24*(2), 187–191. https://doi.org/10.1016/J.COPBIO.2012.08.009

Gilbert, J. A., & Hughes, M. (2011). Gene expression profiling: metatranscriptomics. *Methods in Molecular Biology (Clifton, N.J.), 733*, 195–205. https://doi.org/10.1007/978-1-61779-089-8_14

Gry, J., Black, L., Eriksen, F. D., Pilegaard, K., Plumb, J., Rhodes, M., Sheehan, D., Kiely, M., & Kroon, P. A. (2007). EuroFIR-BASIS—a combined composition and biological activity database for bioactive compounds in plant-based foods. *Trends in Food Science & Technology, 18*(8), 434–444. https://doi.org/10.1016/J.TIFS.2007.05.008

Gustafson, R. (n.d.). Research Guides: Food Science and Nutrition: Databases. Retrieved January 17, 2023, from https://guides.library.ucdavis.edu/food-science-nutrition/databases.

Guttman, D. S., McHardy, A. C., & Schulze-Lefert, P. (2014). Microbial genome-enabled insights into plant–microorganism interactions. *Nature Reviews Genetics, 15*(12), 797–813. https://doi.org/10.1038/nrg3748

Herrero, J., Muffato, M., Beal, K., Fitzgerald, S., Gordon, L., Pignatelli, M., Vilella, A. J., Searle, S. M. J., Amode, R., Brent, S., Spooner, W., Kulesha, E., Yates, A., & Flicek, P. (2016). Ensembl comparative genomics resources. *Database: The Journal of Biological Databases and Curation, 2016*, bav096. https://doi.org/10.1093/DATABASE/BAV096

Holton, T. A., Vijayakumar, V., & Khaldi, N. (2013). Bioinformatics: Current perspectives and future directions for food and nutritional research facilitated by a food-wiki database. *Trends in Food Science & Technology, 34*(1), 5–17. https://doi.org/10.1016/J.TIFS.2013.08.009

Ito, T., Chiba, T., Ozawa, R., Yoshida, M., Hattori, M., & Sakaki, Y. (2001). A comprehensive two-hybrid analysis to explore the yeast protein interactome. *Proceedings of the National Academy of Sciences of the United States of America, 98*(8), 4569–4574. https://doi.org/10.1073/PNAS.061034498

Ivanciuc, O., Schein, C. H., & Braun, W. (2003). SDAP: Database and computational tools for allergenic proteins. *Nucleic Acids Research, 31*(1), 359. https://doi.org/10.1093/NAR/GKG010

Kumar, A., & Chordia, N. (2017). Bioinformatics approaches in food sciences. *Journal of Food: Microbiology, Safety & Hygiene, 2*(2), 1–4. https://doi.org/10.4172/2476-2059.1000E104

Lander, E. S., Linton, L. M., Birren, B., Nusbaum, C., Zody, M. C., Baldwin, J., Devon, K., Dewar, K., Doyle, M., Fitzhugh, W., Funke, R., Gage, D., Harris, K., Heaford, A., Howland, J., Kann,

L., Lehoczky, J., Levine, R., McEwan, P., & Morgan, M. J. (2001). Initial sequencing and analysis of the human genome. *Nature, 409*(6822), 860–921. https://doi.org/10.1038/35057062
Langille, M. G. I., Zaneveld, J., Caporaso, J. G., McDonald, D., Knights, D., Reyes, J. A., Clemente, J. C., Burkepile, D. E., Vega Thurber, R. L., Knight, R., Beiko, R. G., & Huttenhower, C. (2013). Predictive functional profiling of microbial communities using 16S rRNA marker gene sequences. *Nature Biotechnology, 31*(9), 814–821. https://doi.org/10.1038/NBT.2676
Matsunami, H., Montmayeur, J. P., & Buck, L. B. (2000). A family of candidate taste receptors in human and mouse. *Nature, 404*(6778), 601–604. https://doi.org/10.1038/35007072
Max, M., Shanker, Y. G., Huang, L., Rong, M., Liu, Z., Campagne, F., Weinstein, H., Damak, S., & Margolskee, R. F. (2001). Tas1r3, encoding a new candidate taste receptor, is allelic to the sweet responsiveness locus sac. *Nature Genetics, 28*(1), 58–63. https://doi.org/10.1038/NG0501-58
Nagel, G., Szellas, T., Riordan, J. R., Friedrich, T., & Hartung, K. (2001). Non-specific activation of the epithelial sodium channel by the CFTR chloride channel. *EMBO Reports, 2*(3), 249–254. https://doi.org/10.1093/EMBO-REPORTS/KVE045
Notebaart, R. A., van Enckevort, F. H. J., Francke, C., Siezen, R. J., & Teusink, B. (2006). Accelerating the reconstruction of genome-scale metabolic networks. *BMC Bioinformatics, 7*, 296. https://doi.org/10.1186/1471-2105-7-296
Overbeek, R., Olson, R., Pusch, G. D., Olsen, G. J., Davis, J. J., Disz, T., Edwards, R. A., Gerdes, S., Parrello, B., Shukla, M., Vonstein, V., Wattam, A. R., Xia, F., & Stevens, R. (2014). The SEED and the rapid annotation of microbial genomes using subsystems technology (RAST). *Nucleic Acids Research, 42*(Database issue), D206. https://doi.org/10.1093/NAR/GKT1226
Prakash, B., Singh, P. P., Kumar, A., & Gupta, V. (2020). Prospects of omics technologies and bioinformatics approaches in food science. In *Functional and preservative properties of phytochemicals* (pp. 317–340). https://doi.org/10.1016/B978-0-12-818593-3.00010-5
Pridmore, R. D., Crouzillat, D., Walker, C., Foley, S., Zink, R., Zwahlen, M. C., Brüssow, H., Pétiard, V., & Mollet, B. (2000). Genomics, molecular genetics and the food industry. *Journal of Biotechnology, 78*(3), 251–258. https://doi.org/10.1016/S0168-1656(00)00202-9
Priya, S., & Satheeshkumar, P. K. (2020). 5- natural products from plants: recent developments in phytochemicals, phytopharmaceuticals, and plant-based neutraceuticals as anticancer agents. In *Functional and preservative properties of phytochemicals*. Elsevier Inc. http://www.sciencedirect.com:5070/book/9780128185933/functional-and-preservative-properties-of-phytochemicals
Robitaille, G., Tremblay, A., Moineau, S., St-Gelais, D., Vadeboncoeur, C., & Britten, M. (2009). Fat-free yogurt made using a galactose-positive exopolysaccharide-producing recombinant strain of Streptococcus thermophilus. *Journal of Dairy Science, 92*(2), 477–482. https://doi.org/10.3168/JDS.2008-1312
Schmidtke, L. M., Blackman, J. W., Clark, A. C., & Grant-Preece, P. (2013). Wine metabolomics: objective measures of sensory properties of semillon from GC-MS profiles. *Journal of Agricultural and Food Chemistry, 61*(49), 11957–11967. https://doi.org/10.1021/JF403504P
Scrimshaw, N. S. (1997). INFOODS: the international network of food data systems. *The American Journal of Clinical Nutrition, 65*(4 Suppl), 1190S. https://doi.org/10.1093/AJCN/65.4.1190S
Seo, S. W., Yang, J., Min, B. E., Jang, S., Lim, J. H., Lim, H. G., Kim, S. C., Kim, S. Y., Jeong, J. H., & Jung, G. Y. (2013). Synthetic biology: tools to design microbes for the production of chemicals and fuels. *Biotechnology Advances, 31*(6), 811–817. https://doi.org/10.1016/J.BIOTECHADV.2013.03.012
Smid, E. J., & Kleerebezem, M. (2014). Production of aroma compounds in lactic fermentations. *Annual Review of Food Science and Technology, 5*(1), 313–326. https://doi.org/10.1146/ANNUREV-FOOD-030713-092339
Talevi, A., Enrique, A. V., & Bruno-Blanch, L. E. (2012). Anticonvulsant activity of artificial sweeteners: a structural link between sweet-taste receptor T1R3 and brain glutamate receptors. *Bioorganic & Medicinal Chemistry Letters, 22*(12), 4072–4074. https://doi.org/10.1016/J.BMCL.2012.04.076

Talukdar, V., Konar, A., Datta, A., & Choudhury, A. R. (2009). Changing from computing grid to knowledge grid in life-science grid. *Biotechnology Journal, 4*(9), 1244–1252. https://doi.org/10.1002/BIOT.200800073

Tidsall, J. (2013). Beginning Perl for bioinformatic. *Journal of Chemical Information and Modeling, 53*(9).

Tiwari, A., & Sekhar, A. K. T. (2007). Workflow based framework for life science informatics. *Computational Biology and Chemistry, 31*(5–6), 305–319. https://doi.org/10.1016/J.COMPBIOLCHEM.2007.08.009

Ugawa, S., Minami, Y., Guo, W., Saishin, Y., Takatsuji, K., Yamamoto, T., Tohyama, M., & Shimaria, S. (1998). Receptor that leaves a sour taste in the mouth. *Nature, 395*(6702), 555–556. https://doi.org/10.1038/26882

Valdés, A., Cifuentes, A., & León, C. (2017). Foodomics evaluation of bioactive compounds in foods. *TrAC Trends in Analytical Chemistry, 96*, 2–13. https://doi.org/10.1016/J.TRAC.2017.06.004

Whetstine, M. E. C., Drake, M. A., Broadbent, J. R., & McMahon, D. (2006). Enhanced nutty flavor formation in cheddar cheese made with a malty Lactococcus lactis adjunct culture. *Journal of Dairy Science, 89*(9), 3277–3284. https://doi.org/10.3168/JDS.S0022-0302(06)72364-5

Woolf, P., Fu, L., & Basu, A. (2010). Foodwiki: a distributed annotation system for building a nutrition and food database. *The FASEB Journal, 24*(S1), 943.4-943.4. https://doi.org/10.1096/FASEBJ.24.1_SUPPLEMENT.943.4

# Chapter 14
# The Beneficial Impact of Microbes in Food Production, Health, and Sustainability

**Manpreet Kaur, Abdel Rahman Mohammad Said Al-Tawaha, and Arun Karnwal**

**Abstract** Beneficial bacteria have a substantial impact on human health and environmental sustainability. Microbes are extremely important in a variety of food applications. They are crucial to the production of various food items. Since the beginning, multiple bacteria and fungi have been used to make multiple fermented food products and enzymes. Bread, dairy products, alcohol and drinks, enzymes, organic acids, food colours, etc., are prepared using yeast, bacteria, and fungi. These microbes are included to add appealing colour, taste, flavour, and texture and improve the product's marketability. The present chapter will focus on the use of bacteria and yeast in food processing, the production of terpenoids, lipopeptides, and polyphenols, improvement of packaging materials' effectiveness and quality through the application of nanotechnology to ensure food safety.

**Keywords** LPS · Bio-preservatives · Terpenoids · Sustainable

## 14.1 Introduction

Food processing is a branch of the manufacturing business that converts raw animal, plant, and marine ingredients into intermediate or polished, safer-to-eat food products. The objectives of food processing may include:

1. increasing the amount of time that food is wholesome,
2. giving nutrients needed for the body,

M. Kaur
Department of Biotechnology, Chandigarh University, Chandigarh, India

A. R. Mohammad Said Al-Tawaha
Department of Biological Sciences, Al Hussein Bin Talal University, Ma'an, Jordan
e-mail: Abdel-al-tawaha@ahu.edu.jo

A. Karnwal (✉)
Department of Microbiology, Lovely Professional University, Phagwara, Punjab, India



A. Karnwal, A. R. Mohammad Said Al-Tawaha (eds.), *Food Microbial Sustainability*, https://doi.org/10.1007/978-981-99-4784-3_14

3. offering diversity and ease in diet,
4. addition of value.

The definition of food processing is broad, and any unit operation that takes place from the time that raw materials are harvested until they are turned into food items, packaged, and delivered for retail sale may be included (Park et al., 2014). In general, these might be divided into primary and secondary processing. Primary processing, often known as post-harvest or post-slaughter processing, is food preparation for consumption or incorporation into other food products. Secondary processing transforms the prime processed food or component into new food items. It guarantees that food can be utilized for various things (Heldman & Hartel, 1998).

Although microorganisms are essential to sustaining life on the planet, we know very little about the vast microbial species found in different habitats like soils (Goel et al., 2018). Any civilization's ability to endure depends on its capacity for sustainability. When the system is socially supportive, commercially competitive, and environmentally sound, agriculture may be considered sustainable. The dynamics of growth, metabolic activity, and survival of microorganisms in food are influenced by their capacity to colonize the food matrix and develop into a spatial heterogeneity as well as by in situ cell-to-cell ecological interactions that frequently occur in a solid phase. These dynamics also involve stress responses in response to modifications in the physical and chemical conditions in the food microenvironment (Holzapfel, 2002).

Microbes are important because they are the principal dynamic factors in farming systems. Agriculturally important microorganisms control interactions between plants and other pathogenic microflora and the efficiency with which nutrients are available to agricultural plants and soil biodiversity (Singh et al., 2016). Microorganisms contribute completely to the food sector by aiding in the fermentation process that preserves food and dairy products (Gondal et al., 2021). Microorganisms are utilized in fermentation to extend food's shelf life. Since lactic acid bacteria have unique metabolic properties, they are used in many fermentation and preservation processes (Caplice & Fitzgerald, 1999). Apart from extending the shelf life of food and guaranteeing its microbiological safety, fermentation can also improve the digestibility of some foods, such as cassava, by lowering the substrate's toxicity. Lactic acid bacteria are engaged in various fermentation processes in milk, meats, cereals, and vegetables (Rakhmanova et al., 2018). Microorganisms carry out natural fermentation processes. For thousands of years, bread, vinegar, beer, yoghurt, cheese, and wine have been made using yeasts, moulds, bacteria, fruits and vegetables, and fermented fish (Fig. 14.1). One of the earliest techniques for preserving and transforming food is fermentation. This biological procedure keeps food while enhancing its nutritive and olfactory qualities. The starter culture bacteria produce lactic acid to stop the growth of unfavourable microorganisms. Food fermentation products show high economic value and have a vital role in enhancing human health. LAB have had a role in the rise of fermented foods globally, especially those that include probiotics or other bacteria that are good for your health (Soomro et al., 2002).

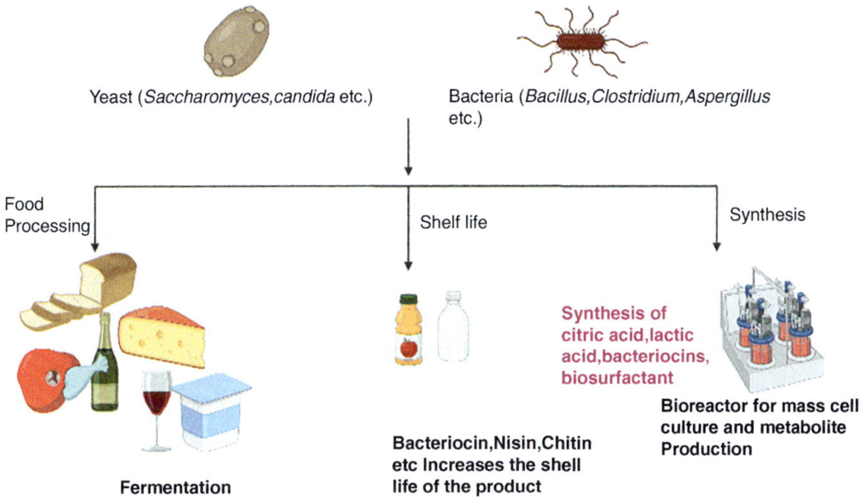


**Fig. 14.1** Role of Microbes (Yeast and Bacteria) in the Food Industry

*Lactococcus, Leuconostoc, Lactobacillus,* and *Pediococcus* are some of the genera involved in fermentation. These bacteria create organic acids, fatty acids, and hydrogen peroxide, among other metabolic products, that have antibacterial properties. LAB produces Class III bacteriocins that hinder the growth of the pathogens involved in food spoilage. According to studies, some lactobacilli delay mycelia's development on the cheese's surface and lower the fungal population in both contaminated cheese and milk. The characteristics of *Lactobacillus plantarum* are known in biotechnology (enzyme systems, vitamins, bacteriocins, antioxidants, antimicrobial activities) and safety and quality aspects in the conventional food system (Zdolec et al., 2018). Various fermented foods are made from animal and plant sources using numerous food-grade bacteria. In addition to having a far longer shelf life than the raw materials from which they are made, foods like matured cheeses, sauerkraut, pickles, and sausages also have scent and flavour characteristics provided by the fermentation-performing organisms. Dairy products that have been fermented are becoming increasingly popular as quick, wholesome, stable, all-natural, and healthy meals. LAB is widely used to produce cultured butter, yoghurt, cheese, and sour cream. Other bacteria, known as secondary microflora, are sometimes introduced to fermented dairy products to produce carbon dioxide, affecting the flavour and changing the finished product's texture (Doyle & Meng, 2006).

Most microbial enzymes used in food processing are sold as enzyme preparations, including the needed enzyme activity, various metabolites of the manufacturing strain, and other ingredients like stabilizers and preservatives. When assessing the safety of an enzyme, the manufacturing strain's safety, specifically its toxigenic potential, continues to be the major factor to be taken into account. Fungi, known as *Aspergillus* spp., can be found in many environments, including almost all soils. These are extremely effective extracellular enzyme makers. Cellulases, pectinases, amylases, etc. that break down complex polysaccharides are produced using them commercially (Mojsov, 2016). Through the production of vitamins, and vital, proteins, improved protein and fibre digestibility, and the degradation of antinutritional components, fermentation increases food's nutritional content. It also offers a source of calories. By lowering harmful substances like aflatoxins and cyanogens and producing antimicrobial substances like lactic acid, bacteriocins, hydrogen peroxide, carbon dioxide, and ethanol that aid in suppressing or eliminating foodborne pathogens, fermentation procedures also improve food safety (Giraffa, 2004).

## 14.2 Factors Affecting Microbial Growth in Food

The main elements that affect microbial development in food can be summed up as follows:

1. Extrinsic factors are connected to the conditions in which the food is kept,
2. Intrinsic factors, which are elements associated with itself,
3. Processing processes,
4. Implicit factors,
5. Aspects relating to the microorganisms themselves, interactions between variables can also have a multifaceted impact on the proliferation of microorganisms in meals.

Typically, pH inhibits development in food by interacting with other factors. Pathogens and other organisms can grow more slowly when certain conditions, like pH, salt, temperature, redox potential, and preservatives, are present. Microorganisms need a few basic nutrients to thrive and maintain their metabolic processes. Depending on the organism, different nutrients have different types and necessary amounts.

Metabiotic associations are fundamental "sequence synergisms," in which the development of one microorganism leads to the development of a second, which can lead to the development of a third, and so on. An exceptional illustration of extended metabiosis is raw milk. The earliest microbes to develop in raw milk are *Lactococcus lactis* and a few coliforms. They produce lactic acid, which fosters the growth of aciduric lactobacilli. When the milk's accumulated acidity prevents lactobacilli from multiplying, oxidative yeasts and moulds start to breed and oxidize the lactic acid, elevating the milk pH and promoting the development of proteolytic bacteria (Preetha & Narayanan, 2020).

## 14.3 Use of Bacteria and Yeast in Food Processing

The food industry's primary goal is to process and preserve the fermented food substrate. Preparing various fermented foods, particularly those that are milk-based, and their shelf life and microbiological safety are ensured by establishing the ideal conditions for fermentation. There have also been reports of the therapeutic benefits of fermented foods.

Most fermented food manufacturing processes are unfavourable to many food-borne microbes and may even be suppressive. The antimicrobial properties of fermentation inhibit pathogens that could be present and bacteria that cause food to spoil. Therefore, potentially dangerous raw materials (such as raw milk and meat) may be employed in manufacturing food with enhanced quality and increased safety using classic food fermentation procedures (Jordan et al., 2014). Fermentation enriches the product by making a variety of tastes, textures, and scents in addition to its nutritional. Lactic acid bacteria show a crucial role in dominating role in the fermentation process (Erten et al., 2014). Single-cell proteins are cited as a source of mixed proteins and are prepared from microbes such as fungi, yeast, bacteria, and algae. Microbial cells are called single-cell proteins and natural protein sources if their cells contain a sizable amount of proteins. Due to the growing population and global protein shortage, microbial biomass is used as food. Single-cell protein has become more significant due to the growing human population and the global shortage of protein-rich microbial biomass for food and feed (Nasseri et al., 2011).

Candida yeast synthesizes extracellular metabolites such as xylitol, erythritol, biosurfactants, ethanol, and exopolysaccharides (Kieliszek et al., 2017). Bacteriocin-producing strains can be added or substituted to the starter cultures to increase the quality and safety of the fermented foods. Numerous LAB bacteriocins are natural, safe, and efficient bacterial pathogens and food-rotting inhibitors. Nisin inhibits the growth of some psychotropic bacteria in cottage cheese, stops the growth of decay lactobacilli in the fermentation of wine and beer, and provides additional protection against clostridial and bacillus spores in canned foods. It also extends the shelf life of milk in warm countries (Soomro et al., 2002).

In the nutrition of cattle and aquaculture, microbial protein is an alternative source of high-quality protein that can take the place of animal protein like fishmeal. Microalgae and algae are employed as food and supplements in the food sector. The two primary processes that use yeast are Baker's yeast and alcohol fermentation, with a worldwide market value for 2019 of up to 9.2 billion euros and a yearly growth projection of 7.9%. The most popular type of mycoprotein, QuornTM, is marketed and distributed in about 15 nations worldwide. They are particularly well suited to mimic the texture and flavour of meat, which accounts for their popularity as a substitute for traditional animal-based goods. Some cyanobacteria and microalgae are the main producers of microbial oil, which can be used in place of vegetable oil in dietary supplements. The abundant fatty acids can take the place of those typically generated from rapeseed, soy, sunflower, and palm oils (Matassa et al., 2016).

Chitin is also commonly used as a preservative in the food industry. Applying the chitin solution to egg cells before inoculation provided a barrier preventing Salmonella from penetrating. Similarly, L. monocytogenes growth on slices of sausages was inhibited by chitosan films. Adding chitin powder at a rate of 0–6% (w/w) to bread prevented the development of *B. cereus* and rope formation. Chitosan and pressure applied together synergistically removed *S. aureus* and *E. coli* from the buffer and regulated microbial growth in apple extract and minced pork in chilled storage (Malinowska-Panczyk et al., 2009).

*S. salivarius, Streptococcus thermophiles,* and *L. bulgaricus* ferment lactose, the milk sugar, and produces lactic acid as a by-product. These bacteria are together known as Lactic Acid Bacteria, or LAB. The bacteria consume the lactose in milk and produce lactic acid as a waste product throughout the feeding process.

A solid mass known as curd is formed by the conversion of the casein by lactic acid. The fermentation of lactose sugar into lactic acid and preservation effects gives yoghurt its jelly texture and flavour. Fermentation also enhances the diet by creating a variety of tastes, textures, and scents (Sandoval-Castilla et al., 2004).

Microbes are used in the production of chocolate. Cacao tree seeds are used to make chocolate. The pods are first fermented with naturally occurring microbes, primarily *Lactobacilli* and *Acetobacter.* Because of the increase in temperature during the fermentation process, the ethanol produced by these microorganisms kills the beans and contributes to the flavour of the chocolate (Kalsoom et al., 2020) (Table 14.1).

In the food sector, microbial enzymes are crucial because they are more stable than enzymes from plants and animals. They can be produced using fermentation processes more rapidly and inexpensively, and because of their high consistency, process adjustment and optimization are simple. The food industry uses amylases extensively for baking, starch liquefaction, brewing, and as digestive aids (van der MJEC et al., 2002). They are frequently employed as flavour enhancers and antistaling agents in the baking industry to enhance the quality of bread. During baking, yeast ferments and produce amylase enzyme. When added to the dough, it enhances the bread's flavour, colour, and toasting properties by converting starch to smaller dextrins, which are then processed into the final product (Blanco et al., 2014) with the help of amylases from *B. amyloliquefaciens*, *B.stearothermophilus,* or *Bacillus licheniformis,* most starch saccharification is accomplished. The starch in the flour is changed into maltose and fermentable sugars by enzymes called glucoseamylases. The yeast fermentation process makes the dough rise. These enzymes are also employed in synthesizing glucose, which results in ethanol formation when fermented with *Saccharomyces cerevisiae* (James et al., 1996).

The ability of yeasts like S. cerevisiae to survive in GI tract and interact antagonistically with GI pathogens such as *Escherichia coli, Salmonella,* and *Shigella* has been defined and demonstrated as one of their probiotic qualities. The industrial manufacturing of alcohol and spirits uses distiller's yeasts. They are often isolated from commercial fermentations of fruit pulps and molasses made from beets or sugar cane. At a semi-industrial scale (Bovill et al., 2001). Kefir yeasts have been

**Table 14.1** Microorganism and their products

| Microorganism | Process | Product | References |
|---|---|---|---|
| *Saccharomyces cerevisiae* | Fermentation | Bread | Heitmann et al. (2018) |
| *Aspergillus oryzae or A. soyae* | Fermentation | Soy sauce | Zhao et al. (2020) |
| *Propionibacterium shermanii* | Fermentation | Cheese | Yerlikaya et al. (2020) |
| *S. thermophilus* | Fermentation | Yoghurt | Wasilewska et al. (2019) |
| *Saccharomyces cerevisiae* and *Zigosaccharomyces rouxii* | Hydrothermal, alkali, acid, or enzymatic pre-treatments | Ethanol | Liguori et al. (2015) |
| *Aspergillus flavus* and *aspergillus tamari* | Submerged fermentation | Ascorbic acid | Mussatto & Roberto, (2005) |
| *Monascus purpureus* strain | Submerged fermentation | Colouring agent | Silbir & Goksungur, (2019) |
| *Rhizopus* spp., *Trichoderma* spp. | Hydrolysis | Single-cell protein | Bekatorou et al. (2015) |
| Penicillium roqueforti | Fermentation | Blue-veined cheeses | Klaenhammer (1993) |
| *L. Lactis* | Fermentation | Kefir | Jay (1996) |
| *L. Acidophilus* | Fermentation | Milk | Lee and Salminen (1995) |
| *Lb. plantarum* | Fermentation | Gari' | Oyewole and Odunfa (1990) |

employed to use whey lactose and produce value-added goods such as biomass, ethanol, lactic acid, etc. (Weatherholtz & Holsing, 1975).

When making sourdough, used as a starter culture to leaven bread, flour, and water are combined with yeast and lactic acid bacteria. Compared to Baker's yeast, sourdough offers significant benefits, including the formation of distinctive flavour and texture and the extension of preservation time through the in situ generation of antibacterial compounds (such as bacteriocins). Commercially, a range of yeast and bacterial blends are employed to create sourdoughs. These sourdoughs serve to condition the dough, extend shelf life, and yield distinctive sensory attributes in bread and baked goods. (Bekatorou et al., 2015). Malolactic fermentation (MLF) is a secondary fermentation in wine often performed by LAB, particularly by *Leuconostoc oenos*, and it typically takes place after alcoholic fermentation by yeast (Liu, 2002). In this metabolic process, the malolactic enzyme catalyses the conversion of L-malic acid to L-lactic acid and $CO_2$ without releasing intermediates. This pathway reduces acidity, making it a vital step in manufacturing wine and cider (Faria-Oliveira et al., 2015). Several stimuli are applied to yeast and LAB cells during fermentation, including high osmotic pressure, hydrostatic pressure, and high ethanol concentrations; unusual tests on the generation of beer under anaerobic conditions, limiting temperature and nutrients were conducted on uninhabited yeast isolates from cachaça fermentation vats in Brazil (Araújo, 2013).

## 14.4 Food Pigments

The food business has set a high priority on developing foods with attractive looks. To make the food more enticing, different colouring agents are added. Microbial pigments must endure severe pH and temperature to meet the standard criteria for the industry (Narsing Rao et al., 2017). At a variety of pH levels, several fungi colours remain stable. The earliest known usage of fungus pigment is in Monascus. Monascus produce the yellow, orange, and purple colours frequently found in Asian cuisine (Dufosse et al., 2014). Food-grade fungi-derived pigments are now readily accessible on the market, including Monascus colours, riboflavin from Ashbya and -carotene from Blakeslea trispora (Dharmaraj et al., 2009). A wide variety of pigments, including melanin, violacein, carotenoids, pyocyanin, actinorhodin, prodigiosin, and zeaxanthin, are also produced by bacteria.

## 14.5 The Sustainable Production of Terpenoids, Lipopeptides, and Polyphenols

### *14.5.1 Terpenoids*

Since microorganisms provide an efficient, ecologically acceptable way to transform low-cost raw materials such as sucrose, glucose, and biomass-derived materials into high-value compounds and fuels, manipulating microbial metabolism has significant benefits (Kampranis & Makris, 2012). One powerful class of secondary metabolites, isoprenoids, and terpenoids contribute more than 50,000 chemicals to the variety of natural product structures. The food and cosmetic industries have shown interest in using this group's members as flavour and scent additions.

The following requirements must be met by microbial hosts to synthesize terpenoids (Fig. 14.2):

1. an enormous metabolic capacity facilitating the effective synthesis of molecules that elicit interest with strong and rapid cell development;
2. well-developed genetic tools and a well-understood metabolic activity;
3. a significant capacity to grow on inexpensive carbon sources (Wang et al., 2018).

The MEP or MVA route, IPP, and DMAPP are required to synthesize terpenoids. Two representative microbial hosts are widely used to produce a variety of terpenoids.

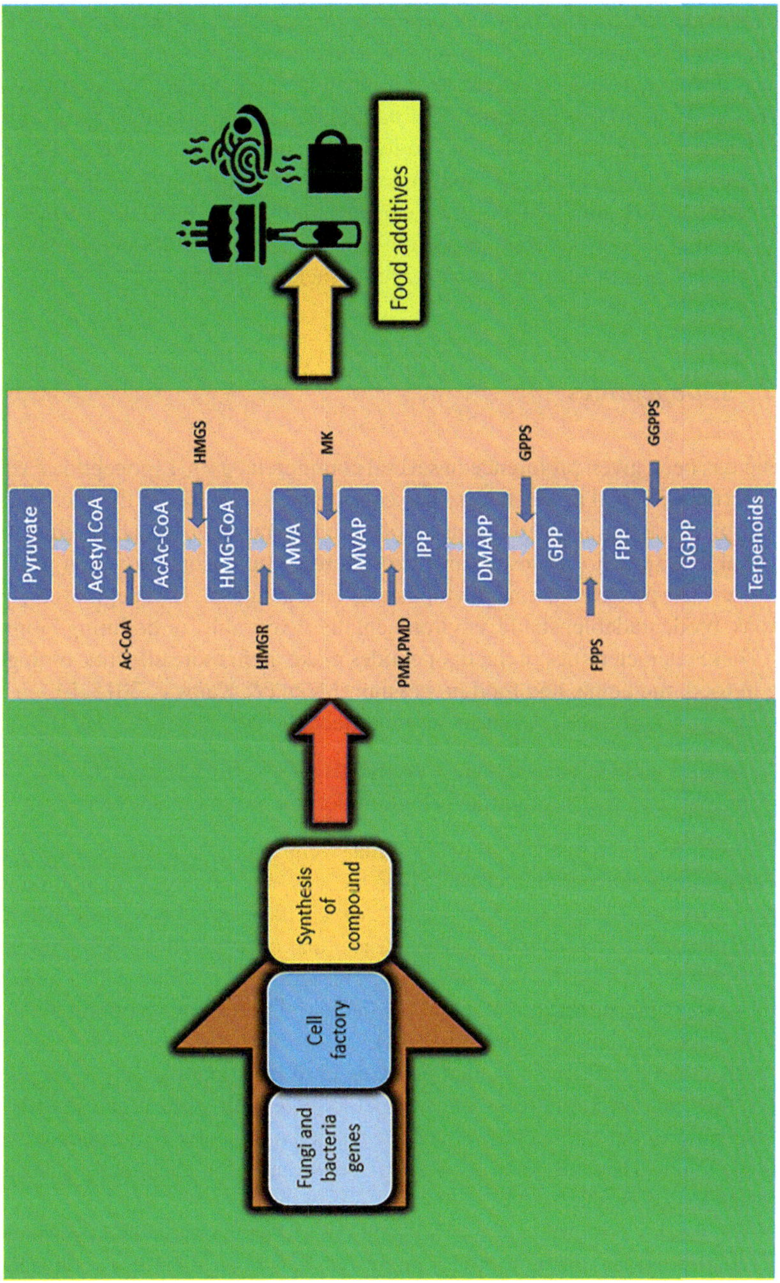


**Fig. 14.2** Biosynthesis of terpenoids

1. *E. coli.*
2. *S. cerevisiae.*

The industry favours *Saccharomyces cerevisiae* over bacteria because it can handle higher osmotic pressure and a lower pH. Currently, yeast strains are being continuously developed and improved to produce large quantities of desired end products (Nevoigt, 2008). The amount of natural terpenoids produced by *E. coli* is relatively low. S. cerevisiae accumulates a lot of ergosterol, which opens up the possibility of using its MVA route to produce terpenoids. Contrary to *E. coli*, *S. cerevisiae* contains redox mechanisms that enable cytochrome P450 to change the terpenoids' skeleton. In synthesizing value-added terpenoids with complex structures, *S. cerevisiae* is superior to *E. coli* (Zhu & Jackson, 2015).

### 14.5.2 Lipopeptides

Surfactin has been given preference in several commercial uses. Lipopeptides' existence in fermented food products was also considered for its uses in the food industry. Additionally, their capacity to cause systemic resistance in plants and their role in bacterial cell spread that results in rhizosphere colonization may open up new avenues for their potential use as phytopharmaceutical agents. Lipopeptides are not poisonous, biodegradable, stable, environmentally favourable, or polluting biomolecules. These characteristics of the lipopeptides make them more effective biologics for use in food, medicine, and food processing (Meena & Kanwar, 2015, b).

Two categories of microbial surfactants are recognized:

1. Lipopeptides and Glycolipids—low molecular weight biosurfactants,
2. Lipoproteins and polysaccharides—high molecular weight biosurfactants.

The emulsifying ability of the low molecular weight molecule can be used in the bioremediation of hydrophobic chemicals in various environments. Several microorganisms produce lipopeptide surfactants, which have structures with different chemical and surface characteristics. The intracellular or extracellular parts of fungi, bacteria, and yeast synthesize the surfactants. The synthesis of value-added metabolites from renewable substrates is one of the industrial processes involving biotechnology products (Kumar & Ngueagni, 2021).

As each molecule of biosurfactant contains both hydrophobic and hydrophilic moieties, they are amphiphilic secondary metabolites with surface and interfacial action. Because of this, BS can lower surface tension or dissolve hydrophobic compounds in water. They are seen as "green" substitutes for chemical surfactants (Biniarz et al., 2017).

Bacillus species primarily release lipopeptides (LPs), a unique class of natural antimicrobial peptides, through NRPS pathways. LPs typically forms connection with the membrane, creating a pore that causes the microbial membrane to become unstable (Zhang et al., 2022). However, on the basis of the kind and structures of these lipopeptides, changes in the mechanism of act against different pathogenic and spoilage microbe have been identified among the described LPs. In addition to improving antibacterial activity, Bacillus spp. can better preserve orange fruit during storage by fabricating (metal particles into LPs (Zhou et al., 2021). Some microorganisms produce biosurfactants that are extensively used in the food industry. *Candida* and *Yarrowia* are among the yeasts used for the production of biosurfactants and the synthesis of emulsifiers. Due to the lack of toxicity or pathogenicity risks, these microorganisms can be used in the food sector. These yeasts include *Yarrowia lipolytica, S. cerevisiae,* and *Kluyveromyces lactis.*

The majority of food industry goods' consistency, texture, dispersion, and scent solubilization are all influenced by emulsification. An emulsifier prevents globule clustering and maintain an aerated environment to stabilize the emulsion. Studies have shown that *Candida valida, Rhodotorula graminis*, the red alga, and bacteria from the *Klebsiella* sp. and *Acinetobacter calcoaceticus* are widely used as emulsifiers. They show better-stabilizing activity than carboxymethyl cellulose and gum arabic (Campos et al., 2013) (Table 14.2).

### *14.5.3 Polyphenols*

Secondary metabolites from plants and polyphenols are mostly present in foods made of plants, including fruits, vegetables, herbs, legumes, spices, and tea. The health advantages of polyphenol are increased by hydrolysis and change from bound to free form caused by microbial fermentation (Sarkar et al., 2022). Fruit

**Table 14.2** Biosurfactant-producing microbes and their function in food processing and preservation

| Type | Function | Application | References |
|---|---|---|---|
| *Candida lipolytica* | Emulsifier | Food fermentation | Barros et al. (2007) |
| *Candida utilis* | Emulsifier | Salad decoration | Shepherd et al. (1995) |
| *Saccharomyces cerevisiae* | Stabilize water–oil | Formation of mayonnaise, cookies, etc. | Barros et al. (2007) |
| *Klebsiella* | Emulsifier | Inhibition of the autooxidation of soybean oil. | Kawaguchi et al. (1996) |
| *Lactobacillus* | Adhesive | Coating agents for food-related utensils | Gudina et al. (2010) |
| *Nesterenkonia* | Emulsifier | Antioxidant activity and protective against *S. aureus* | Kiran et al. (2017) |
| *Bacillus* | Antimicrobial | Shows antimicrobial activity against food-spoiling bacteria | Kourmentza et al. (2021) |
| *Bacillus subtilis (Surfactins)* | Emulsifier | Maintain the texture, stability, and volume and also aid in the emulsification of fat to regulate the accumulation of fat globules | Meena and Kanwar (2015) |

wine's glucosidase production by non-*Saccharomyces cerevisiae* was essential. Alcohols, esters, acids, terpenes, and other volatile compounds may be produced during the hydrolysis of glycoside-bound volatiles in fruit juice by non-*Saccharomyces cerevisiae* with glucosidase activity (Fig. 14.3). It enhanced the fruit wine's flavour and aroma (Li et al., 2022). The naringenin, flavanones, and pinocembrin, formed from the phenylpropanoids p-coumaric acid and cinnamic acid, respectively, were the first plant-derived polyphenols to be synthesized in a microbe. Alternative hosts for the microbial synthesis of polyphenols, including *Corynebacterium glutamicum*, *L. lactis*, and *Streptomyces venezuelae*, have been effectively introduced (Milke et al., 2018) (Table 14.3).

Ascomycetes, deuteromycetes, and basidiomycetes all include laccases, which have great potential for application in the food industry as processing aids.

The usage of laccase includes bioremediation, beverage stability, uses in the banking sector, and a contribution to the enhancement of overall food quality. Despite turbidity, laccase treatment and active filtering significantly boosted colour stability, according to research. Juices treated with laccase have been reported to have significantly less phenolic content and more stable colour. Additionally, it has been discovered that laccase treatment is superior to standard colour and flavour stability methods, such as adding sulphites and ascorbic acid (Brijwani et al., 2010).

During cassava fermentation for gari preparation, it was observed that *L. plantarum* was frequently isolated as a LAB species. Due to the production of many antimicrobial metabolites during fermentation, lactic acid fermentation extends the shelf life of fermented items. The produced compounds such as organic acids,

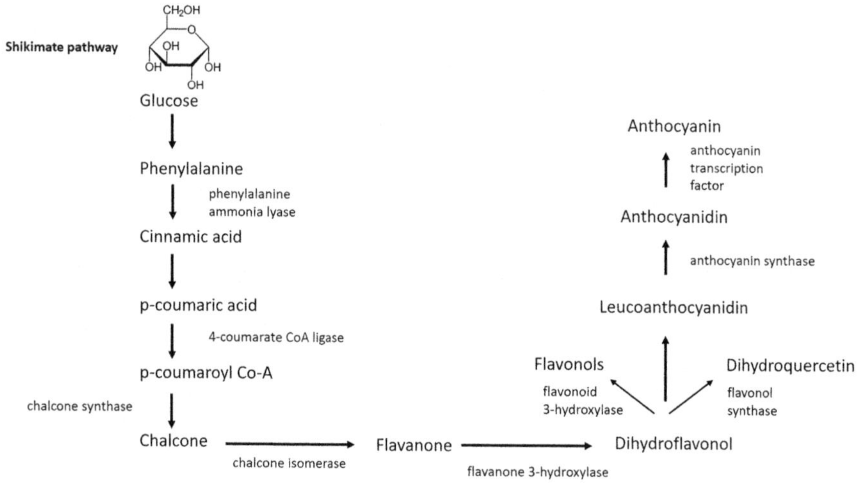


**Fig. 14.3** Biosynthesis of Polyphenols

**Table 14.3** Microorganisms and their polyphenol products

| Product | Organism | References |
|---|---|---|
| Isoflavone genistein | *E. coli* and *S. cerevisiae* | Leonard and Koffas (2007) |
| Resveratrol | *E. coli* and *C. glutamicum* | Heo et al. (2017) |
| Kaempferol, quercetin lucoside | *Bacillus cereus* or *Xanthomonas campestris* | Hyung Ko et al. (2006) |
| Sakuranetin | *E. coli* | Kim et al. (2013) |
| Resveratrol | *Flavobacterium johnsoniae* | Kallscheuer et al. (2016) |
| Resveratrol | *S. cerevisiae* | Li et al. (2016) |
| Ferulic acid | *E. coli* | Kang et al. (2014) |
| Kaempferol | *S. cerevisiae* | Duan et al. (2017) |

generate an acidic environment that is adverse to the growth of harmful and unwanted microbes (Adesulu-Dahunsi et al., 2020). Other antimicrobial bioactive compounds that LAB can produce include hydrogen peroxide, ethanol diacetyl, and bacteriocins. By reducing the antinutrient content of FFB through lactic acid fermentation, the bioavailability of vital dietary minerals in food items is enhanced (O'Sullivan et al., 2010).

The synergistic interaction of LAB and yeast carries out the traditional fermentation of cereals. According to research, LAB's acidifying environment favours yeast growth in fermented items. Yeast, which can supply growth factors, including vitamins and soluble nitrogen molecules, can encourage the growth of LAB. During the fermentation process of cereal for the production of ogi, the association of yeast species (*S. cerevisiae, R. graminis,* etc.) and LAB species (*L. fermentum, L. plantarum,* etc.) have been reported (Teniola et al., 2022). Incorporating phenolic substances such as curcumin, quercetin, and catechin into the foil matrix is one technique to enhance food quality. The use of phenolic compounds in packaging foil has been demonstrated to increase product shelf life and have high antioxidant effects, avoiding oxidation of lipid in fresh ground pork (Gutiérrez-del-Río et al., 2021).

## 14.6 Improvement of Packaging and Quality Through Nanotechnology to Ensure Food Safety

Food packaging's principal objective is to shield the product from its environment. Maintaining the food's quality throughout the product's shelf life is another goal. Additionally, packaging needs to consider legal, commercial, and communication requirement. Product qualities, attributes, and storage and distribution conditions of the individual packaging all affect how long a product will last.

If properly utilized, the growing field of nanotechnology could completely transform the global food processing and packaging industries. From processing and manufacture to handling and storage, till the products reach consumers, food packaging is crucial for protecting the safety and quality of foodstuffs. Food quality degrades because many fresh fruits and vegetables are susceptible to abiotic factors and ethylene (Mustafa & Andreescu, 2020). Recent research has indicated that covering food surfaces with natural or biopolymers can help preserve food. Various beverages, including fruit juices, tea, and coffee, can be clarified, maintained, and encapsulated using chitosan-based matrices, according to a new study (Sridhar et al., 2021). To increase the shelf life of various fruits and vegetables, bionanoencapsulated quercetin (biodegradable poly-D,L-lactide) should be employed, as it has improved the shelf life of tomatoes. The most widely used products made using nanotechnology include Nanogreen Tea, Neosino capsules, Canola Active Oil, Aquanova, and Nutralease (Bratovcic, 2020).

The use of nanoparticles in food has several objectives:

1. using nanoencapsulation to safeguard and enhance the stability of nutritional components, minerals, or nutraceuticals (Fathi et al., 2012).
2. for food product fortification, coating surfaces with protective nanomaterials to provide protective functions (Kiran et al., 2017).
3. fillers to improve the mechanical properties of packaging (Majeed et al., 2013).
4. sensing materials to create in situ food quality and safety monitoring techniques, ideally attached to or included in the packaging (Mills et al., 2012).

Nutraceuticals, viscosifying agents, gelation, nutrition delivery, vitamin fortification, gelation, and nanoencapsulation of tastes are just a few of the food processing techniques that use nanomaterials. The major reasons why food is processed are to preserve the food's integrity and lengthen its shelf life. Enzymes can be employed in a variety of food processing methods to alter food ingredients to enhance flavour, nutritional value, and health advantages, among other things (Asadi & Mousavi, 2006). Due to their assistance in dispersion in food matrices and their larger surface-to-volume ratios when compared to conventional macro-scale support materials, nanomaterials offer superior enzyme support systems. For instance, nano-silicon dioxide particles successfully hydrolysed olive oil, increasing its stability, adaptability, and reusability (Rashidi & Khosravi-Darani, 2011).

Antimicrobial packaging, also referred to as active packaging, comes into contact with the food item or the headspace inside to stop or impede any possible microbial development on food surfaces (Carbone et al., 2016). Many nanoparticles, including silver, copper, chitosan, and metal oxide nanoparticles like titanium or zinc oxide, have demonstrated antibacterial capabilities. Chitin or chitosan addition enhances the weight, toughness, fire resistance, and thermal properties of the polymer matrix (Singh et al., 2017a, b).

Numerous complex technologies have been made possible by nanotechnology and encapsulating methods (Fig. 14.4), allowing the design of devices for application in the food sector. These nanostructured devices could be very advantageous in many areas of food science (Singh et al., 2017a, b). Nanofibers can add additional functional characteristics as well as mechanical strength to preserve foods with delicate textures better. Due to its antioxidant features, green tea has been widely used

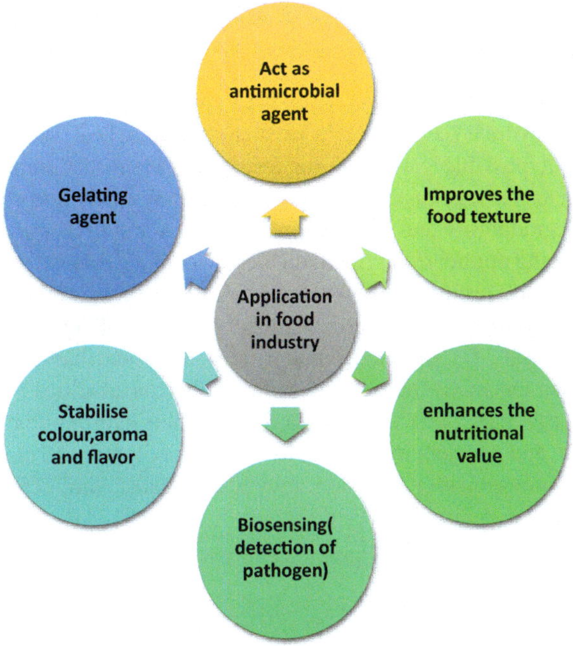


**Fig. 14.4** Application of nanotechnology in the food industry

to make edible food packaging materials due to its antioxidant characteristics. The use of nanomaterial goods as food colour additives, which play a key role in the psychological attractiveness of consumer products, has just received approval. Furthermore, the nanocoatings on various edible food components might serve as a barrier to moisture and gas exchange, transmit flavours, colours, and anti-browning agents, and lengthen the shelf life of manufactured meals (Dwivedi et al., 2018).

New quercetin-starch-based films made on biodegradable chitosan-gelatin have been created. When compared to the control without flavonoid incorporation, it was found that the film containing quercetin-starch improved antioxidant activity. Biodegradable poly-(ester-urethane) film that is loaded with catechin act as an antioxidant agent and is based on a triblock copolymer of poly-(lactic acid) and poly-(caprolactone). This novel material demonstrated effective antioxidant activity, good catechin release, and suitable compost decomposition. A whey protein film synthesized using hydroethanolic extracts of the brown edible macroalgae *Fucus vesiculosus L* act as an active package that can prevent chicken breasts from oxidizing during a 25-day storage period. Under the right environmental conditions, it is possible to trigger the partial breakdown of the milk protein known as lactalbumin by a protease to self-assemble into comparable nanotubes and can be employed in food, nanotechnology, and many more (Rashidi & Khosravi-Darani, 2011).

Microorganism- or bacterial-produced polymers (such as polyhydroxybutyrate, bacterial cellulose, xanthan, curdian, and pullan) are frequently utilized in food packaging.

It is simple to manufacture lactic acid, the PLA's monomer, through the fermentation of a carbohydrate feedstock. Crops such as corn, wheat, molasses, and whey are examples of possible carbohydrate feedstock. Many different bacteria store PHB as carbon and energy reserves. This polyester may easily find commercial uses due to its biodegradability and biocompatibility (PLA and PHB may present a wide range of packaging application potential (Wesley et al., 2014).

A polysaccharide produced from chitin, chitosan is present in the exoskeletons of crustaceans, arthropods, and fungi. Chitosan is a possible nanoparticle for packaging since it is safe for the environment, non-toxic, and has great antibacterial effects. Chitosan enhances the gas and moisture barrier qualities when added to a biodegradable polymer, such as PA films. Due to their hydrophobic nature, hydrogen production, and covalent interaction between biopolymer and chitosan nanoparticles, which slows moisture diffusion, they also enhance antibacterial capabilities and impermeability (Ashfaq et al., 2022).

## 14.7 Conclusion

Any compound with nutritional value is considered food, and when consumed, the body uses it to produce energy and maintain life. Wine, beer, brews, champagne, and other alcoholic beverages are all produced with the help of *Saccharomyces* species. Due to their flavour and health advantages, fermented foods, probiotics, and alcoholic beverages are becoming increasingly popular today. Innovative techniques have been created as a result of technological improvement, opening up a varied series of applications for microorganisms in the food and beverage industries. Numerous routes help produce energy at the cellular level.

Furthermore, positive outcomes have been obtained in using microbes for food preservation, where they may shield the food from moisture, lipids, gases, off-tastes, and aromas. Effective microbial communities that can provide services such as nutrient use efficiency, bioremediation, food packaging quality, food fermentation, and control of phytopathogens at the farming level are well known. They offer excellent vehicle systems to deliver bioactive compounds to the target tissues. Biofilms covering fruits and vegetables are also a hot topic. Bacteria are beneficial in a variety of biological and culinary industries. For these microbial communities to be widely used in the food industry, a comprehensive understanding of the environmental conditions impacting their viability and performance is required. This intriguing technology is widely used in industrial applications, but several problems still need to be addressed and handled, including regulatory concerns and effectiveness. Understanding microbes that offer superior stability, efficacy, and cost-effectiveness must be the focus of future research.

## References

Adesulu-Dahunsi, A. T., Dahunsi, S. O., & Olayanju, A. (2020). Synergistic microbial interactions between lactic acid bacteria and yeasts during production of Nigerian indigenous fermented foods and beverages. *Food control, 110*, 106963.

Araújo, T. M. (2013). *Caracterização bioquímico-molecular de cepas de Saccharomyces cerevisiae isoladas de dornas de fermentação de cachaça Para produção de cervejas [thesis]*. Universidade federal de Ouro Preto.

Asadi, G., & Mousavi, S. M. (2006). Application of nanotechnology in food packaging. In *13th world congress of Food Science & Technology* (pp. 739–739).

Ashfaq, A., Khursheed, N., Fatima, S., Anjum, Z., & Younis, K. (2022). Application of nanotechnology in food packaging: pros and cons. *Journal of Agriculture and Food Research, 7*, 100270.

Barros, F. F. C., Quadros, C. P., Marostica Junior, M. R., & Pastore, G. M. (2007). Surfactina: propriedades quımicas, tecnologicas e funcionais para aplicac¸~oes em alimentos. *Quimica Nova, 30*, 409–414.

Bekatorou, A., Plessas, S., & Mantzourani, I. (2015). Biotechnological exploitation of brewery solid wastes for recovery or production of value-added products. In R. V. Ravishankar (Ed.), *Advances in food biotechnology* (1st ed., pp. 393–414). Wiley.

Biniarz, P., Łukaszewicz, M., & Janek, T. (2017). Screening concepts, characterization and structural analysis of microbial-derived bioactive lipopeptides: a review. *Critical Reviews in Biotechnology, 37*(3), 393–410.

Blanco, C. A., Caballero, I., Barrios, R., & Rojas, A. (2014). Innovations in the brewing industry: light beer. *International Journal of Food Sciences and Nutrition, 65*, 655–660. https://doi.org/10.3109/09637486.2014.893285

Bovill, R., Bew, J., & Robinson, S. (2001). Comparison of selective media for the recovery and enumeration of probiotic yeasts from animal feed. *International Journal of Food Microbiology, 67*, 55–61.

Bratovcic, A. (2020). Nanomaterials in food processing and packaging, its toxicity and food labeling. *Acta scientific nutritional health, 4*(9), 07–13.

Brijwani, K., Rigdon, A., & Vadlani, P. V. (2010). Fungal laccases: production, function, and applications in food processing. *Enzyme research, 2010*, 149748.

Campos, J. M., Montenegro Stamford, T. L., Sarubbo, L. A., de Luna, J. M., Rufino, R. D., & Banat, I. M. (2013). Microbial biosurfactants as additives for food industries. *Biotechnology Progress, 29*(5), 1097–1108.

Caplice, E., & Fitzgerald, G. F. (1999). Food fermentations: role of microorganisms in food production and preservation. *International Journal of Food Microbiology, 50*(1–2), 131–149. https://doi.org/10.1016/S0168-1605(99)00082-3

Carbone, M., Donia, D. T., Sabbatella, G., & Antiochia, R. (2016). Silver nanoparticles in polymeric matrices for fresh food packaging. *Journal of King Saud University-Science, 28*(4), 273.

Dharmaraj, S., Ashokkumar, B., & Dhevendaran, K. (2009). Food-grade pigments from Streptomyces sp. isolated from the marine sponge Callyspongia diffusa. *Food Research International, 42*(4), 487–492.

Doyle, M. P., & Meng, J. I. A. N. G. H. O. N. G. (2006). Bacteria in food and beverage production. *Prokaryotes, 1*, 797–811.

Duan, L., Ding, W., Liu, X., Cheng, X., Cai, J., Hua, E., & Jiang, H. (2017). Biosynthesis and engineering of kaempferol in *Saccharomyces cerevisiae*. *Microbial Cell Factories, 16*(1), 165. https://doi.org/10.1186/s12934-017-0774-x

Dufosse, L., Fouillaud, M., Caro, Y., Mapari, S. A., & Sutthiwong, N. (2014). Filamentous fungi are large-scale pro.

Dwivedi, C., Pandey, I., Misra, V., Giulbudagian, M., Jungnickel, H., Laux, P., & Singh, A. V. (2018). The prospective role of nanobiotechnology in food and food packaging products. *Integrative Food Nutrition Metab, 5*.

Erten, H., Ağirman, B., Gündüz, C. P. B., Çarşanba, E., Sert, S., Bircan, S., & Tangüler, H. (2014). Importance of yeasts and lactic acid bacteria in food processing. In *Food processing: strategies for quality assessment* (pp. 351–378). Springer.

Faria-Oliveira, F., Diniz, R. H., Godoy-Santos, F., Piló, F. B., Mezadri, H., Castro, I. M., & Brandão, R. L. (2015). The role of yeast and lactic acid bacteria in the production of fermented beverages in South America. *Food production and industry*, 107–135.

Fathi, M., Mozafari, M. R., & Mohebbi, M. (2012). Nanoencapsulation of food ingredients using lipid based delivery systems. *Trends in Food Science and Technology, 23*(1), 13–27.

Giraffa, G. (2004). Studying the dynamics of microbial populations during food fermentation. *FEMS Microbiology Reviews, 28*(2), 251–260.

Goel, R., Kumar, V., Suyal, D. C., Narayan, & Soni, R. (2018). Toward the unculturable microbes for sustainable agricultural production. In *Role of rhizospheric microbes in soil* (pp. 107–123). Springer.

Gondal, A. H., Farooq, Q., Hussain, I., & Toor, M. D. (2021). Role of microbes in plant growth and food preservation. *Agrinula: Jurnal Agroteknologi Dan Perkebunan, 4*(2), 106–121.

Gudina, E. J., Teixeira, J. A., & Rodrigues, L. R. (2010). Isolation and functional characterization of a biosurfactant produced by lactobacillus paracasei. *Colloids and Surfaces. B, Biointerfaces, 76*, 298–304.

Gutiérrez-del-Río, I., López-Ibáñez, S., Magadán-Corpas, P., Fernández-Calleja, L., Pérez-Valero, Á., Tuñón-Granda, M., & Lombó, F. (2021). Terpenoids and polyphenols as natural antioxidant agents in food preservation. *Antioxidants, 10*(8), 1264.

Heitmann, M., Zannini, E., & Arendt, E. (2018). Impact of Saccharomyces cerevisiae metabolites produced during fermentation on bread quality parameters: a review. *Critical Reviews in Food Science and Nutrition, 58*(7), 1152–1164.

Heldman, D. R., & Hartel, R. W. (1998). *Principles of food processing*. Springer.

Heo, K. T., Kang, S.-Y., & Hong, Y.-S. (2017). *De novo* biosynthesis of pterostilbene in an *Escherichia coli* strain using a new resveratrol *O*-methyltransferase from *Arabidopsis*. *Microbial Cell Factories, 16*(1), 30.

Holzapfel, W. H. (2002). Appropriate starter culture technologies for small-scale fermentation in developing countries. *International Journal of Food Microbiology, 75*(3), 197–212.

Hyung Ko, J., Gyu Kim, B., & Joong-Hoon, A. (2006). Glycosylation of flavonoids with a glycosyltransferase from *Bacillus cereus*. *FEMS Microbiology Letters, 258*(2), 263–268.

James, J., Simpson, B. K., & Marshall, M. R. (1996). Application of enzymes in food processing. *Critical Reviews in Food Science and Nutrition, 36*, 437–463. https://doi.org/10.1080/10408399609527735

Jay, J. M. (1996). *Modern food microbiology* (5th ed.). Chap man and Hall.

Jordan, K., Dalmasso, M., Zentek, J., Mader, A., Bruggeman, G., Wallace, J., De Medici, D., et al. (2014). Microbes versus microbes: control of pathogens in the food chain. *Journal of the Science of Food and Agriculture, 94*(15), 3079–3089.

Kallscheuer, N., Vogt, M., Stenzel, A., Gätgens, J., Bott, M., & Marienhagen, J. (2016). Construction of a *Corynebacterium glutamicum* platform strain for the production of stilbenes and (2*S*)-flavanones. *Metabolic Engineering, 38*, 47–55.

Kalsoom, M., Rehman, F. U., Shafique, T. A. L. H. A., Junaid, S. A. N. W. A. L., Khalid, N., Adnan, M., & Ali, H. (2020). Biological importance of microbes in agriculture, food and pharmaceutical industry: a review. *Innovare Journal Life Sciences, 8*(6), 1–4.

Kampranis, S. C., & Makris, A. M. (2012). Developing a yeast cell factory for the production of terpenoids. *Computational and Structural Biotechnology Journal, 3*(4), e201210006.

Kang, S.-Y., Lee, J. K., Choi, O., Kim, C. Y., Jang, J.-H., Hwang, B. Y., & Hong, Y.-S. (2014). Biosynthesis of methylated resveratrol analogs through the construction of an artificial biosynthetic pathway in *E. coli*. *BMC Biotechnology, 14*(1), 67.

Kawaguchi, K., Satomi, K., Yokoyama, M., & Ishida, Y. (1996). Antioxidative properties of an extracellular polysaccharide produced by a bacterium Klebsiella sp. isolated from river water. *Nippon Suisan Gakkaishi, 42*, 243–251.

Kieliszek, M., Kot, A. M., Bzducha-Wróbel, A., BŁażejak, S., Gientka, I., & Kurcz, A. (2017). Biotechnological use of candida yeasts in the food industry: a review. *Fungal Biology Reviews, 31*(4), 185–198.

Kim, M.-J., Kim, B.-G., & Ahn, J.-H. (2013). Biosynthesis of bioactive *O*-methylated flavonoids in *Escherichia coli. Applied Microbiology and Biotechnology, 97*(16), 7195–7204.

Kiran, G. S., Priyadharsini, S., Sajayan, A., Priyadharsini, G. B., Poulose, N., & Selvin, J. (2017). Production of lipopeptide biosurfactant by a marine Nesterenkonia sp. and its application in food industry. *Frontiers in Microbiology, 8*, 1138.

Klaenhammer, T. R. (1993). Genetics of bacteriocins produced by starter culture lactic acid bacteria. *FEMS Microbiology Reviews, 12*, 39–86.

Kourmentza, K., Gromada, X., Michael, N., Degraeve, C., Vanier, G., Ravallec, R., & Jauregi, P. (2021). Antimicrobial activity of lipopeptide biosurfactants against foodborne pathogen and food spoilage microorganisms and their cytotoxicity. *Frontiers in Microbiology, 11*, 561060.

Kumar, P. S., & Ngueagni, P. T. (2021). A review on new aspects of lipopeptide biosurfactant: types, production, properties and its application in the bioremediation process. *Journal of Hazardous Materials, 407*, 124827.

Lee, Y.-K., & Salminen, S. (1995). The coming of age of probiotics. *Trends in Food Science and Technology, 6*, 241–245.

Leonard, E., & Koffas, M. A. (2007). Engineering of artificial plant cytochrome P450 enzymes for synthesis of isoflavones by *Escherichia coli. Applied and Environmental Microbiology, 73*(22), 7246–7251. https://doi.org/10.1128/AEM.01411-07

Li, M., Schneider, K., Kristensen, M., Borodina, I., & Nielsen, J. (2016). Engineering yeast for high-level production of stilbenoid antioxidants. *Scientific Reports, 6*, 36827.

Li, S., Bi, P., Sun, N., Gao, Z., Chen, X., & Guo, J. (2022). Characterization of different non-saccharomyces yeasts via mono-fermentation to produce polyphenol-enriched and fragrant kiwi wine. *Food Microbiology, 103*, 103867.

Liguori, R., Soccol, C., de Souza, P., Vandenberghe, L., Woiciechowski, A., & Faraco, V. (2015). Second generation ethanol production from brewers' spent grain. *Energies, 8*(4), 2575–2586.

Liu, S. Q. (2002). Malolactic fermentation in wine–beyond deacidification. *Journal of Applied Microbiology, 92*(4), 589–601.

Majeed, K., Jawaid, M., Hassan, A. A. B. A. A., Bakar, A. A., Khalil, H. A., Salema, A. A., & Inuwa, I. (2013). Potential materials for food packaging from nanoclay/natural fibres filled hybrid composites. *Materials Design, 46*, 391–410.

Malinowska-Panczyk, E., Kolodziejska, I., Murawska, D., & Wolosewicz, G. (2009). The combined effect of moderate pressure and chitosan on Escherichia coli and Staphylococcus aureus cells suspended in a buffer and on natural microflora of apple juice and minced pork. *Food Technology and Biotechnology, 47*(2), 202.

Matassa, S., Boon, N., Pikaar, I., & Verstraete, W. (2016). Microbial protein: future sustainable food supply route with low environmental footprint. *Microbial Biotechnology, 9*(5), 568–575.

Meena, K. R., & Kanwar, S. S. (2015). Lipopeptides as the antifungal and antibacterial agents: applications in food safety and therapeutics. *BioMed research international, 2015*.

Milke, L., Aschenbrenner, J., Marienhagen, J., & Kallscheuer, N. (2018). Production of plant-derived polyphenols in microorganisms: Current state and perspectives. *Applied Microbiology and Biotechnology, 102*(4), 1575–1585.

Mills, A., Lawrie, K., Bardin, J., Apedaile, A., Skinner, G. A., & O'Rourke, C. (2012). An O2 smart plastic film for packaging. *Analyst, 137*(1), 106–112.

Mojsov, K. D. (2016). Aspergillus enzymes for food industries. In *New and future developments in microbial biotechnology and bioengineering* (pp. 215–222). Elsevier.

Mussatto, S. I., & Roberto, I. C. (2005). Acid hydrolysis and fermentation of brewer's spent grain to produce xylitol. *Journal of the Science of Food and Agriculture, 85*(14), 2453–2460.

Mustafa, F., & Andreescu, S. (2020). Nanotechnology-based approaches for food sensing and packaging applications. *RSC Advances, 10*(33), 19309–19336.

Narsing Rao, M. P., Xiao, M., & Li, W. J. (2017). Fungal and bacterial pigments: secondary metabolites with wide applications. *Frontiers in microbiology, 8*, 1113.

Nasseri, A. T., Rasaul, S., Morowvat, M. H., & Ghasemi, Y. (2011). Single cell protein: production and process. *American Journal of Food Technology, 6*(2), 103–116.

Nevoigt, E. (2008). Progress in metabolic engineering of Saccharomyces cerevisiae. *Microbiology and Molecular Biology Reviews, 72*(3), 379–412.

O'Sullivan, L., Murphy, B., McLoughlin, P., Duggan, P., Lawlor, P. G., Hughes, H., & Gardiner, G. E. (2010). Prebiotics from marine macroalgae for human and animal health applications. *Marine Drugs, 8*(7), 2038–2064.

Oyewole, O. B., & Odunfa, S. A. (1990). Characterization and dismonocytogenes in cottage cheese manufactured with a lacticin tribution of lactic acid bacteria in cassava fermentation duringfufu production. *The Journal of Applied Bacteriology, 68*, 145–152.

Park, S. H., Lamsal, B. P., & Balasubramaniam, V. M. (2014). Principles of food processing. *Food processing: principles and applications, 2*, 1.

Preetha, S. S., & Narayanan, R. (2020). Factors influencing the development of microbes in food. *Shanlax International Journal of Arts, Science and Humanities, 7*(3), 57–77.

Rakhmanova, A., Khan, Z. A., & Shah, K. (2018). A mini review fermentation and preservation: role of lactic acid bacteria. *MOJ Food Processing Technology, 6*(5), 414–417.

Rashidi, L., & Khosravi-Darani, K. (2011). The applications of nanotechnology in food industry. *Critical Reviews in Food Science and Nutrition, 51*(8), 723–730.

Sandoval-Castilla, O., Lobato-Calleros, C., Aguirre-Mandujano, E., & Vernon-Carter, E. J. (2004). Microstructure and texture of yogurt as influenced by fat replacers. *International Dairy Journal, 14*(2), 151–159.

Sarkar, P., Abedin, M. M., Singh, S. P., Pandey, A., & Rai, A. K. (2022). Microbial production and transformation of polyphenols. In *Current developments in biotechnology and bioengineering* (pp. 189–208). Elsevier.

Shepherd, R., Rockey, J., Shutherland, I. W., & Roller, S. (1995). Novel bioemulsifiers from microorganisms for use in foods. *Journal of Biotechnology, 40*, 207–217.

Silbir, S., & Goksungur, Y. (2019). Natural red pigment production by *Monascus purpure*us in submerged fermentation systems using a food industry waste: brew spent grain. *Food, 8*(5), 161.

Singh, J. S., Abhilash, P. C., & Gupta, V. K. (2016). Agriculturally important microbes in sustainable food production. *Trends in Biotechnology, 34*(10), 773–775.

Singh, A. V., Hosseinidoust, Z., Park, B. W., Yasa, O., & Sitti, M. (2017a). Microemulsion-based soft bacteria-driven microswimmers for active cargo delivery. *ACS Nano, 11*(10), 9759–9769.

Singh, T., Shukla, S., Kumar, P., Wahla, V., Bajpai, V. K., & Rather, I. A. (2017b). Application of nanotechnology in food science: perception and overview. *Frontiers in Microbiology, 8*, 1501.

Soomro, A. H., Masud, T., & Anwaar, K. (2002). Role of lactic acid bacteria (LAB) in food preservation and human health-a review. *Pakistan Journal of Nutrition, 1*(1), 20–24.

Sridhar, A., Ponnuchamy, M., Kumar, P. S., & Kapoor, A. (2021). Food preservation techniques and nanotechnology for increased shelf life of fruits, vegetables, beverages and spices: a review. *Environmental Chemistry Letters, 19*(2), 1715–1735.

Teniola, O. D., Holzapfel, W. H., & Odunfa, S. A. (2022). Selection, use and the influence of starter cultures in the nutrition and processing improvement of 'ogi'.

van der MJEC, M., van der Veen, B., Uitdehaag, J. C. M., Leemhuis, H., & Dijkhuizen, L. (2002). Properties and applications of starch-converting enzymes of the α-amylase family. *Journal of Biotechnology, 94*, 137–155.

Wang, C., Liwei, M., Park, J. B., Jeong, S. H., Wei, G., Wang, Y., & Kim, S. W. (2018). Microbial platform for terpenoid production: Escherichia coli and yeast. *Frontiers in Microbiology, 9*, 2460.

Wasilewska, E., Zlotkowska, D., & Wroblewska, B. (2019). Yogurt starter cultures of Streptococcus thermophilus and lactobacillus bulgaricus ameliorate symptoms and modulate the immune response in a mouse model of dextran sulfate sodium-induced colitis. *Journal of Dairy Science, 102*(1), 37–53.

Weatherholtz, W. M., & Holsing, G. C. (1975). Evaluation of Torula yeast for use as a food supplement. *Federation Proceedings, 34*, 890.

Wesley, S. J., Raja, P., Raj, A. A., & Tiroutchelvamae, D. (2014). Review on-nanotechnology applications in food packaging and safety. *International Journal of Engineering Research, 3*(11), 645–651.

Yerlikaya, O., Akpinar, A., Saygili, D., & Karagozlu, N. (2020). Incorporation of Propionibacterium shermanii subsp. freudenreichii in probiotic dairy drink production: physicochemical, rheological, microbiological and sensorial properties. *International Journal of Dairy Technology, 73*(2), 392–402.

Zdolec, N., Lorenzo, J. M., & Ray, R. C. (2018). Use of microbes for improving food safety and quality. *BioMed Research International, 2018*, 3902698.

Zhang, B., Xu, L., Ding, J., Wang, M., Ge, R., Zhao, H., & Fan, J. (2022). Natural antimicrobial lipopeptides secreted by bacillus spp. and their application in food preservation, a critical review. *Trends in Food Science Technology, 127*, 26.

Zhao, G., Liu, C., Li, S., Wang, X., & Yao, Y. (2020). Exploring the flavor formation mechanism under osmotic conditions during soy sauce fermentation in aspergillus oryzae by proteomic analysis. *Food Function, 11*(1), 640–648.

Zhou, L., Zhao, X., Li, M., Yan, L., Lu, Y., Jiang, C., et al. (2021). Antibacterial and wound healing-promoting effect of sponge-like chitosan-loaded silver nanoparticles biosynthesized by iturin. *International Journal of Biological Macromol, 181*, 1183.

Zhu, Q., & Jackson, E. N. (2015). Metabolic engineering of *Yarrowia lipolytica* for industrial applications. *Current Opinion in Biotechnology, 36*, 65–72. https://doi.org/10.1016/j.copbio.2015.08.010

# Chapter 15
# Microbiome as a Key Player in Sustainable Food and Food Safety

**Khushboo, Inderpal Devgon, and Arun Karnwal**

**Abstract** Microorganisms have a significant influence on human health and environmental sustainability. Microorganisms that cause food poisoning are a severe problem for consumers, the food industry, and regulatory bodies. Other hazardous microbes ruin the food, shorten its shelf life, and thus result in significant monetary losses. Despite recent developments in food processing technology, the potential for disease transmission through food still exists. Most physical methods used to preserve commercial food items are combined with chemical preservatives. Consumer demand for alternative natural antimicrobial agents is rising, and the food sector must keep up with this trend. One of the most crucial components of sustainability is growing and canning your own food. Traditionally, families handed on their knowledge of food preservation to succeeding generations as a way of life. Consumers face a knowledge gap while attempting to preserve their produce. This chapter will briefly explore the role of different microbes in food preservation. It describes the many microorganisms and techniques crucial for long-term food preservation.

**Keywords** Food · Preservation · Microorganisms · Sustainable · Beneficial

## 15.1 Introduction

Foods are consumed for their nutritional value and are made of different organic ingredients. Foods contain organic substances such as proteins, fats, moisture, carbohydrates, and minerals derived from either plants or animals. Foods turn perishable due to physical, chemical, or microbial factors. Food's nutritional value, aroma,

Khushboo · I. Devgon
Department of Microbiology, School of Bioengineering and Biosciences, Lovely Professional University, Phagwara, India

A. Karnwal (✉)
Department of Microbiology, Lovely Professional University, Phagwara, Punjab, India


A. Karnwal, A. R. Mohammad Said Al-Tawaha (eds.), *Food Microbial Sustainability*, https://doi.org/10.1007/978-981-99-4784-3_15

color, consistency, and taste can all be affected by item spoiling (Amit et al., 2017). Thus, many food items must be preserved to maintain the quality of food for a more extended period. The practices or strategies used to manage both internal and external factors that might cause food spoiling is referred to as food preservation. The primary objective of food preservation is to increase the food's shelf life while maintaining its flavor, color, texture, and inherent nutritional advantages.

Preservation of food has been a long-standing practice that started when prehistoric humans recognized the need to keep food after slaughtering a large animal that they could not all consume. Learning to preserve food was the first and most crucial step in the development of civilization. The basic techniques for food preservation have been used by numerous cultures across time and space (Nummer, 2002). Traditional food preservation techniques are often used worldwide, including pickling, drying, freezing, chilling, pasteurization, and other chemical preservation techniques. Only a few instances of scientific study and advancements impacting the creation of new and existing technologies include irradiation, high-pressure technology, and hurdle technology (Blum, 2012; Freedman, 2011). Food preservation processing has developed into a highly interdisciplinary topic since it encompasses procedures related to the growing, harvesting, processing, packaging, and distribution of food.

Consequently, it would be helpful to utilize an integrated method to preserve food during the food production and processing stages. The worldwide market for processed food products is currently above $7 trillion and encourages growth over time (WFM, 2013). The development of the food processing industry in many nations is primarily due to rapid industrialization and globalization. Food processing is a valuable part of the manufacturing sector in developing countries, and its contribution to national GDP rises with the level of national income (Wilkinson & Roch, 2006), according to a review of the UNIDO Industrial Statistics Database (2005).

The microorganisms that affect food may be categorized into two broad categories: those that cause spoiling and degradation and those that man uses to preserve food. The soluble components of food are used by bacteria, yeasts, and molds, which are tiny plant life. They also produce enzymes that cause the disintegration of food tissue, which makes many of the contents liquid and accessible to the microorganisms. These natural processes, which occur constantly, are primarily responsible for transforming organic matter into materials that improve soil fertility. These little forms of life cause the degradation of leaves, wood, grass, etc. Microorganisms are frequently acknowledged as having an impact on food degradation. However, by helping in fermentation of the food and dairy products for preservation, they also contribute a productive involvement to the food industry (Kar, 2014). Microorganisms are utilized in the fermentation process to extend food's shelf life. Due to their distinctive metabolic features, lactic acid bacteria (LAB) are utilized in various fermentation and preservation processes. Milk, meats, cereals, and veggies are a few examples. Producing bacteriocins is LAB's unique characteristic. LABs are used for preservation since it has antimicrobial properties. It serves as a "starter culture" for the fermentation of milk, producing acid as a result.

This chapter examines several physiological, chemical, and biological causes of food decomposition and the classification of food products. Here, the fundamentals and developments of several simple and advanced food preservation methods—which are said to prevent food from spoiling and provide a longer shelf life—along with their workings, conditions for application, benefits, and drawbacks—are covered. This article also discusses the trend of processed and preserved foods on the international market. This chapter presents a thorough knowledge that might be very helpful to researchers, technicians, and industry management to develop efficient and integrated food preservation techniques and ensure food safety.

## 15.2 Food Categorization

Foods may be widely categorized based on their shelf lives, uses nutrient content, and processing methods (Fig. 15.1). In Table 15.1, many food types are compiled, and in the sections that follow, they are briefly reviewed.

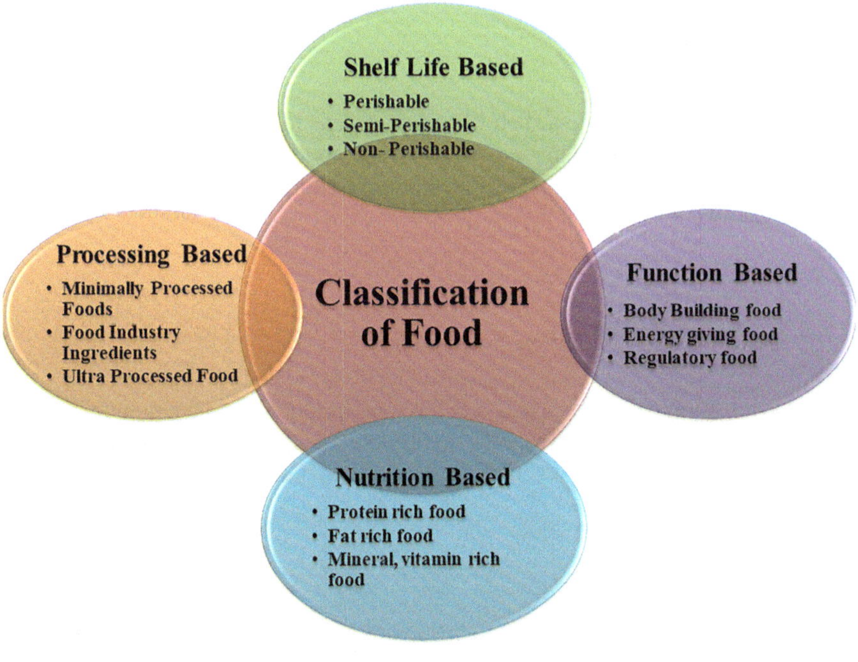


**Fig. 15.1** Categorization of food

**Table 15.1** Categorization of foods based on functions and nutrients

| Categories | Function | Food items |
|---|---|---|
| Function-based | Body building foods | Pulses, milk, fish, meat, nuts, vegetables etc. |
| | Energy-giving foods | Dry fruits, cereals, butter, sugars, oils, etc. |
| | Regulatory foods | Water, raw vegetables, fruits, and beverages |
| | Protective foods | Fresh vegetables, fruits, milk beverages, etc. |
| Nutrition based | Carbohydrate | Starchy vegetable, wheat, rice, etc. |
| | Proteins | Fish, meat, milk, nuts, soybean |
| | Fats | Edible oil, butter, fish |
| | Vitamins or minerals | Fruits and vegetables |

## 15.3 Categorization of Foods Based on Shelf Life

Food gradually loses its color, texture, taste, nutritional content, and edibility due to the natural spoiling process. People have a higher chance of getting sick and, in some circumstances, dying if they eat spoiled food items (Steele, 2004). Depending on how quickly they spoil, foods can be classified as nonperishable, semi-perishable, or perishable.

### *15.3.1 Perishable*

Foods are defined as having a shelf life of a few days to around 3 weeks. Perishable food items include milk, dairy products, meats, poultry, eggs, and shellfish. Food items may spoil immediately if particular preservation measures are not understood (Doyle, 2009).

### *15.3.2 Semi-Perishable*

Under ideal storage circumstances, certain food products can be stored for about 6 months. They are classified as semi-perishable foods. Some examples of semi-perishable foods are fruits, vegetables, cheeses, and potatoes (Chopra, 2005).

### *15.3.3 Nonperishable*

Nonperishable food items are natural and processed foods with an eternal shelf life. These foods have an extended shelf life of many years or more. Nonperishable foods include dry beans, almonds, sugar, canned fruits, mayonnaise, and peanut butter (Monteiro et al., 2010).

## 15.4 Using Nutrients and Functions to Classify Foods

Foods may be grouped into the following categories based on how they help the body function: (a) foods that build and repair the body; (b) foods that provide energy; (c) regulatory foods; and (d) foods that are protective. Foods can be categorized as (a) carbohydrate-containing foods, (b) protein-containing foods, (c) fat-containing meals, or (d) vitamin- and mineral-rich foods, depending on their nutritional value. Table 15.1 lists several food products per their nutritional content and uses.

## 15.5 Food Groups Depending on the Degree and Use of Processing

The food industries utilize various food processing methods to transform fresh ingredients into food items. Based on the level and purpose of food processing, foods may be divided into three main categories: (a) minimally processed foods, (b) processed foods, and (c) ultra-processed food products (Barbosa-Cánovas et al., 2005). Table 15.2 provides a classification of foods depending on the level and use of processing.

## 15.6 Mechanism of Food Spoilage

Food deterioration causes a decrease in the nutritional values and digestibility of food. Food deterioration and food safety are interrelated (Steele, 2004). The early stages of food deterioration can be detected through a change in color, aroma, taste, or texture or even by looking at the item itself. Several physical, microbial, or chemical mechanisms can cause food to deteriorate. These processes are not necessarily mutually exclusive since the degradation brought on by one process may stimulate another. Temperature, pH, air, nutrition, and different chemicals are the primary factors that cause food deterioration.

**Table 15.2** Categorization of food based on food processes

| Food | Purpose of processing | Examples |
|---|---|---|
| Minimally processed food | To make complete food readily available and safe | Vacuum-packed fruit, vegetables, frozen beans, cereals and pulses, frozen fish, pork, meat, pasteurized milk, eggs, yogurt, tea, coffee, non-salted nuts, dry fruits, and beans |
| Processed food or ingredients | Production of ingredients used in the meal preparation | Raw noodles and pasta, butter, milk cream, cheese, sweeteners, and food industry ingredients, like fructose syrups |
| Ultra-processed food | To produce ready-to-eat food with a longer shelf life | Chocolate bars, protein bars, ice cream, sugared fruits, soft drinks; pre-prepared meat, nuggets, burgers; salted, pickled, smoked meat and fish; vegetables bottled in brine; fish canned in oil; breads, biscuits, chips cakes, and pastries |

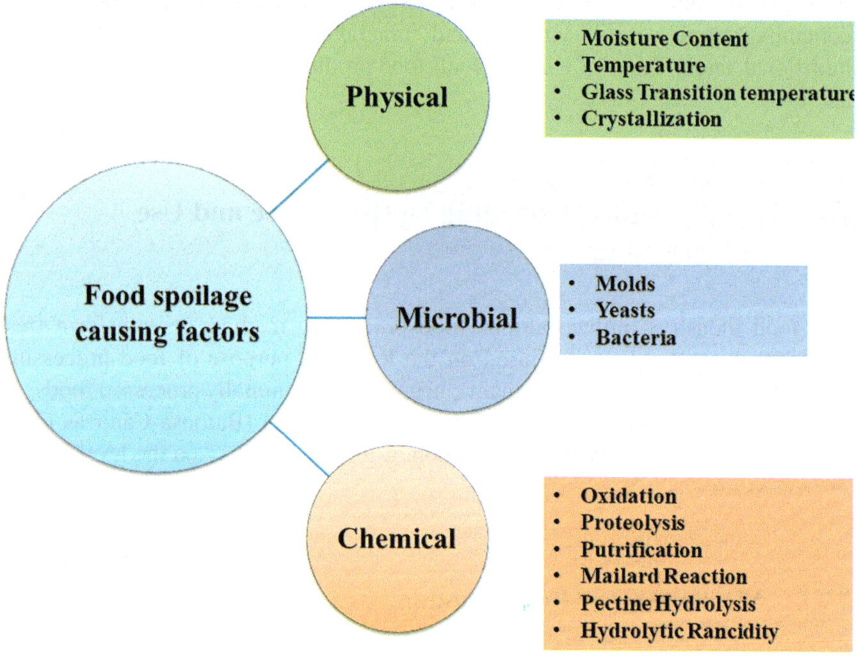


**Fig. 15.2** Food spoilage causing factors

In Fig. 15.2, many food rotting parameters are displayed, and the following sections briefly discuss each.

### 15.6.1 *Physical Degradation*

Food that has undergone physical modification or instability is referred to as the physical degradation of food. Gain or loss of moisture, movement of moisture between different components, and the physical separation of materials or components are a few examples of physical degradation (Rahman, 2007a). The primary factors affecting physical degradation are relative humidity, temperature, transient glass temperature, the structure of the crystal, and crystallization.

### 15.6.2 *Moisture Level*

The change in the content of water in various food items is a principal reason for their degradation. It could take the shape of water gain, water loss, or water movement (Fabunmi et al., 2015). A food item's water activity (aw) directly correlates with the food's moisture transfer (Kader et al., 1989). The ratio of the vapor pressure of water in a system to the vapor pressure of pure water at the same temperature is used to represent the thermodynamic attribute known as water activity (aw) (White & Cakebread, 1966; Balasubramanian & Viswanathan, 2010). It is also possible to substitute the equilibrium humidity level at the same temperature for pure water vapor pressure. With increasing temperature, water activity in food items decreases. At room temperature, foods typically have a water activity of 1, but at 20 °C and 40 °C, the values are 0.82 and 0.68, respectively (Karmas et al., 1992).

### 15.6.3 *Temperature*

The temperature has a significant impact on how quickly fruits and vegetables deteriorate. Post-harvest vitality can be encouraged in suitable temperatures while ripening is delayed. Additionally, relative humidity and airflow around fruit and vegetables are needed for slow ripening of them. These ideal circumstances are sometimes referred to as modified atmospheres (MA). The metabolism of the goods is typically hampered by temperature, which also affects how quickly MA is attained (Levine & Slade, 1988). Low temperatures may have undesirable impacts on foods that are susceptible to freeze damage. When food products are partially frozen at lower temperatures, their cells shatter, causing harm. Most tropical vegetables and fruits are sensitive to the effects of cold. When the temperature is between 5 °C and 15 °C, this often happens before the food product begins to freeze. Due to the glass transition phenomenon, dry food items stored in highly humid environments change state (Levine & Slade, 1986).

### 15.6.4 *Crystal Formation and Growth*

Food deterioration can also be a result of freezing. Foods that are slowly frozen or frozen repeatedly suffer significantly due to crystal development. They have significant extracellular ice accumulation. These foods are more stable than processed meals that are frozen slowly because rapid freezing creates ice inside the food cells (Reid, 1990). During freezing cycles, emulsifiers and other water-binding agents can minimize the growth of large ice crystals (Levine & Slade, 1988). Sugar can crystallize in foods with high sugar content when moisture accumulates or the temperature rises. As a result, sugar appears gray or white and rises to the surface from the inside. Sugar crystallization causes sugar biscuits to stolon and causes graininess in sweets and ice creams. Fructose or starch can be added to sugar solutions to prevent crystallization. Time is also a critical factor in the sugar crystallization process of food products above the appropriate glass transition temperature (Roos & Karel, 1991).

### 15.6.5 *Food Spoilage by Microbes*

Food rotting that results from the action of microorganisms is known as microbial spoilage. Additionally, it is the primary source of foodborne illnesses. Different microbes frequently damage perishable foods. By modifying storage temperature, lowering water activity, lowering pH, adding preservatives, and utilizing suitable packaging, the development of most bacteria may be delayed or stopped (Tianli et al., 2014).

## 15.7 Microorganisms That Cause Food to Deteriorate

Molds, yeasts, and bacteria are the three main kinds of microorganisms that cause food to deteriorate. Many microorganisms have an impact on food under their active states.

## 15.8 Some Causes of Microbial Spoilage

Both internal and extrinsic variables may impact food microbial deterioration (Jay, 2000). The projected shelf life of foods is determined by their inherent qualities, which also impact the kind and pace of microbial deterioration. The main intrinsic factors contributing to food degradation include endogenous enzymes, substrates, light sensitivity, and oxygen (in't Veld, 1996). These characteristics can be modified during the formulation of the food product to regulate food quality and safety (Barnwal et al., 2010). Water activity, pH, oxidation–reduction potential, and nutritional content are intrinsic variables of food decomposition. The presence of other microorganisms, relative humidity, temperature, and their activity are extrinsic causes of food spoiling (Jay, 2000).

### *15.8.1 Chemical Degradation of Food*

Foods inherently undergo chemical and biological processes, which provide unappealing sensory outcomes in food items. Fresh foods may experience minor quality changes as a result of (a) the growth of microbes and their metabolism, which alters pH, (b) harmful substances, and/or (c) the lipids oxidation and fat pigmentation, which produces unfavorable odors and colors (in't Veld, 1996). Microbial activities and chemical deterioration are connected. However, chemical processes such as oxidation rely solely on temperature changes.

### *15.8.2 Oxidation*

Amino acids form ammonia and some organic acid in the presence of oxygen. This is the primary spoiling reaction in refrigerators containing fresh meat and fish (Pitt & Hocking, 2009). The process through which unsaturated fats (lipids) react with oxygen is known as lipid oxidation or "rancidification" (Enfors, 2008). Food items may experience color change, off-flavor, and toxic chemical production. Metal oxides can function as a catalyst for rancidification, and light exposure speeds up the process. Following this process, carbonyl molecules are created, which give food its sour flavor (Criado et al., 2005).

### 15.8.3 Proteolysis

The peptide and iso-peptide linkages of protein are hydrolyzed in a constrained and extremely specific way at the time of proteolysis. It is a widespread and irreversible post-translational alteration. Various protease enzymes are necessary for the complete occurrence (Rogers & Overall, 2013). In numerous regulatory processes, various specialist proteases are crucial. Additionally, diseased and normal circumstances are linked to specialized proteolytic events (Igarashi et al., 2007). Nitrogen-containing foods usually cause this response. Proteolysis is the process by which proteins are broken down into smaller amino acids.

### 15.8.4 Putrefaction

In a sequence of anaerobic processes known as putrefaction, amino acids convert to a complex mixture of some unpleasant sulfur compounds, organic acids, and amines like hydrogen sulfide and mercaptans. The necessity of bacteria throughout the process makes this a biological phenomenon. Protein putrefaction also produces ammonia phenols and indole in addition to amino acids (Van Boekel, 2008). The majority of these substances smell unpleasant. At temperatures higher than 15 °C, putrefaction is reasonably prevalent in meats and other protein-rich meals. In these foods, microbial activity is facilitated by the high temperature.

## 15.9 Microbes' Role in the Food Industry

Microbes play a significant role in the food sector. They handle the distribution of various foods, and they frequently cause food poisoning, which results in intoxication and disease (Solms, 1969). The most frequent times that microorganisms contaminate food are when it is transported from the field to the distribution facility when it is processed, packaged, transported, collected, and just before it is used. Food poisoning has been compared to three (3) gram-positive (+ve) rods, even though the majority of pathogenic or disease-causing bacteria are gram-negative (−ve): A toxin produced by *Clostridium botulinum* is known as botulinum toxin. Compared to conventional crops, they spread very fast and use less area (Igarashi et al., 2007). Compared to conventional crops, they spread very fast and use less area. They use low-cost agricultural and industrial waste products such petroleum oil, methanol, ethanol, sugar, molasses, and paper mill waste. They aid in product recycling, which lowers emissions, which is the secondary benefit. They cultivate a lot of fruit. In a single day, many tonnes of protein are generated in a yeast growing medium weighing 1000 pounds. This is 25–50 times more than maize and around 10–15 times more than soybeans (Pitt & Hocking, 2009). The cells' high protein content is the reason for this. Protein concentration varies between 40% and 50% in yeast cells between 20% and 40% in algal cells (Rogers & Overall, 2013).

Natural fermentation processes are carried out by microorganisms, and for thousands of years, bread, vinegar, beer, yogurt, cheese, and wine have all been made using yeasts, molds, and bacteria as well as fruits, vegetables, and fermented fish. The earliest techniques for preserving and transforming food are fermentation (Solms, 1969). This biological procedure preserves food while improving its nutritive and organoleptic qualities. Microbes are used in both the fermentation and production of both alcoholic beverages and dairy products. Bread, buttermilk, cheese, sour cream, and yogurt are the most popular dairy products made by bacteria. Soil organisms are often the main participants in soil biota food chains, contributing to many different food chains and acting as each other's nutrition providers (Panda, 2003).

## 15.10 Role of Microbes in Food Preservation

Numerous chemicals have been used by the food industry for a long time to prevent the growth of germs that cause food degradation. By lowering the activity of both pathogenic and spoilage bacteria, their use helps to extend the shelf life of foods. Sometimes they show a negative impact on the organoleptic and sensory characteristics of particular foods (Desrosier & Singh, 2014). Bio-preservatives have been developed to lessen the requirement for these preservation, processing, and manufacturing techniques. Bio-preservatives are substances that are produced naturally by plants, animals, and bacteria and extend the shelf life of food. These substances reduce or even get rid of dangerous organisms in food while improving functioning and quality.

Many of these chemicals are antimicrobials that also function as antioxidants, damage cell membranes, and obstruct the routes of bacteria that synthesize bioactive molecules. However, consumers' interest in natural preservative chemicals is rising as a result of the adverse impacts of synthetic food additives on human health and the environment (Hoff & Castro, 1969). Demand for safe food items manufactured with natural ingredients has increased along with consumer knowledge of the potential health risks associated with consuming meals prepared with synthetic chemicals. These bioactive substances, which are derived from living things such as plants, animals, and helpful bacteria, have been praised as possible substitutes for artificial food additives. Many of these substances damage cell membranes and obstruct the routes used by biosynthetic bacteria. They also serve as antioxidants and antimicrobials.

It has been argued that bioactive substances derived from plants, animals, and helpful microbes might replace synthetic food additives (Rahman, 2007a, b). Microorganisms are found in a wide range of food products and can have an impact on both the quantity and quality of food. The environment provided by food products is ideal for microorganism growth. Food degradation or deterioration may result from the proliferation of microorganisms. Similarly, when bacteria are found in food, they may cause stomach related problems mainly food poisoning. Modern civilization is concerned about plastics. Plastic recycling is the most challenging aspect of its utilization. Plastics provide a serious environmental risk since they cannot be recycled or biodegraded. Biodegradable materials are actively being investigated for use in food procuction due to their tremendous potential (Rahman, 2007a, b).

These substances can take the place of biodegradable plastics, reducing the harm to the environment that they cause. On the other hand, biodegradable materials contain inferior biopolymers with other useful components and are less susceptible to water and gas invasion. One of the most crucial aspects of food safety is the packaging. By applying nanotechnology to boost the effectiveness and caliber of packaging materials, food safety may be enhanced. However, they are only partially effective in preventing the flow of $O_2$, $H_2O$, and $CO_2$. The use of polymers containing nanoparticles can increase the permeability of food containers. Clay nanoparticle-infused packaging, for instance, has outstanding mechanical, radiative, thermal, and barrier properties that block the passage of $O_2$, $CO_2$, and moisture. Additionally, they support food preservation, color and taste retention, and microbial growth inhibition (Doyle, 2009). As the demand for meat increased, livestock industrial farming was established to increase animal production. Intense animal management and enhanced animal development were required to boost production. To meet the criteria, antibiotic containing animal feed was used. Despite the fact that antibiotics effectively stop the spread of infectious diseases, improper use of in-feed antibiotics has resulted in a number of issues including environmental pollution from animal waste, resistant to antibiotics, antibiotic contaminants in livestock products, and low or no disease resistance in livestock. As a result, to address the problems brought on by excessive in-feed antibiotic usage, natural antibacterial substances must be developed.

## 15.11 Role of Microbes in Protection and Storage of Foods

### *15.11.1 Fruit*

Microorganisms need specific concentrations of oxygen, acid content, ammonia, and glucose metabolism to survive and thrive, just like all other living things. Since most fruits are acidic, the principal pests that attack them are molds and yeasts that are acid-tolerant and survive in acidic conditions. Fruit bacterial degradation is just a minor issue. These microbes degrade pectin, starch, and other carbohydrates, causing the fruit to soften, discolor, and become undesirable. Even during the earliest stages of degradation, it becomes unpleasant. Various methods can be used to prevent fruit storage decay. Cold storage is practically always used to preserve fruits and fruit products. Low temperatures inhibit both microbial growth and organic fruit processes. Fruits, especially citrus fruits, have waxy or fungicidal outer coverings. Various fruits have also been covered with fungicidal wrappings (Rodriguez & Mesler, 1985).

### 15.11.2 Fruit Juices

If no preservation method is used, the many microorganisms found in freshly made fruit juices will rapidly grow. Mycoderma typically forms a white to gray wrinkled scum on the fruit juice's surface, destroying the fruit acids and sugars and resulting in musty flavors. One way to stop this surface development is to isolate the juice from the air because the scum needs access to oxygen in order to grow. Fruit juices that are not protected from the air may develop a cotton sort of mold on their surface. Fresh fruit juices encourage yeast growth, which breaks down the carbohydrates into both carbon dioxide and alcohol. This method turns the juices into wines under controlled circumstances, but in open containers, a lot of acetic acid is produced (Rahman, 2007a, b). Fruit juices quickly become unsuitable for eating as a result of this fermentation and acetification. The juices can be preserved through pasteurization, a sterile filtration process, or freezing. The method that is most frequently used is pasteurization process, in which the juice is heated either in-bottle or in bulk before being bottled. Due to variations in acid concentration, amount of sugar, and initial microbial contamination, the temperature and procedure used for each juice are varied. Fermentation is stopped by freezing, but it begins as soon as the beverage is defrosted.

### 15.11.3 Vegetables

Fruits and fresh vegetables have diverse compositions and personalities, and they naturally have various surface bacteria and rotting processes. Since most veggies are not acidic, they may support a wide variety of bacteria and germs that do not thrive on fruits, both on the surface and within the tissue. Vegetables that have bacteria and mold growing on them degrade the pectic substance, causing the tissue to lose shape and become liquid and squishy. The majorities of veggies sweat and significantly increase in temperature when stacked together (Rodriguez-Gonzalez et al., 2015). The product spoils within a few hours due to the warm, humid atmosphere that is excellent for the growth of numerous microorganisms. Fresh veggies should not be stored in big containers or piled up; instead, they should be handled quickly and kept as cold as possible. Any storage area should allow for adequate air circulation. The soluble sugars in the vegetables are converted by certain bacteria into lactic acid, which gives the final product its sour flavor and odor. Fresh, cooked, and thawed frozen foods all experience this souring, which should be taken into consideration when determining whether the food is safe to eat.

## 15.11.4 *Frozen Foods*

The manufacturing and consumption of frozen goods, such as ice cream, beef, fish, chicken, eggs, fruits, and vegetables, has expanded over the past few years. This has rekindled interest in how microorganisms react to low temperatures, specifically 32°. Fahrenheit has not been shown to be low enough to stop the growth of several bacteria, yeasts, and molds over extended periods of storage that can alter the foods' appearances, taste, textures, and overall suitability for human consumption (Drake et al., 2008). From the perspective of microbial deterioration, storage at temperatures over 15 F is not regarded as safe neither is this temperature low enough to prevent unfavorable enzymatic changes in some items. The freezing rate, the product nature, acidification, type and size of packaging container, pack airtightness, subfreezing storage temperature, storage period, initial contamination, etc. all affect the rate of eradication of bacteria and the predominant types that survive (Walter, 1991). During the initial periods of freezing, less resilient microbial forms are killed. Following that, the degree of destruction is moderate. Some microbial types may continue to exist for a long time. In the frozen food sector, incorrect treatment of the product either before freezing or after thawing is the main cause of deterioration. Before being frozen, vegetables are scalded to deactivate the enzymes already present in the tissue. The heat from this procedure also kills several germs.

Vegetables that are left out before being scalded or frozen run the risk of developing germs that will soften the tissue, release lactic acid, and have other undesirable effects. Although frozen food items are not sterile, their microbiological content can be reduced by using them right away after thawing, maintaining subfreezing temperatures at which microbial development is known not to occur, and using them in a hygienic manner during preparation and handling. Before being frozen, fruits are often packaged in syrup or dry sugar (Karel & Lund, 2003). Both in the material that will be frozen and in the material that has been thawed, yeast growth begins fast. In exposed areas, mold development is to be expected, and unfrozen material darkens and develops off smells and aromas as a result of the interaction between airborne microorganisms and oxygen. Just freshly caught fish must be utilized when handling fish that will be frozen since decomposition sets in quickly (Ohlsson & Bengtsson, 2002). Despite having a generally acceptable appearance when frozen, items with poor original quality will have unpleasant smells and aromas when defrost. Starting with high-quality ingredients, handling them quickly under hygienic circumstances, storing and transporting them at correct freezing temperatures, and using them quickly after thawing are all important to secure and maintain good quality in frozen food products.

### 15.11.5 Canned Food Products

In the canning preservation foods are packed in tightly sealed containers, the enzymes are inhibited and the microorganisms are died by the action of heat. Certain microorganisms can survive and grow in the product when inappropriate processing techniques are used. Some bacteria create gas, which causes the can ends to expand. Other kinds of spoilage are referred to as "flat sour," as they create acid without gas. Food in cans with spoilage indicators or an off flavor or odor should be thrown away. Molds cannot develop in the lack of oxygen, yet badly treated fruits may ferment and become spoiled due to yeasts (Berk, 2018). Molds cannot develop in the lack of oxygen, yet badly treated fruits may ferment and become spoiled due to yeasts. Canns swelling happen as a result of the production of carbon dioxide at the time of yeast fermentation process. For the different food products, the processing time and temperature have been figured out in commercial practice until there is generally minimal spoiling. For the different food products, the processing time and temperature have been figured out in commercial practice until there is generally minimal spoiling (Rayaguru & Routray, 2010).

### 15.11.6 Eggs

Microorganisms are mostly to blame for the rotting and degradation of fresh, stored, and processed eggs. Large amounts of microorganisms are a sign of inappropriate handling and frequently of unclean settings. While some eggs have microorganisms in them when they are first deposited, most eggs are uninfected. Lysozyme, a substance found in the egg white that destroys bacteria, is undoubtedly responsible for a large portion of the egg's lack of contamination. A larger loss from bacterial spoiling is to be anticipated if the flock is handled in a way that results in a high percentage of "dirties" or if the eggs are not treated properly (Leniger & Beverloo, 2012). Green whites, digested whites, white rots, and black rots are the main forms of damaged eggs that are discovered during and after storage. These kinds often exhibit clear chemical degradation and have a lot of bacterial infection. Due to their high sulfur and nitrogen levels, microorganisms that break down eggs produce particularly unpleasant smells and aromas. Bacterial growth is another reason why eggs smell musty. In practical practice, almost all diseased eggs are found by candling, ensuring that few contaminated eggs are taken by consumers.

In storage, the presence of molds on the surface of eggs, giving them an unpleasant esthetic, but this problem may be prevented with careful management. Very little issue should arise if the moisture content is adequately managed and chance of contamination and packaging materials are removed (Syamaladevi et al., 2016). Unless the egg is allowed to stay in storage for a long period, at which point the mold will pierce the shell and produce darkening, such molding, has no impact on the egg's nutritional value. In order to manufacture high-quality goods, egg-breaking operations must adhere to strict hygienic standards. Since a single musty egg might accidently be included in a batch via inattentive examination, extreme caution should be taken before breaking to exclude all bad eggs. Because of how prevalent this musty flavor or odor is, every effort should be made to keep such ingredients out of the commercial pack. Before adding the eggs to the main batch in the mixer, they should first be cracked, looked at, and smelled separately. When handling frozen eggs, speed is key (Salvato et al., 2003). Until they are ready to be broken, the fresh eggs should be maintained in cold storage. As soon as they are broken, they need to be churned, then quickly placed into the final containers and stored in a freezer. There is a potential that bacteria will grow in the egg mass, producing unpleasant smells and aromas as well as an overall reduction in quality.

### *15.11.7 Milk*

Milk is a very perishable drink item, principally because it offers the perfect environment for the microorganism growth and development. One well-known instance of this action is milk souring. Some microorganisms produce lactic acid by utilizing the sugar present in milk. This lactic acid is produced under normal circumstances at room temperature relatively quickly (Agrahar-Murugkar & Jha, 2010). When a specific level of acidity is reached, casein is transformed into its insoluble state, which leads to the formation of curd. A rennet-like enzyme is produced by bacteria named *Bacillus subtilis*, and it causes sweet milk coagulation of milk before eventually being almost entirely digested. The development of undesired microorganisms frequently results in the production of very offensive aromas, soapy or bitter tastes, as well as a variety of odd colors or gums in the milk. Cream may develop yeast growths that cause it to "gas" or froth up. Milk may be preserved in a nutritious and delectable state for a significantly longer period of time thanks to pasteurization, which is often used to eliminate the germs (Kutz, 2007). It goes without saying that every hygienic precaution should be followed while handling milk and cold storage should always be utilized to slow the growth of microorganisms.

### *15.11.8 Meat*

When meats are handled quickly, hygienically, and with the appropriate cold storage techniques, generally minimal microbial spoiling should happen (Jangam et al., 2010). Ham souring or "bone souring," a bacterial infection near to the bone that has caused significant trouble, is unquestionably caused by the animal heat being slowly removed or by the brine not being absorbed properly. During slaughter, damage to the tissue and bones may be a significant factor. It's also common to encounter soured beef. Fresh meat may develop a mold growth under specific humidity, cold storage, and handling circumstances, however, this is usually of little importance unless it detracts from the product's look. Bacteria and molds may create putrefaction under circumstances of inappropriate handling and storage, making the product unsuitable for human consumption. In order to inhibit bacterial action, delay enzyme activity, and lessen the potential of rancidity, methods including salting, smoking, drying, and freezing are used (Delong, 2006).

### 15.11.9 Carbon Hydrate Resources

Due to the insufficient amount of moisture many bacteria, yeasts, and molds are grow in sugars, starches, or syrups. Conversely, some food spoiling bacteria can occasionally be found in carbs. If the keeping quality of the canned food items is not protected with extreme care, the carbohydrates may induce spoiling when added (Baker, 1997). The term "thermophilic" refers to this particular kind of bacteria, which means "heat loving." The ideal temperature for growth is around 131 °F, which is far higher than the temperatures that are tolerated by the majority of bacterial species. This group's spores are extremely heat-resistant and difficult to destroy during the routine processing of food products (Sequeira-Munoz et al., 2006). This group's spores are extremely heat-resistant and difficult to destroy during the routine processing of food products. This genus of bacteria commonly contaminates processed foods due to improper chilling after processing or storage in very heated environments. By removing food spoilage germs from sugar, sugar producers are benefiting the food industry. Molds and yeasts cannot survive the current procedures used to make sugar and syrup. If they found, they are a result of improper handling and storage contamination.

When yeasts are used to make candy and other confections, several issues arise. These yeasts may flourish in environments with extremely high concentrations of sugar. The candy foams or bursts open under the pressure of the $CO_2$ they release (Bhat et al., 2012). Soft drinks represent a significant portion of the country's food budget, and each year, the production of these drinks requires several hundred tons of sugar and other components that provide energy. In many cases, the availability of sugar makes a product a possible environment for the development of several microorganisms including bacteria, molds and yeasts. If precautions aren't taken to prevent their entry, spoiling frequently happens (Sagar & Kumar, 2010). More frequently than molds, yeasts and bacteria cause drinks to deteriorate. Due to the slower turnover of the stock, deterioration, which manifests as cloudiness and a deposit of sediment at the bottom of the bottle, happens more frequently in the winter than in the summer. The production of drinks is now generally free from spoiling because to ongoing advancements in bottling equipment, ingredient management, and bottling process techniques.

## 15.12 Microorganisms Are Used in Food Processing and Preservation

The importance of microorganisms' role in food preservation is usually overlooked since the role that they play in food spoilage is so highly stressed. Although controlled fermentation is one of the oldest methods ever used by humans to preserve food, many large corporations rely on it to keep food products fresh or alter their flavor.

### *15.12.1 Beer and Wine*

The production of wine from the grape juice and other fruits was without a doubt the earliest application of microorganisms by man. Everywhere fruit was supplied; this natural process was used to extend the shelf life of fruit products. Today, the United States produces more than 10,000,000 barrels of wine annually. The outer layer and stems of most other fruits, as well as the stems of grapes, naturally contain yeast. When the berries are crushed, the yeasts in the juice quickly grow, converting the sugar into carbon dioxide and alcohol (Chopra, 2005). By carefully monitoring the fermentation, it is possible to prevent the growth of almost all other microbes in the product. By sealing the product from air penetration and letting the yeast fermentation to produce carbon dioxide consuming the entire $O_2$ out of the container. This method can eliminate the bulk of slime or surface-growing organisms. Champagne is produced by keeping the bottle closed and pumping carbon dioxide to the liquid, which results in pressures of up to 120 pounds per square inch. To achieve a clean, healthy fermentation, the majority of wine industries inoculates the must or juice with a pure culture of wine yeast. Due to this, wine yeast has an advantage over other microorganisms and is prevented from engaging in an unfavorable type of fermentation (Clark et al., 2009).

The majority of wine yeasts thrive in acidic conditions and at temperatures a little lower than those that are most favorable for the number of bacteria to cultivate. Treatment with sulfur dioxide or pasteurization must are frequently used in place of low temperatures (Karel & Lund, 2003). Yeast can tolerate the few concentrations of sulfur dioxide that would normally kill the majority of spoilage organisms. Once all the sugar in the must has been fermented to generate alcohol, the yeasts die from a lack of food and the alcohol's lethal effects. Usually wine contain up to 14% alcohol by volume, but if juice contains very high content of sugars, spontaneous fermentation may be able to achieve 16 or 17%.

Casks should be tightly sealed to keep them protected from the air in order to prevent wine from being spoiled. In bottled wine, pasteurization can also eliminate organisms that cause spoilage. Beer is made by the yeast fermentation of malted grain extracts. Beer fermentation converts carbohydrates to alcohol and carbon dioxide, just like wine fermentation does; however, beer generates far less alcohol overall. One product where the carbon dioxide is tightly compressed is beer. This helps to preserve the goods while also making it more palatable. Due to the beer's low acidity, lactic acid bacteria could potentially cause it to ferment if the proper precautions are not taken. To maintain its quality while being stored in bottles or cans, the majority of contemporary beers are pasteurized (Rogers & Overall, 2013).

### 15.12.2 *Vinegar*

Similar to making beer or wine, the first step in making vinegar is the production of alcohol. The alcohol that results from the fermentation of mashes, molasses, or other sugary materials can be used to make white or distilled vinegar. Fermented cider is used to produce cider vinegar (Cattelan et al., 2013). Following fermentation, acetic acid is produced from the alcohol by a microbial fermentation-induced oxidation process. While yeast fermentation happens without the presence of air, the conversion of alcohol to acetic acid can only happen in the presence of air. Yeast fermentation takes place without air, whereas the conversion of alcohol to acetic acid can happen only in the presence of air. Unlike the generator procedure, which allows the alcohol solution to drip through towers loaded with sawdust, sand, jute, or other filler material, the old barrel method allows the acetic acid bacteria to develop on the alcohol solution's surface. The utilization of microbes in these two rather straightforward procedures results in the yearly production of thousands of gallons of vinegar (Bhat et al., 2012).

### 15.12.3 *Yeast and Bread*

Yeast-risen bread has been a mainstay of the diet for countless generations. Yeast is given the opportunity to establish a culture in a flour and water dough. The dough rises and becomes lighter as a result of the yeast's growth and release of carbon dioxide. The taste of the bread is also influenced by yeast or certain types of microorganisms. Some types of bread encourage the growth of the lactic acid bacteria, creating distinct flavors (Jangam et al., 2010). During baking, the alcohol produced by the fermentation that killed the microorganisms is pushed off the bread. Under rare conditions, the proliferation of bacteria can cause bread to become ropy, moldy, or intensely pigmented. With the express purpose of cultivating and harvesting a yeast crop, a sizable business has sprung up due to the usage of this type of yeast as food for humans and as animal feed. The only microorganisms that man consumes in this general category of yeast are ingested for their minerals and vitamins (Sagar & Kumar, 2010).

### 15.12.4 Fermented Products

The making of the so-called fermented foods makes use of a specific bacteria's capacity to convert carbohydrates to lactic acid (Kristensen & Purslow, 2001). These consist of silage-like animal feeds, pickles, sauerkraut, and other condiments. Lactic acid is produced throughout these activities as a result of carbohydrates. The preservation has a dual purpose because (1) the acidic environment has a selective effect that prevents microorganisms from spoiling and (2) the majority of the food for microorganisms has been consumed during the fermentation process, leaving no more material to support significant microbial populations. Since salt has a curing effect and these lactic acid bacteria are salt resistant, it is used in practically all fermented foods. Before the product is utilized as food, the salt and acid are soaked out of the cucumbers after they have fermented in brines potent enough just to prevent the majority of spoilage microorganisms. Pickled olives are likewise made using this general procedure (Kutz, 2007). When it comes to sauerkraut, the amount of salt is just approximately 2%, and the final fermented product is what is eaten. Microbial activity gives sauerkraut its pleasantly acidic taste. If at all feasible, the procedure in all three circumstances needs to be carried out without the presence of air to avoid the development of scum. For the manufacturing of pickles, sauerkraut, and pickled olives in 1937, about 8,000,000 bushels of cucumbers, 132,000 tons of cabbage, and 22,000 tons of olives were utilized, totaling a market worth of over $7,800,000 for the raw materials. Although very little, if any, salt is used, the employment of microorganisms in the formation of silage is relatively comparable.

### 15.12.5 Dairy Products

Microorganisms are widely used in the production of a variety of dairy products, including butter, yogurt, cheese, and fermented dairy items. A variety of microorganisms and various processes are used to produce dairy products with distinctive aromas and tastes. Milk is soured by bacteria in order to form cheese, and ripening is achieved by encouraging the growth of additional bacteria in pressed cheese. Bacterial digestion of the curd produces unique flavors, which is what results in this (Mizuta et al., 1999). Molds work to produce various kinds of cheese, such as Camembert, Brie, and Roquefort, and their growth results in distinct tastes. Sour cream butter is made by using churning cream that has undergone bacterial ripening, which has led to the production of lactic acid. The fat globules have more easily solidified throughout the churning, and they take on some of the savory flavors that the bacteria produced during the fermentation stage. Unwanted microorganisms may produce bitter or off flavors in the cream if proper handling methods, including as pasteurization, cooling, and the use of starters, are not employed. Microorganisms are used to produce a variety of sour milk drinks, such as cultured buttermilk, *acidophilus* and *bulgaricus* milk, kefir, and koumiss. The latter two, which include a significant amount of alcohol, are produced when bacteria and yeasts collaborate.

## 15.13 Conclusion

Learning how to preserve food was one of the most revolutionary discoveries of human civilization because it enabled people to settle down in one place and start communities. Extending the shelf lives of foods without reducing their original nutritional value is important and challenging. Food is a perishable, organic commodity that can spoil as a result of microbial, physical, and chemical activities. In the past, cooling, freezing, drying, heating, and fermenting were just a few of the traditional techniques for extending the shelf life and preserving food while retaining its nutritional value and original texture. Preservation methods have evolved and become more contemporary throughout time in response to increasing needs. The newest developments in food preservation include pulsed electric field effect, high-pressure food preservation, and irradiation. As food additives and preservatives, other chemical reagents have also been developed. The use of chemical preservatives and additives in food products is, however, raising concerns due to potential health risks.

## References

Agrahar-Murugkar, D., & Jha, K. (2010). Effect of drying on nutritional and functional quality and electrophoretic pattern of soyflour from sprouted soybean (Glycine max). *Journal of Food Science and Technology, 47*(5), 482–487.

Amit, S. K., Uddin, M., Rahman, R., Islam, S. M., & Khan, M. S. (2017). A review on mechanisms and commercial aspects of food preservation and processing. *Agriculture & Food Security, 6*(1), 1–22.

Baker, C. G. (1997). *Industrial drying of foods*. Springer Science & Business Media.

Balasubramanian, S., & Viswanathan, R. (2010). Influence of moisture content on physical properties of minor millets. *Journal of Food Science and Technology, 47*(3), 279–284.

Barbosa-Cánovas, G. V., Altunakar, B., & Mejía-Lorío, D. J. (2005). *Freezing of fruits and vegetables: an agribusiness alternative for rural and semi-rural areas* (Vol. 158). Food & Agriculture org.

Barnwal, P., Singh, K. K., Mridula, D., Kumar, R., & Rehal, J. (2010). Effect of moisture content and residence time on dehulling of flaxseed. *Journal of Food Science and Technology, 47*(6), 662–667.

Berk, Z. (2018). *Food process engineering and technology*. Academic press.

Bhat, R., Alias, A. K., & Paliyath, G. (2012). *Progress in food preservation*. Wiley.

Blum, D. (2012). *Food that lasts forever* (Vol. 12). TIME magazine.

Cattelan, M. G., de Castilhos, M. B. M., Sales, P. J. P., & Hoffmann, F. L. (2013). Antibacterial activity of oregano essential oil against foodborne pathogens. *Nutrition & Food Science, 43*(2), 169–174.

Chopra, P. (2005). *Food & Nutrition Education*. APH Publishing.
Clark, S., Costello, M., Drake, M., & Bodyfelt, F. (2009). *The sensory evaluation of dairy products*. Springer Science & Business Media.
Criado, M. V., Pinto, V. E. F., Badessari, A., & Cabral, D. (2005). Conditions that regulate the growth of moulds inoculated into bottled mineral water. *International Journal of Food Microbiology, 99*(3), 343–349.
DeLong, D. (2006). *How to dry foods: the most complete guide to drying foods at home*. Penguin.
Desrosier N.W., Singh R.P., (2014). Food preservation. Encyclopaedia Britannica Inc. https://www.britannica.com/topic/food-preservation. Cited 4 May 2017.
Doyle, M. P. (2009). *Compendium of the microbiological spoilage of foods and beverages*. Springer Science & Business Media.
Drake, M. A., Drake, S., Bodyfelt, F. W., Clark, S., & Costello, M. (2008). *The sensory evaluation of dairy products* (2nd ed.). Springer.
Enfors, S.-O. (2008). *Food microbiology*. KTH-Biotechnology.
Fabunmi, O. A., Osunde, Z. D., Alabadan, B. A., & Jigam, A. A. (2015). Influence of moisture content and temperature interaction on mechanical properties of DESMA (novella pentadesma) SEED. *Journal of Advances in Food Science & Technology, 2*(2), 81–85.
Freedman, D. H. (2011). The bright, hi-tech future of food preservation, in discover magazine.
Hoff, J. E., & Castro, M. D. (1969). Chemical composition of potato cell wall. *Journal of Agricultural and Food Chemistry, 17*(6), 1328–1331.
Igarashi, Y., Eroshkin, A., Gramatikova, S., Gramatikoff, K., Zhang, Y., Smith, J. W., Osterman, A. L., & Godzik, A. (2007). CutDB: a proteolytic event database. *Nucleic Acids Research, 35*(suppl_1), D546–D549.
in't Veld, J. H. H. (1996). Microbial and biochemical spoilage of foods: an overview. *International Journal of Food Microbiology, 33*(1), 1–18.
Jangam, S. V., Law, C. L., & Mjumder, A. S. (2010). *Drying of foods, vegetables and fruits* (Vol. 1, 1st ed.).
Jay, J. M. (2000). *Modern food microbiology* (6th ed.). Aspen Publishers.
Kader, A. A., Zagory, D., Kerbel, E. L., & Wang, C. Y. (1989). Modified atmosphere packaging of fruits and vegetables. *Critical Reviews in Food Science & Nutrition, 28*(1), 1–30.
Kar, B. K. (2014). Multi-stakeholder partnership in nutrition: an experience from Bangladesh. *Indian Journal of Community Health, 26*(Supp 1), 15–21.
Karel, M., & Lund, D. B. (2003). *Physical principles of food preservation: Revised and expanded*. CRC Press.
Karmas, R., Pilar Buera, M., & Karel, M. (1992). Effect of glass transition on rates of nonenzymic browning in food systems. *Journal of Agricultural and Food Chemistry, 40*(5), 873–879.
Kristensen, L., & Purslow, P. P. (2001). The effect of processing temperature and addition of mono- and di-valent salts on the heme-nonheme-iron ratio in meat. *Food Chemistry, 73*(4), 433–439.
Kutz, M. (2007). *Handbook of farm, dairy, and food machinery*. William Andrew.
Leniger, H. A., & Beverloo, W. A. (2012). *Food process engineering*. Springer Science & Business Media.
Levine, H., & Slade, L. (1986). A polymer physico-chemical approach to the study of commercial starch hydrolysis products (SHPs). *Carbohydrate Polymers, 6*(3), 213–244.
Levine, H., & Slade, L. (1988). Principles of "cryostabilization" technology from structure/property relationships of carbohydrate/water systems: a review. *Cryo-Letters, 9*(1), 21–63.
Mizuta, S., Yamada, Y., Miyagi, T., & Yoshinaka, R. (1999). Histological changes in collagen related to textural development of prawn meat during heat processing. *Journal of Food Science, 64*(6), 991–995.
Monteiro, C. A., Levy, R. B., Claro, R. M., Castro, I. R. R. D., & Cannon, G. (2010). A new classification of foods based on the extent and purpose of their processing. *Cadernos De Saude Publica, 26*, 2039–2049.
Nummer, B. A., & Brian, A. (2002). *Historical origins of food preservation*. National Center for Home Food Preservation. University of Illinois Extension.

Ohlsson, T., & Bengtsson, N. (2002). *Minimal processing technologies in the food industries*. Elsevier.
Panda, H. (2003). *Herbal foods and its medicinal values*. NIIR Project Consultancy Services.
Pitt, J. I., & Hocking, A. D. (2009). *Fungi and food spoilage* (Vol. 519, p. 388). Springer.
Rahman, M. S. (2007a). *Handbook of food preservation*. CRC press.
Rahman, M. S. (2007b). Food preservation: overview (chapter 1). In M. S. Rahman (Ed.), *Handbook of food preservation* (2nd ed.). Taylor & Francis Group, LLC.
Rayaguru, K., & Routray, W. (2010). Effect of drying conditions on drying kinetics and quality of aromatic Pandanus amaryllifolius leaves. *Journal of Food Science and Technology, 47*(6), 668–673.
REID, D. (1990). Optimizing the quality of frozen foods: understanding and minimizing the effects of each stage in the production of frozen foods, particulary blanching and frozen storage, can optimize their quality. *Food technology (Chicago), 44*(7).
Rodriguez, F., & Mesler, R. (1985). Some drops don't splash. *Journal of Colloid and Interface Science, 106*(2), 347–352.
Rodriguez-Gonzalez, O., Buckow, R., Koutchma, T., & Balasubramaniam, V. M. (2015). Energy requirements for alternative food processing technologies—principles, assumptions, and evaluation of efficiency. *Comprehensive Reviews in Food Science and Food Safety, 14*(5), 536–554.
Rogers, L. D., & Overall, C. M. (2013). Proteolytic post-translational modification of proteins: Proteomic tools and methodology. *Molecular & Cellular Proteomics, 12*(12), 3532–3542.
Roos, Y., & Karel, M. (1991). Plasticizing effect of water on thermal behavior and crystallization of amorphous food models. *Journal of Food Science, 56*(1), 38–43.
Sagar, V. R., & Kumar, S. P. (2010). Recent advances in drying and dehydration of fruits and vegetables: a review. *Journal of Food Science and Technology, 47*(1), 15–26.
Salvato, J. A., Nemerow, N. L., & Agardy, F. J. (2003). *Environmental engineering* (5th ed.). Wiley.
Sequeira-Munoz, A., Chevalier, D., LeBail, A., Ramaswamy, H. S., & Simpson, B. K. (2006). Physicochemical changes induced in carp (Cyprinus carpio) fillets by high pressure processing at low temperature. *Innovative Food Science & Emerging Technologies, 7*(1–2), 13–18.
Solms, J. (1969). Taste of amino acids, peptides, and proteins. *Journal of Agricultural and Food Chemistry, 17*(4), 686–688.
Steele, R. (2004). *Understanding and measuring the shelf-life of food*. Woodhead Publishing.
Syamaladevi, R. M., Tang, J., Villa-Rojas, R., Sablani, S., Carter, B., & Campbell, G. (2016). Influence of water activity on thermal resistance of microorganisms in low-moisture foods: a review. *Comprehensive Reviews in Food Science and Food Safety, 15*(2), 353–370.
Tianli, Y., Jiangbo, Z., & Yahong, Y. (2014). Spoilage by Alicyclobacillus bacteria in juice and beverage products: Chemical, physical, and combined control methods. *Comprehensive Reviews in Food Science and Food Safety, 13*(5), 771–797.
Van Boekel, M. A. (2008). Kinetic modeling of food quality: a critical review. *Comprehensive Reviews in Food Science and Food Safety, 7*(1), 144–158.
Walter, R. H. (1991). *Analytical and graphical methods for pectin* (pp. 189–225). Academic Press.
White, G. W., & Cakebread, S. H. (1966). The glassy state in certain sugar-containing food products. *International Journal of Food Science & Technology, 1*(1), 73–82.
Wilkinson, J., & Roch, R. (2006). Agri-processing and developing countries.
World Food Market Overview Marketing Essay. 2013. http://www.ukessays.com/essays/marketing/world-food-market-overviewmarketingessay.php?cref=1.

The manufacturer's authorised representative in the EU is Springer Nature Customer Service Centre GmbH, Europaplatz 3, 69115 Heidelberg, Germany. If you have any concerns regarding our products, please contact ProductSafety@springernature.com

Printed and bound by CPI Group (UK) Ltd, Croydon, CR0 4YY
01/10/2026
02225872-0011